METEOROLOGICAL MONOGRAPHS

VOLUME 32   NOVEMBER 2004   NUMBER 54

NORTHEAST SNOWSTORMS

VOLUME I: OVERVIEW

Paul J. Kocin
Louis W. Uccellini

American Meteorological Society
45 Beacon Street, Boston, Massachusetts 02108

Credits for photographs on the dust jacket

Front cover
Richmond Hill, New York City, 27 Dec 1947 (New York Historical Society).

Back cover
Top row: (left) Skiers at Radio City Music Hall, Feb 2003 (courtesy of Rob Gardiner, www.nyclondon.com); (middle) Hamilton St, 30 Jan 1966 (*Blizzards and Snowstrorm of Washington, D.C.,* Historical Enterprises, 1993, p. 63); (right) Snarled traffic on the Northern State Pkwy, 12 Feb 1983 (*Great Blizzards of New York City,* Historical Enterprises, 1994, p. 73).
Upper-middle row: (left) Cars buried in snow near the Capitol, 16 Feb 1958 (*Washington Weather: The Weather Sourcebook for the D.C. Area,* Historical Enterprises, 2002); (middle) New York City, 6 Apr 1982 (photo reprinted courtesy of the *New York Times*); (right) Snowplows on the Long Island Expressway, 11 Feb 1983 (photo reprinted courtesy of *Newsday*).
Lower-middle row: (left) Student walks home from school, 13 Jan 1964 (courtesy of *Washington Weather: The Weather Sourcebook for the D.C. Area,* Historical Enterprises, 2002); (middle) Walking across the Memorial Bridge, 7 Feb 1967 (*Washington Weather: The Weather Sourcebook for the D.C. Area,* Historical Enterprises, 2002); (right) After the snowstorm of 15 Feb 1958 (*Washington Weather: The Weather Sourcebook for the D.C. Area,* Historical Enterprises, 2002).
Bottom row: (left) Times Square, 7 Feb 1967 (reprinted courtesy of the *New York Times*); (middle) Knickerbocker Snowstorm, 28 Jan 1922 (*Washington Weather: The Weather Sourcebook for the D.C. Area,* Historical Enterprises, 2002); (right) A hotel at Rockaway Beach, 1920 (*Great Blizzards of New York City,* Historical Enterprises, 1994, p. 3).

ISBN 1-878220-64-0
ISSN 0065-9401

Support for this monograph has been provided by the National Oceanic and Atmospheric Administration's National Weather Service and the National Centers for Environmental Prediction.

Published by the American Meteorological Society
45 Beacon St., Boston, MA 02108

For a catalog and ordering information for AMS Books, which include meteorological and historical monographs, see www.ametsoc.org/pubs/books.

Printed in the United States of America
by Allen Press, Inc., Lawrence, KS

Awakening to deep snow, Richmond Hill, New York City, 27 Dec 1947 (from New York Historical Society).

TABLE OF CONTENTS

Chapter 8. Summary, Forecast Advances, and a Northeast Snowfall Impact Scale (NESIS)

PREFACE

Saturday, 8 February 1969, was not the bitterly cold day that often serves as a harbinger of a severe winter snowstorm on New York's Long Island. Instead, it was generally sunny and cool, about 40°F(5°C). The weather forecast that morning, though, provided a little bit of encouragement to a 13-year-old snow fanatic: "Tomorrow will be cloudy with rain OR SNOW likely."

I[1] knew from experience that the probability of a big snowstorm on Long Island was not very great because of its close proximity to the relatively warm Atlantic Ocean. I had already felt the crushing disappointment when potentially major snowstorms became rainstorms, and sure days off from school became days like all the rest. My pessimism was reinforced by the local weather reports on Saturday evening, indicating southeasterly winds, a bad sign for impending snowstorms. As I went to bed that night, I prepared myself for the usual disappointment, since outdoor temperatures held well above freezing and the wind continued to blow from the ocean.

When I awakened the following morning, a strong wind was whistling through the air-conditioning unit that extended outside my room. This sound was not enough to arouse me until I suddenly realized that the glow emanating from the translucent shutters covering my window was unusually bright for early morning. I shot out of bed and flung open the shutters to discover that the world outside was white. There wasn't very much accumulation on the ground yet, maybe an inch, but the snow was falling fast and flying horizontally from the northeast. A busy day lay ahead of me!

While my family still slept, I ran from window to window to see how the storm was changing the environment outside. I turned the radio on, only to become annoyed at the weather forecasters' insistences that the snow would soon end. After a while, confidence in the original forecasts eroded as the storm grew more and more intense.

By late morning, my family awoke in astonishment as they gazed out the window, wondering what to do now that they knew the day would be spent inside. The snow had picked up in intensity, falling at perhaps 1 or 2 inches (2.5 to 5 cm) per hour. Savoring every moment, I felt victorious when the weather forecasters finally succumbed to the storm and admitted that they didn't know how much more would fall or when the storm would end. Then, the inevitable comparisons with earlier blizzards began to pour forth from the radio and TV.

The day wore on and snow continued to fall heavily. The wind increased, making it difficult to determine how much of the swirling snow outside was due to falling snow versus snow blowing off the roof or from the ground. Drifting snow eventually began to cover several windows, making it more difficult to observe what was obviously a great storm. As the drifts grew, it became impossible to open the front door, and by evening, the cars parked on the street appeared as white swells in the sea of snow that covered the street. Tremendous gusts of wind whistled through the house and actually seemed to shake the foundation.

The snow finally ended in the middle of the night. About 18 hours had passed since the time the early forecasts indicated that the storm would cease. By morning, the only sound heard was that of an overworked snowplow. Radio reports now documented the extent of the storm. The entire New York City metropolitan area was paralyzed. As several hundred cars became stranded, scores of motorists spent a harrowing night on the Tappan Zee Bridge north of New York City. Many other motorists met similar fates on the myriad of roadways throughout the metropolitan area that were littered with stranded cars. Thousands more were stranded at airports that remained closed anywhere from 1 to 2 days following the storm. Food was airlifted by helicopter to Kennedy Airport to feed the weary travelers. Schools closed on Monday and many didn't reopen until Thursday, Friday, or even the following Monday. Snowclearing equipment was slow to be deployed in parts of the New York City area and couldn't keep up with the storm. Some city streets remained unplowed many days after the storm, creating a public furor. Dozens of deaths and several hundred injuries were attributed to the storm while millions of dollars were lost due to delayed or lost business.

This particular snowstorm, and other similar storms during the late 1950s and 1960s, provided the motivation for the monograph that follows. Questions concerning how these storms develop, what weather patterns provide clues that foretell such events, and what factors delineate snow/no snow situations, have challenged forecasters and researchers alike. This study is our attempt at describing a phenomenon that has stirred our curiosity, and provided countless hours of speculation, entertainment, and, for those numerous false alarms, profound disappointment. The disappointments (i.e., those storms that changed to rain or veered harmlessly out to sea) will not be highlighted here. The greatest snowstorms to affect the northeastern coast of the United States during the 1950s, 1960s, 1970s, 1980s, 1990s, and the beginning of the 21st century will be explored in the following chapters.

----

[1] Paul J. Kocin

ACKNOWLEDGMENTS

First we wish to acknowledge and thank the National Weather Service, the Office of Oceanic and Atmospheric Research, and The Weather Channel for their support throughout this effort to update and upgrade the Snowstorm Monograph.

We have benefited from the support of many individuals who have all made important contributions to the successful completion of this 8-year project. From our days at the NASA/Goddard Space Flight Center, we thank those who encouraged us and nurtured the original monograph, published in 1990, that served as a basis for this updated version: Dr. David Atlas, Dr. Joanne Simpson, Kelly Pecnick, Mary des Jardins, Lafayette Long, Bill Skillman, Dr. Franco Einaudi, Dr. Robert Atlas, and Dr. Ken Spengler.

Many meteorologists from the National Center for Environmental Prediction's Hydrometeorological Prediction Center and The Weather Channel also helped with analyses, including Dan Graf, Dan Petersen, Nicole Vanderzon, Jon Van Ausdall, Greg DeVoir, Jon Flatley, and many others. We also thank Robert Kistler from the Environmental Modeling Center for helping with the reanalysis process used to produce the maps in volume II, which was supported through the NOAA Office of Global Programs, and specifically Dr. Ming Ji.

We especially single out Keith Brill from NCEP's Hydrometeorological Prediction Center and Michael N. Baker from the Scientific Application International Corporation for attending to the reanalysis and the many steps required in production of the final figures in volume II. Volume II would not have been possible without their expert follow through. We'd also like to thank Mike Halpert and Wesley Ebisuzaki of the Climate Prediction Center for their help in the production of the DVD. We thank Rafael Ameller of StormCenter Communications, Inc., for his expertise in production of many schematics and other figures.

We thank Ray Ban, Stu Ostro, and Tawnya Carter from The Weather Channel for their continued support, and John Stremikis from the University of Wisconsin Extension for walking us through the new electronic world of manuscript editing and production that shaped both volumes. We also thank Mary Des Jardins for her development of the GEneral Meteorological PAcKage (GEMPAK), which was used extensively to create the figures for all of the case studies and related figures produced in this monograph. We also thank Brian Doty from the Center for Ocean Land Atmosphere Studies (COLA) for the use of GrADS for accessing the data from the DVD in volume II and providing easy display capabilities from that gridded dataset.

Chris Velden of Space Science and Engineering Center at the University of Wisconsin—Madison has always provided ready access to high-quality satellite data and imagery. Thomas Karl and his able staff at the National Climatic Data Center were always quick and willing to respond to our seemingly endless requests for data. The staff of the NOAA Central Library in Maryland was also an incredibly valuable resource in acquiring data for the many analyses. Photographs were provided by Jon Nese, the New York Historical Society, the Library of Congress, Kevin Ambrose, Rob Gardiner, Keith Stanley, Eric Pence, Bill Swartwout, and many others.

Thomas Karl, Lance Bosart, and Peter Ray offered reviews and many suggestions that greatly helped us refine the final version of the monograph.

Last but not least, we thank the able staff at the American Meteorological Society Headquarters, who helped shepherd this two-volume monograph through the editorial process to produce the final product, especially Gretchen Needham and Stuart Muench.

Chapter 1

INTRODUCTION

The winter of 2002/03 was a reminder that winter storms can have a powerful impact on the northeast United States. Snow fell repeatedly from Halloween past April Fools' Day, capped off by the record-setting snowfall during the Presidents' Day holiday of 16–17 February 2003. Following several winters whose mild temperatures and light or nonexistent snowfall might make one believe that severe winters might be headed for extinction, the harsh winter of 2002/03 was a throwback to many of the famous and infamous winters and winter storms that characterized the 20th century.

The 20th century has now come and gone, leaving a legacy of memorable winters, snowstorms, and blizzards in the northeast United States. For anyone growing up in this region during the late 1950s and 1960s,[1] one could count on at least one "real good" snowstorm each winter. As the 1960s ended, so did the apparent barrage of winter storms that many Northeast residents had become accustomed to. By the early 1990s, snowy winters in the northeast United States seemed to be a relic of a different climate regime. Coastal storms, with wind and blinding snow, occurred during the 1970s and 1980s, but only rarely. The winter of 1977/78 marked the last "great" winter to affect the East Coast, and an updated version of this book was more likely to appear as a Historical Monograph rather than as a Meteorological Monograph. However, nature had several surprises left up its sleeve with the approach of the 21st century.

During the early and middle 1990s, snow fell in quantities rarely, if ever, observed in this part of the country. The winters of 1992/93 and 1993/94 established new seasonal snowfall records throughout New York, Pennsylvania, and New England, many of which were later broken during the winter of 1995/96. Virtually all seasonal snowfall records in many Northeast cities, some that have stood since the 19th century, were broken. When we talk about the snows of yesteryear, storms such as the March 1993 "Superstorm" and the January 1996 "Blizzard of '96" will become the benchmarks with which storms in the 21st century will be compared, just as such legendary snows as the "Blizzard of 1888," New York City's "big snow of 1947," the "Appalachian Storm" of November 1950, the snowstorms of the 1960/

61 winter, and the New England blizzard of February 1978 have served in that capacity.

The snowstorms of the 1990s and during the winter of 2002/03 will become the memories for millions of people, especially children, who will remember these storms well into this century. Even though snow "droughts" dominated most winters since 1990, those few winters during which the snow would not stop falling will likely lead future "old timers" to talk about how "it doesn't snow like it used to back when I was a kid near the turn of the century." It is also likely that the severe winters of the early and middle 1990s and 2002/03 will inspire a new generation of future meteorologists and amateur weather observers from those children who have now been drawn to one of the wonders of the atmosphere, the great Northeast winter snowstorm. These emotions are felt by many who lived through the 6–8 January Blizzard of 1996. Following the blizzard, columnist Russell Baker (1996) captured the spirit of the moment in an essay that appeared in the *New York Times*:

What a Lovely Blizzard

"Where are the snows of yesteryear?" asked the poet. We can now tell him. The weekend blizzard, like the snows of yesteryear, not only was wonderfully copious but also, like those storms of childhood, brought adventure for almost everybody.

This was not universally welcomed. There is something alarming at first in being bullied by nature, in being told that our airplanes are suddenly unable to take us to the next stop down the road toward success and the well-ordered life.

Discovering that our powerful automobiles are suddenly useless, that a horse and buggy would be more useful just now is alarming. We who have tamed the atom and populated the once-lonely farmland with suburbs reachable only because our cars can go wherever we wish are suddenly mocked by our powerlessness.

Some people can't take that. In the Piedmont hills where I enjoyed the blizzard, people with four-wheel drive cars took their machines onto the highways despite terrifying driving conditions. Some may have desperately needed bread and milk, but many, I suspect, were simply boasting

---

[1] Both authors grew up on Long Island, New York, during the 1950s and 1960s.

to nature and neighbors of their unstoppable machine power.

Sudden destruction of the normal order had its virtues this weekend. What after all is so delightful about being strapped rigidly into the normal order?

Thomas Mann in "The Magic Mountain" observed that time slows down when one escapes the normal order and that life becomes more interesting and, because time has slowed, longer. Fred Allen defined the normal order as a "treadmill to oblivion."

This weekend there was not much normal order. You could bow to the majesty of nature and enjoy it, or hate it and sputter.

A friend of mine who had planned to drive a rental truck loaded with furniture to a new house in North Carolina this weekend had to give up after the first 30 miles. Fortunately, he was near a warm hearth where he could lodge until the weather settled down. This island of security happened to be his mother-in-law's house.

She was delighted to have him and her daughter pay an unscheduled visit, and they were cheered by the prospect of settling in for a cozy few days until the normal order resumed. The blizzard had elbowed them off a treadmill and provided a gentle adventure.

Adventures are rare these days, even the gentle kind. Everybody seems so busy making money or worrying about being fired, or "downsized," to use the corporate euphemism, that we spend most of our lives staring at the bottom line.

The bottom line—what a depressing metaphor with its suggestion of the grave, that final accounting before lawyers and heirs throw an iron wreath on Mr. Success and divvy up the bounty.

Some spoilsport record-keeper can probably produce figures proving it is silly to call this weekend's blizzard in the East one of the great snows of yesteryear, and more statistics to prove that the snows of yesteryear were not worth raving about anyhow. Where would we be without statistics to save us from life's romantic pleasures?

I distinctly remember the great snow of—well, never mind the year. I was 11 years old and mesmerized by a girl named Gladys. All I remember now is her hair. It was dark and long and thick and hung down in ringlets.

The snow of that yesteryear was the loveliest I ever saw. Never had there been better sledding. We boys sledded night after night until we knew it was past suppertime, then, soaking wet, went home to scoldings for being late. I persisted in coming late despite the scoldings, for I had found that Gladys of the glorious ringlets, who had until then been utterly unapproachable, loved sledding downhill at breakneck speed.

And ah, the beauty part: She had no sled, so had to be

content with piggyback rides with the boys. After a few test runs, she must have decided that I had the fastest sled of all. The woman was mine!

The snow melted too soon. Sledding stopped. I was moved to another city far away and never saw Gladys again. The memory of that snow has stayed with me ever since. They haven't made them like that for years, but this weekend they came close.

While heavy snowstorms are clearly responsible for major disruptions affecting many millions of individuals, a great snowstorm can become a defining event in people's lives. This book is dedicated to those who are mesmerized, for whatever the reason, by the beauty and the incredible spectacle of nature that transforms the order of everyday life into an "adventure" in a sea of snow. The aesthetic appeal of snowstorms is shared by many, who, overlooking the discomforts and hardships such storms may yield, find the transformation of familiar surroundings into spectacular landscapes of white and gray an exhilarating experience. The following excerpt of a poem by John Greenleaf Whittier captures the strong emotions brought forth by these great snow- and windstorms:

Snowbound

The sun that brief December day
Rose cheerless over hills of gray,
And, darkly circled, gave at noon
A sadder light than waning moon.
Slow tracing down the thickening sky
Its mute and ominous prophecy,
A portent seeming less than threat,
It sank from sight before it set.
A chill no coat, however stout,
Of homespun stuff could quite shut out,
A hard, dull bitterness of cold,
That checked, mid-vein, the circling race
Of life-blood in the sharpened face,
The coming of the snow-storm told.
The wind blew east; we heard the roar
Of Ocean on his wintry shore,
And felt the strong pulse throbbing there
Beat with low rhythm our inland air.

Meanwhile we did our nightly chores,
Brought in the wood from out of doors,
Littered the stalls, and from the mows
Raked down the herd's-grass for the cows:
Heard the horse whinnying for his corn;
And, sharply clashing horn on horn,
Impatient down the stanchion rows
The cattle shake their walnut bows;
While, peering from his early perch
Upon the scaffold's pile of birch,
The cock his crested helmet bent
And down his querulous challenge sent.
Unwarmed by any sunset light

The gray day darkened into night,
A night made hoary with the swarm
And whirl-dance of the blinding storm,
As zigzag, wavering to and fro,
Crossed and recrossed the winged snow;
And ere the early bedtime came
The white draft piled the window-frame,
And through the glass the clothes-line posts
Looked in like tall and sheeted ghosts.

So all night long the storm roared on:
The morning broke without a sun;
In tiny spherule traced with lines
Of Nature's geometric signs,
In starry flake, and pellicle
All day the hoary meteor fell;
And, when the second morning shone,
We looked upon a world unknown,
On nothing we could call our own.
Around the glistening wonder bent
The blue walls of the firmament,
No cloud above, no earth below,-
A universe of sky and snow!
The old familiar sights of ours
Took marvelous shapes;
Strange domes and towers
Rose up where sty or corn-crib stood,
Or garden-wall, or belt of wood;
A smooth white mound the brush-pile showed,
A fenceless drift what once was road.

1. Impact

Episodes of heavy snowfall are cause for concerns that are larger in scope than the mere discomfort and inconvenience of shoveling the driveway or walks. The nation's complex infrastructure of highways, city streets, and local roads challenge the Department of Transportation, state agencies, and municipal governments to effectively combat snowfall to allow safe transit and maintain commerce (McKelvey 1995). Most people are unaware of the significant efforts, in terms of both planning and expense, put forward by local and state agencies devoted to snow removal and highway maintenance.

There are few statistics *directly* relating cause and effect between automobile accidents, injuries, and deaths and snow- and ice storms. Statistics are kept relating automobile deaths and injuries *for* certain roadway and weather conditions that show a significant number of weather-related crashes occur with snow and ice. For example, an average of more than 85,000 crashes occurred each year (for the period 1995–2001), nationwide, resulting in at least one injury, when the pavement was reported as either snowy and slushy or icy (L. Goodwin, Mitretek Systems, Inc., unpublished manuscript). In addition, an average of more than 46,000 crashes resulting in at least one injury occurred per year (1995–

2001) when snow was reported as "falling." Fatality statistics are kept by the Fatality Analysis and Reporting System (FARS; information online at http://www-fars.nhtsa.dot.gov/) and include reports of road conditions in fatal traffic accidents, with separate observations termed snowy, slushy, and ice and show that snowy or ice-covered roads may have been factors in 10,164 fatalities between 1994 and 2001, an average of 1270 fatalities per year.

The combined effects of heavy snow, high winds, and cold temperatures can be particularly debilitating and pose a serious threat to safety in a heavily populated and highly industrialized environment. Such is the case along the northeast coast of the United States, a region that extends roughly from Virginia to Maine and includes the densely populated metropolitan centers of Washington, D.C.; Baltimore, Maryland; Philadelphia, Pennsylvania; New York, New York; Boston, Massachusetts; and dozens of smaller urban areas. This region is commonly referred to as the "Northeast urban corridor" (see Fig. 1-1) and is home to nearly 50 million people. Here, heavy snowfall associated with intense coastal cyclones, often called nor'easters, may maroon millions of people at home, at work, or in transit; severely disrupt human services and commerce; and endanger the lives of those who venture outdoors. Snowstorms have their greatest impact on transportation, being especially disruptive to automotive travel and to aviation. Snowstorms have a particularly debilitating effect on aviation by causing widespread delays, airport closings, and occasionally contributing to serious airline accidents.[2]

These snowstorms can also have a long-term impact on the nation's economy. Examples include the snowstorms of March 1993 and January 1996, which have resulted in economic losses in the billions of dollars (NCDC 2003) that rival those associated with major hurricanes. In both of these instances, state and local resources were unable to keep pace with the enormous expenses incurred during each storm, and the Federal Emergency Management Agency (FEMA) responded with numerous disaster declarations, allowing federal funds to be used in disaster relief. The Department of Commerce measured a downturn in the economy following the March 1993 Superstorm. Some studies,[3] based on economic indicators that are heavily weighted by employment statistics, have also suggested that a major snowstorm in the heavily populated Northeast significantly influences the regional and the national economies, since a major storm temporarily puts mil-

---

[2] The 14 January 1982 Air Florida crash into the 14th St. Bridge in Washington, D.C., occurred during a heavy snowstorm. Other mishaps, such as the 22 March 1992 crash at New York's John F. Kennedy International Airport and an aborted takeoff on 16 February 1996 at New York's LaGuardia Airport, are examples of snowfall's threat to aviation.

[3] See the Liscio Report, January 31, 1993, Vol. 3, No. 3; February 2, 1994, Vol. 3, No. 4; and January 9, 1996, Vol. 5, No. 1.

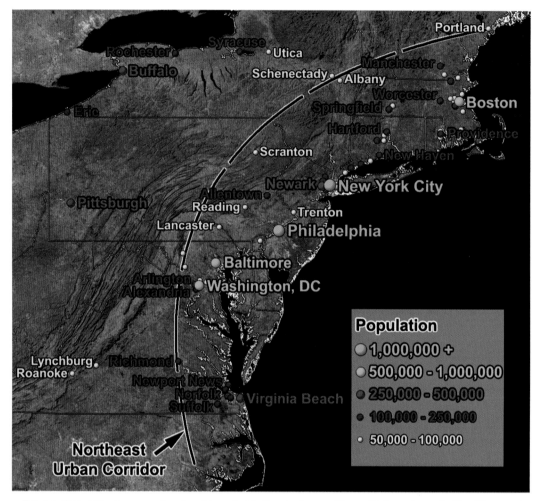

FIG. 1-1. This region is commonly referred to as the Northeast urban corridor and is home to nearly 50 million people. Here, heavy snowfall associated with intense coastal cyclones known as "nor'easters" may maroon millions of people at home, work, or in transit; severely disrupt human services and commerce; and endanger the lives of those who venture outdoors.

lions of people out of work. Retail sales and housing activity are also affected by heavy snows and severe cold. These recent reports have suggested that the nation's economic strength was significantly weakened following the major snowstorms in February 1978, March 1993, and January and February 1994. During the harsh winter of 1977/78, the economy slowed from a 9% growth rate at either end of the winter season to only 1% during the winter itself. Once severe weather conditions eased, the economy rebounded significantly.

The occurrence of snowstorms is certainly not limited to the northeast United States. Accordingly, there is an increasing literature on the climatology and factors responsible for heavy snowfall for other regions of the nation that are severely impacted by winter weather, including the Front Range of the Rocky Mountains (Dunn 1987, 1988, 1992; Howard and Tollerud 1988; Rasmussen et al. 1992; Mahoney et al. 1995; Snook and Pielke 1995; Wesley et al. 1995), the eastern and south-

eastern United States (Suckling 1991; Maglaras et al. 1995; Gurka et al. 1995; Keeter et al. 1995; Mote et al. 1997), the central United States (Younkin 1968; Marwitz and Toth 1993; Rauber et al. 1994; Gyakum 1987; Weisman 1996), the western United States (Carpenter 1993; Ferber et al. 1993), and the Great Lakes (Niziol 1987; Niziol et al. 1995; Hjelmfelt 1990, 1992; Byrd et al. 1991). However, there is little question that the impact of heavy snowstorms in the Northeast urban corridor is huge, since, as will later be shown, the numbers of people affected by these storms can be staggering.

2. The challenge of forecasting Northeast snowstorms

Northeast snowstorms result from a complex interaction of physical processes that acts to organize a widespread region of ascent, entrain large amounts of water vapor necessary to produce heavy amounts of precipitation, and supply a source of cold air that enables precipitation to reach the ground as snow. These physical mechanisms are associated with cyclonic and anticyclonic weather systems that traverse the middle latitudes and interact with the unique topography of the eastern United States and adjacent waters. Topographical factors that contribute to these weather systems include land–ocean temperature and frictional contrasts; the influences of the Atlantic Ocean, Gulf of Mexico, and Great Lakes in supplying moisture and on airmass modification; the position of the Gulf Stream; and the effects of the Appalachian Mountains on the low-level temperature and wind fields. This combination of factors favors the northeastern United States and its offshore waters as important sites for cyclogenesis (Petterssen 1956; Klein and Winston 1958; Reitan 1974; Colucci 1976; Sanders and Gyakum 1980; Hayden 1981; Roebber 1984) that can occasionally result in heavy snowfall in the Northeast urban corridor.

The ability of weather forecasters to recognize and account for the many elements that influence the development and evolution of major snowstorms is crucial for the accurate prediction of heavy snowfall in the northeastern United States. Throughout much of the 20th century, attempts to forecast such storms have yielded mixed results, at best. Prior to 1950, reliance on hourly surface weather observations and an increasing amount of upper-air data derived from synoptic-scale rawinsonde balloon releases were the only means for forecasting, which could be useful for cases involving preexisting storms that followed easily predictable paths. However, suddenly developing storm systems were often not predicted because the significant changes in the prestorm environment could only be inferred from observing the "synoptic scale" evolution of upper-level features and approximating their interaction with lower-tropospheric processes.

As examples, the severity of two of the most debilitating snowstorms in New York City's history, the March 1888 blizzard (Kocin 1983) and the city's record 26.4-in. (66 cm) snow of December 1947, was virtually unforecast, as described in later chapters. Even the advent of upper-air observing systems, satellite data, the advances made in computer technology, and the development of numerical weather prediction models during the latter half of the 20th century provided no guarantee of accurate forecasts for many important storms, including the "Lindsay Storm" of February 1969, and the February 1979 Presidents' Day storm (Bosart 1981; Uccellini et al. 1984, 1985, 1987; Atlas 1987). The public's lack of confidence in forecasters' abilities to predict such storms was, to some degree, sustained by continued forecast deficiencies for major storm systems. These factors contributed to a public perception that these weather systems were either unpredictable or that forecasters did not know what they were doing, or both.

In the 1970s and 1980s, operational and research numerical weather forecasts started having more success in many heavy snow situations, as demonstrated by simulations of the intense New England snowstorm of February 1978 (Brown and Olson 1978), the spring blizzard of April 1982 (Kaplan et al. 1982), the "Megalopolitan" snowstorm of February 1983 (Sanders and Bosart 1985), and a convective snow event in March 1984 (Kocin et al. 1985). The failure of the operational forecast models to adequately predict the February 1979 Presidents' Day storm served as a catalyst for research efforts that led to improvements in the physical representation of boundary layer and diabatic effects in numerical model predictions. In more recent years, the quality of forecasts of major cyclone events has improved markedly in the 24–48-h range, with useful prediction of storm systems also made out to 3–5 days in advance (Uccellini et al. 1999). These advances are described in more detail in chapter 8 and have culminated in dramatic improvements in the forecasts of the timing, location, and intensity of cyclogenesis since the late 1980s. As a result, "surprise" snowstorms that affect large areas are becoming a relatively rare event. Nevertheless, the modern, model-based forecast process is not infallible, as demonstrated by the failure of many numerical prediction models to predict a major East Coast snowstorm on 24–25 January 2000 (Buizza and Chessa 2002; Zhang et al. 2002; Bosart 2003).

Even with several examples of successful predictions noted above, even good forecasts of major snowstorms were not completely accurate. Forecasts that can be described as successes, such as the operational numerical model forecast of the 11–12 February 1983 Megalopolitan snowstorm, the March 1993 Superstorm, and the January 1996 Blizzard of '96, contained flaws that were significant. In the 1983 and 1996 storms, a major snowfall was predicted by forecasters several days in advance for the middle Atlantic states, clearly a success. However, it was not clear in the early forecasts of each storm whether the northward extent of excessive snowfall amounts would reach New York City and southern New England, a failure for both cases. In the case of the March 1993 Superstorm, the prediction of heavy snow and a major cyclone up to 5 days in advance was a remarkable success. However, the rapid development of the cyclone over the Gulf of Mexico was underforecast, while the East Coast development was overforecast, leading to a significant underestimation of the storm's strength over the Gulf of Mexico (Uccellini et al. 1995). This forecast error minimized the expected impact of the storm along the Gulf coast, especially in the state of Florida, where damage was severe and dozens of lives were lost. More recently, a snowstorm in December 2000 was accurately predicted for New York City and much of New Jersey up to 5 days in advance, with heavy snowfall amounts forecast 36 h before the snow began falling. This success was tempered, however, by the forecast of significant snowfall that never materialized over Washington, D.C.; Baltimore; and portions of central Pennsylvania, which were also warned of heavy snow. Therefore, even relatively exceptional model forecasts still may contain seemingly small errors that can have a profound impact on the accuracy of weather forecasts that are released to the general population.

3. Basis for a snowstorm monograph

In view of the continuing need to improve our understanding of these storm systems, this two-volume monograph provides an analysis of the horizontal and vertical structures and evolution of many of the most crippling snowstorms to affect the heavily populated Northeast region during the 20th century. These snowstorms are examined from historical, climatological, synoptic, and dynamical perspectives.

It is not the intent of this monograph to introduce detailed dynamical interpretations of each cyclone event, as can be found in specialized diagnostic and modeling studies of individual cases. The Genesis of Atlantic Lows Experiment (GALE; Dirks et al. 1988), the Canadian Atlantic Storms Program (CASP; Stewart et al. 1987), and the Experiment on Rapidly Intensifying Cyclones over the Atlantic (ERICA; Hadlock and Kreitzberg 1988) have served as excellent sources of specialized studies of cyclogenesis that address the relative importance of the various processes that influence cyclone development along the east coast of the United States.

Brandes and Spar (1971) attempted to define the necessary conditions for heavy snowfall along the East Coast using a composite approach. They found that compositing failed to resolve antecedent patterns 12–24 h prior to heavy snowfalls. Apparently, the synoptic- and mesoscale features in the geopotential height and wind fields that are important factors in the initial stages of many snowstorms were overly smoothed by combining many storm systems characterized by significant case-to-case variability. This eliminates important signals that could be used to specify those conditions that lead to Northeast snowstorms. Therefore, the approach used in this monograph is based on sequences of conventional weather charts and detailed descriptions of individual storms. This method enables us to identify patterns in the surface, lower-tropospheric, and upper-tropospheric fields that precede and accompany the development of heavy snowfall, to develop schematic diagrams that capture the essential processes that contribute to these storms, and to document the case-to-case variability that frustrates efforts to specify required conditions or standard scenarios that are applicable to all storms.

The interactions of the varied mechanisms and processes that produce these storms span the spectrum of time and length scales from the nucleation of cloud particles to planetary wave dynamics. The temporal and spatial resolution of the operational weather analyses from which this study is derived limits us to describing only those features that operate over periods of several hours to several days and lengths of 10^2–10^4 km [the meso-, synoptic, and planetary scales; see Orlanski (1975)]. Details on the physical principles in snow formation can be found in Houze (1993) and other texts on cloud physics, and will not be addressed here.

The monograph is designed to provide a foundation for researchers, students, and weather observers interested in investigating the processes that interact to produce major winter storms and to serve as an easily referenced guide for forecasters concerned with predicting heavy snowfall along the northeastern coast of the United States. In the following pages, material from the original version of this monograph (Kocin and Uccellini 1990) and review articles on rapid cyclogenesis and the evolution of the forecast process (Uccellini 1990; Uccellini et al. 1999) have been retained and updated.

But there is much that is new. The monograph is composed of two volumes. In volume I, the chapter on climatology has been expanded to cover seasonal snowfall and relationships with climatic variables such as the El Niño–Southern Oscillation (ENSO) and the North Atlantic Oscillation (NAO). New chapters have also been added, one to address the mesoscale character of snowfall distributions, a continuing source of interest and frustration to forecasters unable to diagnose and predict phenomena that are still poorly understood. An additional chapter distinguishes differences between major snowstorms and storms that involve a change from snow to rain within the Northeast urban corridor, as well as storms marked by moderate snowfall, and ice storms. Also included are discussions of the scientific and technological issues that contribute to the increased understanding and general ability to predict the snowfall that accompanies winter storms. New cases that have occurred since the publication of the original monograph have also been included in volume II to provide a synoptic climatology for snowstorms that have occurred over a 50-year period. A Northeast Snowfall Impact Scale (NESIS) has also been devised to provide a comparative assessment of individual storms that can be communicated to the public.

In volume I, a climatological review of seasonal snowfall during the 20th century, an overview of moderate to heavy snow events during the latter half of the 20th century, and their relationships to climatic variability are provided in chapter 2. A description of the meteorological characteristics of 30 major snowstorms during the latter half of the 20th century follows in chapters 3 and 4, derived from conventional analyses of sea level pressure, geopotential height, wind, temperature, and other parameters. Chapter 5 focuses on cases characterized as "near misses"—those storms that start as snow but either change to rain, ice pellets and freezing rain—or "moderate" snowstorms, whose snowfall distribution is not as widespread nor heavy as in those cases summarized in chapters 3 and 4. Mesoscale aspects of the snowfall distribution associated with heavy snow events, both in terms of the detail apparent in major snow events, and of the isolated heavy snow event, are described in chapter 6. Chapter 7 provides an overview of the physical processes that contribute to these storms. Chapter 8 summarizes the revolutionary changes that have occurred during the past 100 years in forecasting winter storms and concludes with discussion of the Northeast Snowfall Impact Scale. A compilation of seasonal snow statistics for more than 30 sites has been included in the appendix.

In volume II, detailed descriptions of several historical cases, dating as far back as the 18th century through 1950, are provided in chapter 9. A comprehensive series of weather map analyses for each of the 30 storms that occurred between 1950 and 2000 and summarized in chapters 3 and 4 of volume I is presented in chapter 10, as well as an analysis of the February 2003 Presidents' Day Snowstorm II and another snowstorm in December 2003. Brief case histories and map analyses of numerous interior and moderate snowstorms and ice storms summarized in chapter 5 of volume I are also provided in chapter 11. The digital data on standard pressure levels and related maps used in volume II are also provided on a DVD that is included with volume II. A final chapter concludes with analyses of early and late season snow events.

Chapter 2

CLIMATOLOGY

Northeast snowstorms are best understood and appreciated in the context of the climatological range of conditions experienced in the northeast United States. In this chapter, the climatology of seasonal snowfall and significant snowstorms in the northeast United States is described. The distribution of seasonal snowfall during the 20th century is also discussed, with emphasis placed on the periodic and episodic nature of heavy snowfall associated with major snowstorms. The possible relationships between climate variability related to the Southern Oscillation [SO; or El Niño–Southern Oscillation (ENSO)], the North Atlantic Oscillation (NAO), and global warming are also discussed.

The distribution of moderate (10 cm or 4 in. and greater) and heavy snow (25 cm or 10 in. and greater) events within the Northeast urban corridor is examined during the latter half of the 20th century, with a focus on its locations, frequency, and months of occurrence. An examination of these events provides a basis for the selection of the most significant storms that have impacted the Northeast urban corridor. A discussion of snowfall measurement issues that influence the interpretation of all snowfall amounts presented throughout the monograph follows.

1. Snowfall measurement issues

Observations and summaries of snowfall presented in this chapter, as well as throughout this monograph, are affected by the difficulties inherent in measuring snowfall. The issues associated with snowfall measurement are condensed from Doesken and Judson (1996), who discuss the many difficulties in taking accurate and consistent snow measurements. This discussion is presented to help analyze some of the difficulties in assessing snowfall and to keep the snowfall analyses presented throughout this monograph in some perspective.

The measurement of snow is far from an exact procedure, being influenced by where, how often, and on what surfaces the measurements are made. Snowfall is defined in Doesken and Judson (1996) as the depth of new snow that has fallen and accumulated in a given period, and is measured daily at a specified time of observation using a ruler and a snowboard,[1] in a location protected from the wind and representative of the area (i.e., not in a drift).

National Weather Service policy[2] has directed that snowfall measurements can be taken up to four times daily at 6-hourly intervals in order to differentiate between *snowfall* and *snow cover*. If a snowboard is not available, wooden decks and grassy surfaces are usually good alternatives although the observer must be careful to subtract the airspace between the grass and the lowest layer of snow. The measurement is read to the nearest 0.1 in. and is recorded unless blowing, drifting, melting, or settling has occurred. When these conditions occur, the observer must use judgment to determine snowfall and pay heed and consider the following precautions:

a) Measure snowfall where the effects of blowing and drifting are minimized. In situations where wind-blown snow cannot be avoided, an average of several representative measurements should be employed.

b) Snow often melts as it lands on the ground. If no accumulation is ever noted even on grassy surfaces, snowfall should be measured as a "trace." If snow partially melts, and some accumulation takes place, the daily snowfall should be recorded as the greatest accumulation of new snow observed at any time during the day. If snow accumulates, melts, and accumulates again, the snowfall is the sum of each new accumulation.

c) Snow, especially deep snow with low water content, also settles and compacts as it lies on the ground. According to Doesken and Judson (1996), the daily snowfall should be the maximum accumulation of new snow observed at any time during the day. If one uses frequent snow observations of a cleared surface to measure snowfall, say several times per day, or hourly, the total snowfall can exceed the daily snowfall, sometimes significantly during long-duration events.

An excellent example of how differing measurement techniques can affect a final snow measurement occurred during the 6–8 January 1996 "Blizzard of '96." This long-duration snowfall event affected the Northeast urban corridor, beginning as all snow, followed by a changeover in some locations to ice pel-

---

[1] A snowboard is usually a solid, painted piece of wood, approximately 2 ft on a side, that is laid on the ground.

[2] Snow measurement guidelines for National Weather Service Cooperative Observers are available from the National Weather Service, Office of Climate, Weather and Water Services, May 1997.

lets, and then back to snow. During this event, some measurements differed by a factor of 2 to 1 [some sites measured 15 in. (37 cm) while nearby sites reported 30 in. (75 cm)]. If an observer waited to take a measurement until the snow ended, some of the snow had probably settled due to the storm's long duration and high winds. Therefore, this measurement would likely underestimate the total snowfall, while accurately measuring the snow depth. Alternatively, some observers measured the snowfall at differing intervals, clearing off their boards to begin a new measurement each time, and measuring a final snow depth. These observers reported significant discrepancies between the summed measurements and the final snow depths. An observer taking hourly measurements would have a snowfall tally much greater than snow depth at the end of the event. Which reported amount represents an accurate measure of snowfall?

In most cases of relatively short-term snowfall events, frequent measurements may closely approximate the final snow depth if settling and melting are minimized. However, for the greater and longer-lasting events, which are perhaps the most noteworthy cases, the weight of the snow can lead to compaction, especially if a changeover to sleet or rain occurs sometime during the event. The complexity of issues surrounding snowfall measurement suggests that a standard, such as the 6-hourly measurements instituted by the National Weather Service, be applied uniformly so a consistent dataset is obtained.

d) It was a recommendation of Doesken and Judson (1996) that if snow changes to another form of precipitation, snow measurements be taken at time of changeover, when snow depth is at a maximum. If sleet accumulates afterward, it should be measured as new "snow" and if it changes back to snow, new measurements should begin at the time of changeover.

e) Summer hail should *not* be reported as snowfall to help distinguish early and late season snow and ice pellet events from thunderstorm-generated hail, when possible (hail occurs throughout the year and during the cold months is difficult to distinguish from other forms of frozen precipitation).

Snowfall observations used throughout this monograph were obtained from records of first-order and co-operative observer networks available through the National Climatic Data Center (NCDC). According to Doesken and Judson (1996), both sources of data are plagued by inconsistent measurement techniques and intervals. Therefore, exact amounts should be accepted with a certain degree of caution. In general, analyses of snowfall for most storms exhibit surprisingly consistent snowfall patterns, lending to confidence that in spite of possible irregularities, the measurements still have considerable value.

TABLE 2-1. (a) 10 snowiest and (b) 10 least snowy seasons of the 20th century for an ensemble of 18 sites depicted in Fig. 2-2.

(a) 10 snowiest		(b) 10 least snowy	
Season	Avg (in.)	Season	Avg (in.)
1995/96	80.3	1994/95	14.8
1966/67	58.5	1918/19	14.9
1977/78	56.2	1912/13	14.9
1947/48	55.3	1991/92	16.4
1960/61	53.8	1988/89	16.6
1963/64	51.8	1936/37	18.2
1906/07	51.8	1990/91	18.6
1993/94	51.6	1931/32	19.2
1915/16	50.6	1997/98	19.3
1933/34	48.8	1972/73	19.6

2. Seasonal snowfall

The mean distribution of seasonal snowfall over the northeastern United States is largely dependent upon latitude, with snow totals ranging from as little as 6 in. (15 cm) in the southeastern corner of Virginia to greater than 100 in. (250 cm) across sections of central and northern New England, New York, and West Virginia (Fig. 2-1). The impact of elevation on temperature affects the distribution of snowfall, which is demonstrated by the snowfall maxima over more mountainous inland terrain. The increase in snowfall at higher latitudes is skewed from southwest to northeast by the Atlantic Ocean's moderating influence on temperature patterns along and near the coastline.

Within the heavily populated urban corridor adjacent to the Atlantic Ocean, mean seasonal snowfall ranges from less than 10 in. (25 cm) in the Tidewater region of Virginia; to 10–20 in. (25–50 cm) across Virginia through central and eastern Maryland, Delaware, and southern New Jersey; 20–30 in. (50–75 cm) throughout the area between Washington, D.C., and New York City; and 30–40 in. (75–100 cm) in southern New England. A summary of seasonal snowfall statistics for numerous locations within the northeast United States is provided in the appendix.

a. Seasonal snowfall in the Northeast urban corridor

To provide a representative overview of seasonal snowfall in the Northeast urban corridor, 18 sites (Fig. 2-1) are selected to represent the variations in seasonal snowfall that occurred during the 20th century. These sites are selected because all have maintained snowfall records dating back at least to the early 20th century. The mean seasonal snowfall patterns for these 18 sites show distinct maxima and minima during the 20th century (Fig. 2-2, top), with the mean snowfall for the 18 sites approaching 34 in. (85 cm).

For the combined 18 sites, the 10 snowiest and least snowy seasons of the 20th century are provided in Table 2-1. The snowiest seasons appear to occur primarily during the second half of the 20th century, with the

RELIEF MAP OF THE NORTHEAST
UNITED STATES

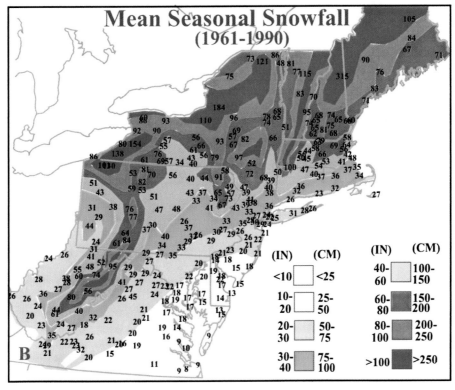

FIG. 2-1. (top) Relief map of the northeast United States. (bottom) Mean seasonal snowfall (Oct–May 1960/61–1989/90) for the northeastern United States. Values are site averages in inches and contour intervals are given in inches and centimeters.

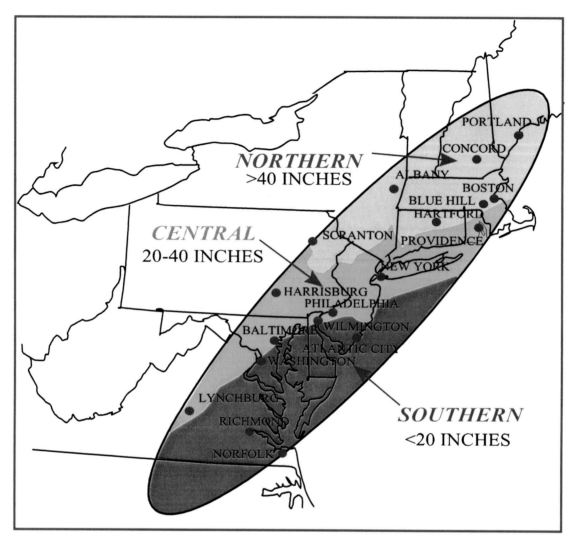

FIG. 2-2. Eighteen sites used to compute mean seasonal snowfall for the Northeast urban corridor. Northern, central, and southern regions are used to demarcate differences in seasonal snowfall. Northern region represents mean snowfall greater than 40 in. (100 cm), central region represents mean snowfall between 20 and 40 in. (50 and 100 cm), and southern region represents mean snowfall less than 20 in. (50 cm).

winter of 1995/96 greatly exceeding all other seasons with a mean of 80 in. (200 cm). The greatest mean seasonal snowfalls (greater than 50 in. or 125 cm) occurred during the winters of 1906/07, 1915/16, 1947/48, 1960/61, 1963/64, 1966/67, 1977/78, and 1993/94. The least mean seasonal snowfalls with means of less than 20 in. (50 cm) occurred in 1912/13, 1918/19, 1936/37, 1972/73, 1988/89, 1990/91, 1991/92, 1994/95, and 1997/98. The least snowy season occurred in 1994/95, with an average of just under 15 in. (37 cm). *One remarkable result from this analysis is that the snowiest and least snowy seasons of the 20th century occurred in consecutive years!* The winter of 1995/96 had the most, while the winter of 1994/95 had the least. A pattern of *increasing interannual variability* occurred throughout the 1990s and into the early 21st century,

including two snowy winters (2000/2001 and 2002/2003) surrounded by several very snow-deficient winters (1997/1998, 1998/1999, 1999/2000, and 2001/2002). Whether this increasing variability is related to the potential effects of global warming (i.e., Houghton et al. 2001) is beyond the scope of this study.

A 5-yr running mean was applied to this record to help distinguish any dominant trends and is shown in Fig. 2-3 (bottom). In this figure, the dominant snowy period of the 20th century occurred during the late 1950s through the 1960s. Other snowy periods included the late 1900s, the middle 1910s through the early 1920s, the mid-1930s, the late 1940s, the late 1970s, and the mid-1990s. The dominant snowfall minima occurred in the late 1920s and early 1930s, the early 1950s, and the late 1980s and early 1990s.

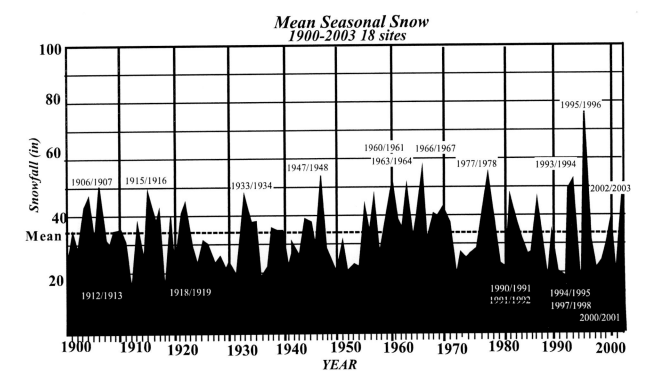

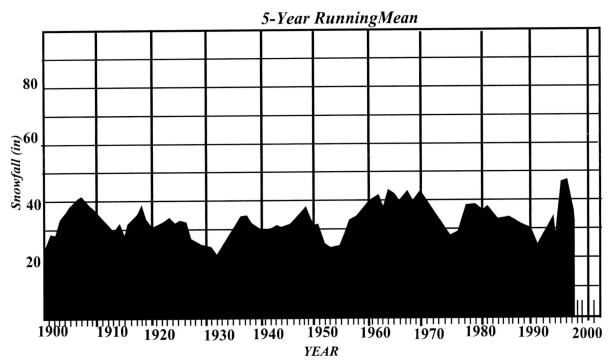

FIG. 2-3. (top) Mean seasonal snowfall for an ensemble of 18 sites during the 20th century. (bottom) The 5-yr running mean seasonal snowfall for the same sites as in Fig. 2-2.

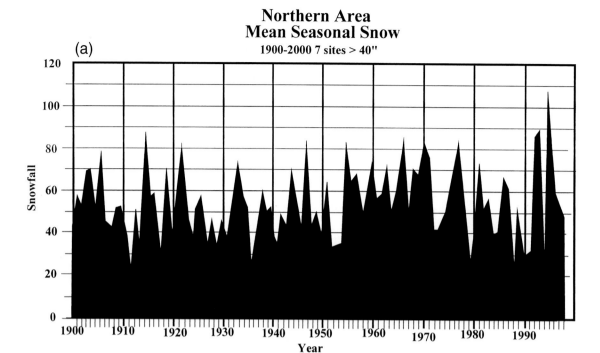

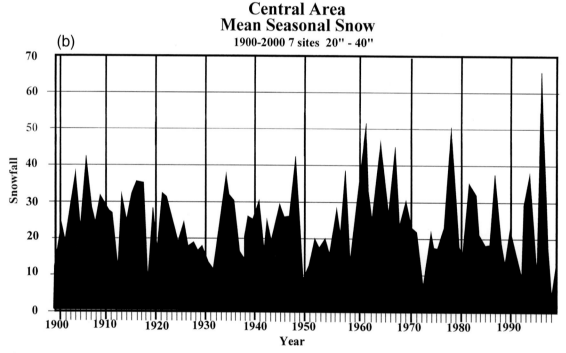

Fig. 2-4. (a) Mean seasonal snowfall for the northern region composed of seven sites depicted in Fig. 2-2, averaging >40 in. (>100 cm) per season, during the 20th century. (b) Mean seasonal snowfall for the central region composed of seven sites depicted in Fig. 2-2, averaging 20–40 in. (50–100 cm) per season, during the 20th century.

1) REGIONAL VARIABILITY

The 18 sites are subdivided into three areas (Fig. 2-2) to help distinguish intraregional variations in seasonal snowfall. The northern area is represented by seven sites

that average greater than 40 in. (100 cm) seasonally and include central New England, interior New York, and northeastern Pennsylvania (Fig. 2-4a). The central domain is represented by seven sites that average between 20 and 40 in. (50 and 100 cm) and includes southern

Southern Area
Mean Seasonal Snow
1900-2000 4 sites < 20"

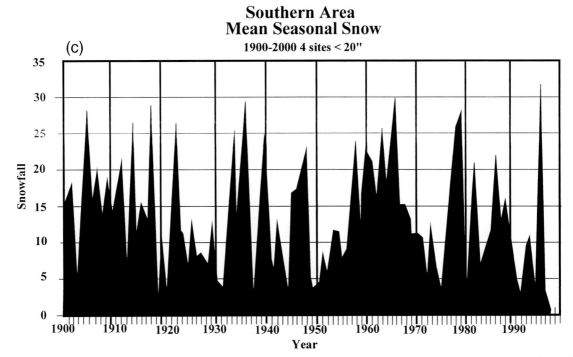

FIG. 2-4. (*Continued*) (c) Mean seasonal snowfall for the southern region composed of four sites depicted in Fig. 2-2, averaging >40 in. (<50 cm) per season, during the 20th century.

New England, eastern Pennsylvania, southeastern New York, northern New Jersey, northern Virginia, and northern Maryland (Fig. 2-4b). The southern domain is represented by four sites that average less than 20 in. (50 cm; Fig. 2-4c) and includes eastern Virginia, southeastern Maryland, southern Delaware, and southern New Jersey.

The 100-yr mean snowfall for the three regions are as follows. In the northern domain, the mean snowfall is 54.2 in. (137 cm), while the middle region's mean is 24.7 in. (63 cm) and the southern domain's mean is 13.6 in. (35 cm). Distinct maxima and minima characterize all three records, but the middle and southern domains present a more "spiky" appearance, due to the influence of fewer snowfall events overall, which can dominate the snowfall record.

The variations in all three regions show some similar patterns to each other, such as the tendency toward snowy winters in the middle 1930s, the late 1940s, the late 1950s through the 1960s, and the century maximum during 1995/96, as well as some of the snow droughts. However, some intraregional differences complicate an examination of 18 sites across a domain that extends from Virginia to Maine. Note that the interannual variability increases from north to south since warmer temperatures in the southern region can preclude much snowfall in some seasons from those in which colder temperatures may permit more snowfall. For example, many snow events across New England might only be liquid events over the middle and southern domains. The winters of 1970/71 and 1993/94 are two seasons

that saw excessive snowfall across New England but only rain or ice across portions of the middle and southern regions. In another example, a relatively snowy winter across the southern domain may be due to a few snow events whose influence did not extend into the more northern domains. Such was the case during the winters of 1978/79 and 1979/80, when the primary storm track was suppressed far enough south that Virginia and Maryland received heavy snows while the more northern domains experienced relatively modest amounts.

2) CONTRIBUTION OF MODERATE AND HEAVY SNOW EVENTS TO SEASONAL SNOWFALL TOTALS

In this section, the focus shifts from a review of seasonal snowfall to an assessment of the snowfall events that contribute to snowy winters. Beginning primarily in October–December and ending in March–May for most areas in the urban Northeast, each winter season is composed of a highly variable number of snowfall events, which can be "moderate" (defined here as more than 4 in. or 10 cm) or "heavy" (defined here as more than 10 in. or 25 cm).

To present a more general climatology of the coastal region between Virginia and Maine, 21 regions (Fig. 2-5) are selected to catalog the monthly and seasonal distributions of moderate and heavy snow events for the winter seasons of 1949/50–1998/99.[3]

---

[3] These regions are based on subdivisions found in *Climatological Data,* the monthly publication used to obtain the snowfall reports.

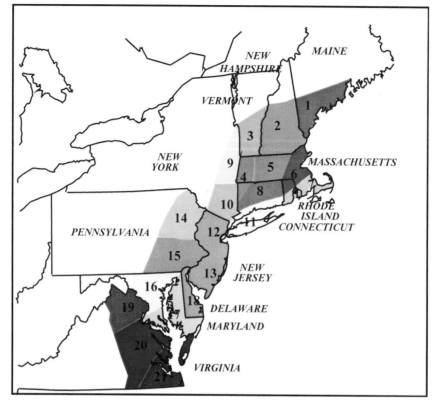

FIG. 2-5. Locations of 21 regions selected to catalog the monthly and seasonal distributions of heavy snow occurrences described in chapter 2.

TABLE 2-2. Total number of snowfall events exceeding 10 and 25 cm (by region) for the winter seasons 1949/50–1998/99, including the seasonal average of 10-cm events and the average time interval in years between 25-cm events [in parentheses are similar values for the period 1954/55–1984/85; see Kocin and Uccellini (1990)].

Region No.	Location	Total No. of events > 10 cm	Seasonal avg	Total No. of events > 25 cm	Avg time interval (yr) between 25-cm events
1	Coastal ME	312 (203)	6.2 (6.8)	51 (34)	1.0 (0.8)
2	S NE	338 (216)	6.7 (7.2)	59 (41)	0.8 (0.7)
3	S VT	334 (220)	6.6 (7.3)	63 (41)	0.8 (0.7)
4	W MA, NW CT	296 (194)	5.9 (6.5)	60 (38)	0.8 (0.8)
5	Central MA	228 (148)	4.5 (4.9)	57 (38)	0.9 (0.8)
6	E MA, N RI	197 (125)	3.9 (4.2)	47 (31)	1.1 (1.0)
7	SE MA, S RI	122 (79)	2.4 (2.6)	30 (18)	1.7 (1.7)
8	Central CT	159 (103)	3.1 (3.4)	26 (19)	1.9 (1.6)
9	E NY	246 (160)	4.9 (5.3)	52 (33)	1.0 (0.9)
10	SE NY	178 (109)	3.5 (3.6)	33 (24)	1.5 (1.3)
11	NYC, Long Island, and S CT	116 (75)	2.3 (2.5)	18 (15)	2.8 (2.0)
12	N NJ	126 (83)	2.5 (2.8)	24 (18)	2.1 (1.7)
13	S NJ	71 (51)	1.4 (1.7)	15 (11)	3.3 (2.7)
14	NE PA	160 (100)	3.2 (3.3)	35 (23)	1.4 (1.3)
15	SE PA	106 (72)	2.1 (2.4)	23 (16)	2.2 (1.9)
16	Central MD	99 (67)	2.0 (2.2)	16 (11)	3.1 (2.7)
17	SE MD	71 (49)	1.4 (1.6)	10 (6)	5.0 (5.0)
18	DE	74 (50)	1.4 (1.7)	13 (7)	3.8 (4.3)
19	N VA	98 (63)	1.9 (2.1)	21 (17)	2.3 (2.0)
20	East-central VA	78 (53)	1.6 (1.8)	11 (7)	4.5 (4.3)
21	SE VA	39 (29)	0.8 (1.0)	4 (3)	12.5 (10.0)

50-year TOTAL NUMBER OF 10-CM EVENTS BY MONTH
(1949/50-1998/99)

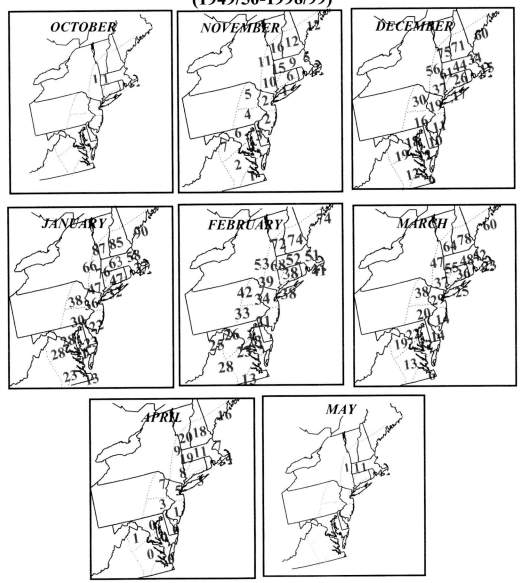

FIG. 2-6 The 50-yr total number of snowfall events exceeding 10 cm (by region and month) for the winter seasons between 1949/50 and 1998/99.

An occurrence of moderate or heavy snowfall is noted for each region in Fig. 2-5 whenever it was determined that at least half the area received snow amounts in excess of the threshold values. The following two sections define the temporal and spatial distributions of moderate to heavy snow events within the Northeast region.

(i) Moderate snow events

During a 50-season period from 1949/50 through 1998/99, the number of storms that have produced snowfall accumulations exceeding 4 in. (10 cm) (Fig. 2-6; Table 2-2) ranges from 39 in southeastern Virginia, an average of less than one event per season, to more than 300 in sections of central New England, an average of over 6 events per season. Within the most densely populated regions of the Northeast urban corridor from northern Virginia to southern New England,[4] approximately 70–120 events have occurred, averaging from 1.5 to 2.5 occurrences per season. Farther north and

---

[4] An area comprising regions 7, 11, 12 13, 15, 16, 17, and 19.

inland,[5] the number of moderate and heavy snow events increases from approximately 150 to 290 across northeastern Pennsylvania, eastern New York, Connecticut, and Massachusetts, averaging 3–6 per season. Over coastal Maine, southern Vermont, and New Hampshire (regions 1–3), the number of events exceeds 310, averaging about 6 or 7 events per season.

Storms capable of producing snowfall in excess of 4 in. (10 cm) typically occur during December, January, February, and March (Fig. 2-6). In October, only 2 of the 21 regions shown in Fig. 2-5 have experienced a general snowfall exceeding 10 cm during the second half of the 20th century.[6] Accumulations of 10 cm or greater have also been observed in isolated areas of the more heavily populated Northeast urban corridor as early as 10 October (1979) (see volume II, chapter 12). In November, episodes of general snowfall exceeding 10 cm occur sporadically, with a frequency as great as once every 3–4 yr over inland sections of southern and central New England. Significant early and late season snowstorms are discussed in chapter 12 of volume II.

By December, all regions can experience snow events exceeding 10 cm. The far northern sections of the domain receive moderate events with nearly equal likelihood in any of the months from December through March although there is a slightly greater tendency in January. Across extreme southern New England and much of the coastal region of the middle Atlantic states, the frequency of moderate to heavy snowfalls reaches a peak in January and February. From Virginia to southern New England, the frequency of moderate to heavy snow events diminishes in March, although there is a slightly greater tendency in the northern domain for snowstorms to occur in December than in March. Conversely, there is a slightly greater tendency over the middle domain for snowstorms to occur in March than in December. In all areas, the incidence of winter-type storms declines quickly after March.[7] Other less widespread occurrences have occurred in hilly terrain or in northern New England as late as late May, and even into June. Only the highest peaks have seen snow in July and August.

In summary, there is a general increase in the number of events characterized by moderate snowfalls as one moves northeastward along the urban corridor with a significant increase in the rate of change as one moves from southeastern New York into New England (Fig. 2-7, top). In general, the mean number of 10-cm events

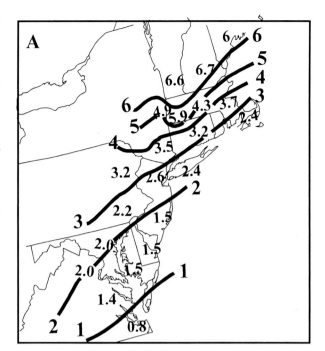

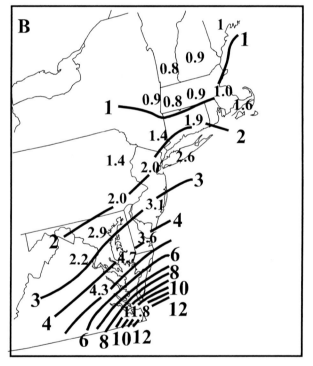

FIG. 2-7. (top) Mean number of greater than 4-in. (10 cm) snowfall events per season for each of 21 regions. (bottom) Mean number of years between 10-in. (25 cm) snowfall events for each of 21 regions.

---

[5] An area comprising regions 4, 5, 6, 8, 9, 10, and 14.

[6] These two regions, one in eastern New York and the other in western Massachusetts, experienced a uniform moderate snowfall on 3–4 October 1987 (see volume II, chapter 12), which is especially unusual since it not only occurred in October but also during the first few days of the month (see Bosart and Sanders 1991).

[7] An unusually heavy snowfall was observed over portions of interior New York State and southern New England on 9 May 1977, one of the latest such occurrences in the sample (see volume II, chapter 12).

increases from an average of fewer than one event per season over the Tidewater region of Virginia, to 1.5–2.5 events from the Washington to New York corridor, to about 3–4 events in the Boston area, and to 6–7 events in interior New York and New England.

50-year TOTAL NUMBER OF 25-CM EVENTS BY MONTH
(1949/50–1998/99)

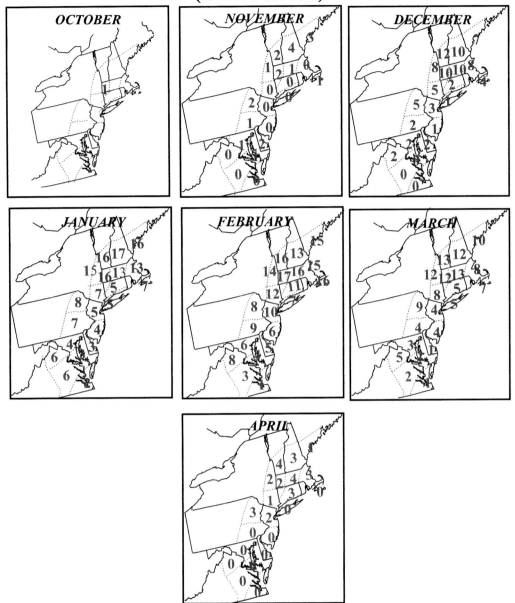

FIG. 2-8. The 50-yr total number of snowfall events exceeding 25 cm (by region and month) for the winter seasons between 1949/50 and 1998/99.

(ii) Heavy snow events

In contrast to the seasonal occurrence of moderate snow events, storms that deposit 10 in. (25 cm) or more are relatively rare events (Fig. 2-7, bottom). Only 4 such storms have occurred over the 50-yr period in southeastern Virginia, an average of only 1 every 12 years, with 10–16 events over the coastal regions from Virginia to southern New Jersey (Table 2-2), approximately 1 every 3–6 yr. The heavily populated urban corridor from northern Virginia through extreme southern New

England experienced 16–30 major snowfalls, approximately 1 every 2–3 yr. These are relatively infrequent events that appear to maintain a uniform distribution along a southwest-to-northeast axis from Washington, D.C., to New York City and Long Island. The inland areas of northeastern Pennsylvania, eastern New York, and southern New England received 33–57 heavy snows, approximately 1 every 1–2 yr, while 51–63 events were noted across northern New England, from near to slightly greater than 1 every year, occurring with

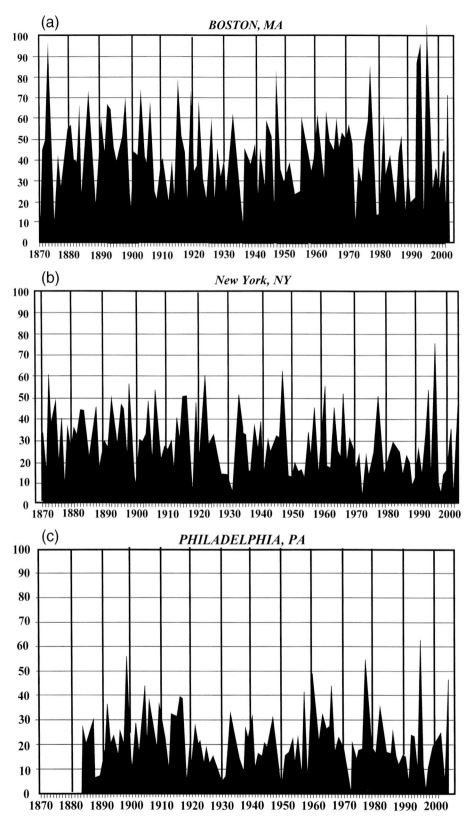

FIG. 2-9. Seasonal snowfall for (a) 1871–1999, for Boston; (b) 1869–1999, for New York City; and (c) 1884–1999, for Philadelphia.

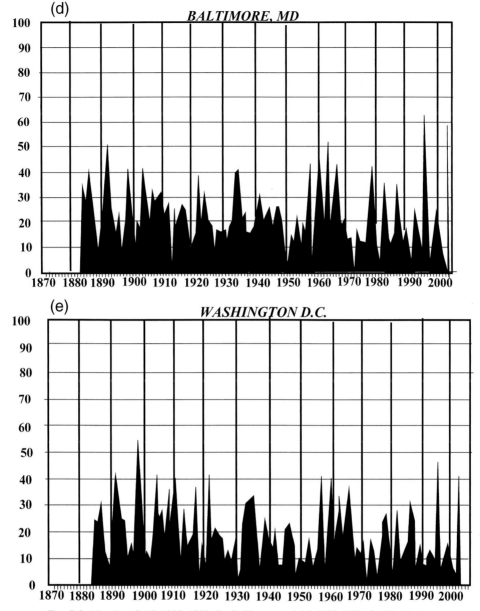

FIG. 2-9. (*Continued*) (d) 1885–1999, for Baltimore; and (e) 1885–1999, for Washington.

a regularity that is no surprise to residents of the interior Northeast and New Englanders.

These major snowstorms typically occur from December through March (Fig. 2-8), with very few events in November or April. For much of southern New England and the middle Atlantic states, the frequency of storms with snowfall exceeding 10 in. (25 cm) shows a significant maximum during February. This contrasts with the nearly equal likelihood of 4-in. (10 cm) amounts in January and February (Fig. 2-10). Thus, the numbers shown in Fig. 2-8 support the notion that February is the "big month" for the "big snow" along the Northeast coast, at least during the 50-yr period ending with the winter season of 1998/99.

3) SNOWFALL FOR BOSTON, NEW YORK, PHILADELPHIA, BALTIMORE, AND WASHINGTON, D.C

In the five largest cities of the Northeast urban corridor, snowfall records have been kept for more than a century (Figures 2-9a–e; also see chapter 9). To help assess any trends in seasonal snowfall and illustrate the year-to-year variability that characterizes seasonal snowfall, 30-yr (1971–2000) means are compared to means computed over several other intervals (Table 2-3) during the history of record for each city.

In Boston, Massachusetts, seasonal snowfall has varied from a minimum of 9 in. (23 cm; 1936/37) to a

TABLE 2-3. Seasonal snowfall means (in.) for selected time intervals for Boston, New York, Philadelphia, Baltimore, and Washington.

Time intervals	Boston	New York	Phila-delphia	Baltimore	Washing-ton D.C.
1870–1999	42.5	28.6	21.7	21.2	18.2
1900–99	41.3	27.0	21.9	20.5	19.1
1870–99	47.8	33.6	22.8	25.3	24.7
1900–49	40.6	29.2	21.8	22.2	19.5
1950–99	42.0	24.7	21.7	18.6	16.9
1971–2000	41.6	22.0	20.0	18.0	14.7
Max	107.6	75.6	63.1	62.5	54.9
	(1995/96)	(1995/96)	(1995/96)	(1995/96)	(1898/99)
Min	9.0	2.8	T	1.2	0.1
	(1936/37)	(1972/73)	(1972/73)	(1972/73)	(1972/73, 1997/98)

TABLE 2-4. Mean seasonal snowfall (in.) by decade. Minima and maxima are italicized.

Decade intervals	Boston	New York	Phila-delphia	Baltimore	Washing-ton D.C.
1870–79	45.4	32.3	—	—	—
1880–89	45.0	32.7	21.8	26.6	22.7
1890–99	*52.6*	*35.8*	23.8	24.6	*25.7*
1900–09	39.8	28.9	25.3	25.0	23.5
1910–19	39.1	30.6	*27.6*	21.2	20.9
1920–29	43.2	29.5	17.5	19.6	17.2
1930–39	37.0	24.4	17.5	22.8	18.5
1940–49	43.9	32.8	21.1	23.0	17.5
1950–59	37.1	20.2	*15.2*	*14.7*	*12.8*
1960–69	49.4	32.0	29.1	*32.2*	24.8
1970–79	44.6	22.5	21.7	17.8	13.6
1980–89	*32.8*	*19.9*	20.4	18.5	17.1
1990–99	49.7	24.4	18.3	17.7	12.9

maximum of 107.6 in. (273 cm) during the record winter of 1995/96. The 30-yr mean (1971–2000) of 41.6 in. (106 cm) is comparable to 50-yr averages computed both for the first and for the last half of the 20th century but is 6 in. (13 cm) less than averages computed during the last 30 yr of the 19th century (Table 2-3).

In New York, New York, seasonal snowfall has varied from 2.8 in. (7 cm; 1972/73) to 75.6 in. (192 cm; 1995/96). For the 130-yr period from 1870 to the end of the 20th century, significant changes are observed in the snowfall record. The most recent 30-yr average shows a mean of 22.0 in. (56 cm), which is less than the 50-yr mean for the last half of the 20th century and is 7 in. (18 cm) less than the 50-yr average over the first half of the 20th century and greater than 11 in. (28 cm) less than averages computed during the last 30 yr of the 19th century. Therefore, seasonal snowfall appears to have decreased significantly in New York City within the last 50 yr.

In Philadelphia, Pennsylvania, mean seasonal snowfall appears to have remained relatively constant since records were started in the winter of 1884/85 with an average of 20–23 in. (50–58 cm), including a 30-yr (1971–2000) mean of 20.0 in. (51 cm), showing a decline during this period.

In Baltimore, Maryland, the 1971–2000 average is 18.0 in. (46 cm), less than the first half of the 20th century (22.2 in.; 57 cm), but significantly less than the mean during the late 19th century (25.3 in.; 64 cm).

The snowfall records for Washington, D.C., appear to show some of the same tendencies as of those from New York City with a sharp decrease in seasonal snowfall during the last half of the 20th century. In Washington, the 1971–2000 average is 14.7 in. (37 cm), while the means during the late 19th and first half of the 20th century are 24.7 and 19.5 in. (63 and 50 cm), respectively.

In Boston, New York, Philadelphia, Baltimore, and Washington, an examination of mean seasonal snowfall during each decade of the 20th century (Table 2-4, bottom) shows that, in general, seasonal snowfall during

the late 19th century was greater than current 30-yr means by as much as 50%. In New York City, Baltimore, and Washington, the first half of the 20th century has experienced greater average seasonal snowfall than the second half, while Philadelphia and Boston's means have remained nearly the same or dropped in the late 20th century.

(i) Prior to 1950

As discussed above, the period from the late 1880s through the middle 1920s experienced greater than "normal" snowfall compared to the 30-yr means constructed during the late 20th century. For example, more than 50% to nearly 75% of the winters in the five cities experienced above "current normal" snowfall (Boston, 21 winters; New York, 30; Philadelphia, 27; Baltimore, 29; and Washington, 29) during 40 winters between 1884/85 and 1923/24 (see chapter 9).[8]

Following this relatively long and extended snowy period (which still had its share of relatively snow-free winters), much of the late 1920s and 1930s was characterized by subnormal snowfall. In particular, nearly all winters between 1923/24 and 1931/32 exhibited below normal snowfall, especially in Philadelphia, Baltimore, and Washington. Even though this period was relatively snow free, a few significant snowstorms did occur. The late 1920s and the early 1930s exhibit some of the lowest seasonal totals for any 5-yr period this century with New York City averaging less than 12 in. (30 cm) and Philadelphia only averaging a little over 9 in. (23 cm). Midway through the 1930s, seasonal snowfall increased throughout much of the region with the winters of 1933/34 and 1934/35 exhibiting above nor-

---

[8] During this snowy period, some great historic storms occurred, most notably the March 1888 "Blizzard of '88," the cold wave and blizzard of February 1899, and the January 1922 "Knickerbocker Storm" (see volume II, chapter 9).

mal amounts[9]. The late 1930s through the 1940s experienced subnormal snowfall, especially from New Jersey southward. However, the late 1940s saw a return to above normal snowfall in New York and Boston, highlighted by New York City's "big snow" of December 1947 (volume II, chapter 9).

(ii) 1950–2000

During the second half of the 20th century, subnormal snowfall was common across the region early in the period. The winters of 1949/50–1954/55 saw a snow drought across the entire region. During this period, seasonal means were only 30 in. (75 cm) in Boston, 14 in. (35 cm) in New York, 12 in. (30 cm) in Philadelphia, 11 in. (28 cm) in Baltimore, and 9 in. (23 cm) in Washington. In Philadelphia, Baltimore, and Washington, the decade of the 1950s was the least snowy this century (Table 2-4). The 1950s first snowy regime did not occur until March 1956, when a succession of late season snowstorms occurred. The most memorable winter of the 1950s occurred during 1957/58 when many snowstorms occurred throughout the winter and made it one of the most severe during the late 20th century for much of the region.

The late 1950s through the 1960s were characterized by especially severe winters, possibly the longest spate of severe winters since the 1890s, which were marked by several memorable snowstorms. The decade of the 1960s was the snowiest of the century for all cities except New York City. In particular, the winters of 1959/60, 1960/61, 1963/64, 1965/66, 1966/67, and 1968/69 were all major snow-producing seasons. Over a 10-season period between 1957/58 and 1966/67, Boston averaged 51 in. (130 cm), New York 33 in. (84 cm), Philadelphia 30 in. (76 cm), Baltimore 33 in. (84 cm), and Washington 27 in. (69 cm). These figures are 25% to nearly double current 30-yr means. Many of the major snowstorms described in the following two chapters and in volume II occurred during this period. Following the winter of 1968/69, New England continued to experience significant snowfall through the winter of 1971/72. However, much of the rest of the Northeast urban corridor experienced a snowfall drought during the early and middle 1970s with few major storms occurring until the winter of 1977/78. During this lull, the winter of 1972/73 was remarkably free of snow in the major metropolitan areas, with seasonal totals ranging from a trace at Philadelphia to 10 in. (25 cm) at Boston. New England continued to experience some major snow events but New York, Philadelphia, Baltimore, and Washington each only averaged between 10 and 17 in. (25 and 43 cm) per season between 1970/71 and 1976/77. This snow drought ended with a return to cold conditions

during the winter of 1976/77, but much of the snow fell mainly in northern domains. By the late 1970s, cold and heavy snows returned to much of the region, especially during the winter of 1977/78, which rivaled any during the late 1950s or 1960s. January and February 1978 saw several major snowstorms and the 1970s ended with the stormy winter of 1978/79.

The winters of the 1980s into the early 1990s were generally mild with few snowstorms. This period is one of the least snowy and most protracted of the 20th century. For example, seasonal snowfall averaged only 31 in. (78 cm) in Boston, 18 in. (45 cm) in New York, 16 in. (40 cm) in Philadelphia, 15 in. (38 cm) in Baltimore, and 14 in. (35 cm) in Washington from 1983/84 through 1991/1992, approximately 75%–90% of normal. There were, however, some notable exceptions during this long period of mild, snow-depressed winters. The early half of the 1980s saw some of the coldest winter outbreaks of the last half of the 20th century, especially in January 1982, December 1983, and January 1985, but major snowstorms did not occur in conjunction with these outbreaks. One particularly crippling snowstorm occurred in February 1983 and a series of snowstorms affected the Northeast throughout the winter of 1986/87. Despite these active seasons, the 1980s were the least snowy decade of the century in New York City and Boston (Table 2-4). Following this period, few snowstorms of any consequence occurred from the late 1980s to the early 1990s.

The snow drought of the 1980s and early 1990s ended abruptly with the winter of 1992/93, when a series of major snowstorms began to affect much of the region from Pennsylvania and New York into New England. This winter was highlighted by a major interior snowstorm in December 1992 and the March 1993 "Superstorm." The 12–14 March 1993 Superstorm produced the most widespread distribution of heavy snowfall over the eastern United States this century (Kocin et al. 1995). During the following winter of 1993/94, record snowfall again occurred across many portions of the Northeast with numerous snowstorms from Pennsylvania and New York into New England and many freezing rain and ice pellet events in the middle Atlantic states.

Following the quiet winter of 1994/95, the winter of 1995/96 will probably be remembered as the snowiest of the 20th century for much of the region between Virginia and Massachusetts. New seasonal snowfall records were set in Boston, New York, Philadelphia, Baltimore, and many other locations with snowfall exceeding 2–4 times the normal seasonal totals. In some cases, new seasonal records far exceeded the old standards. For example, Boston reported 107.6 in. (273 cm), breaking the modern record of 96.3 in. (245 cm) set in 1993/94 and the historical record of 96.4 in. (245 cm) in 1873/74. Providence, Rhode Island, measured 106.1 in. (269 cm); the previous record was 75.6 in. (192 cm) during 1947/48. Hartford, Connecticut, measured 115.2

---

[9] This period was marked by some of the coldest winter weather of the 20th century, in December 1933 and February 1934, and some significant snowstorms, particularly during February 1934.

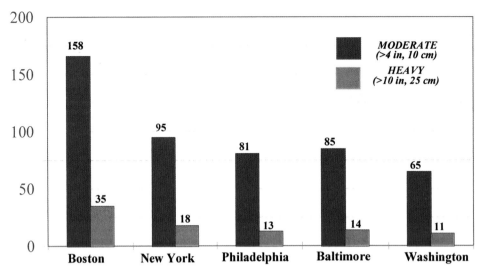

FIG. 2-10. Total number of moderate (>4 in.; 10 cm) and heavy (> 10 in.; 25 cm) snowfall events (by city) for the winter seasons 1949/50–1998/99 in Boston, New York City, Philadelphia, Baltimore, and Washington. Data were obtained from the following locations: Boston (Logan International Airport), New York City (Central Park), Baltimore (Baltimore–Washington International Airport), and Washington (National Airport).

in. (293 cm); the previous record was 84.9 in. in 1993/94. New York City's Central Park recorded 75.6 in. (192 cm), while the previous record was 63.2 in. in 1947/48; Newark, New Jersey measured 78.4 in. (199 cm), the most seasonal snow there since record keeping began in 1843.

Following the record-setting winter of 1995/96, seasonal snowfall again reverted to significantly subnormal conditions to finish out the century (see tables in the appendix), highlighted by the winter of 1997/98, one of the least snowiest this century (Table 2-1). As the 21st century began, extreme interannual variability continued. For example, Baltimore experienced its third least snowy winter (since 1883/84) with 2.3 in. (6 cm) in 2001/02 while experiencing its second snowiest winter the following season with 58.1 in. (148 cm) in 2002/03. In general, low seasonal snowfall totals occurred in 1998/99, 1999/2000 (except for a 2-week period in January 2000), and 2001/02 (one of the five least snowy winters in Boston, New York, Philadelphia, Baltimore, and Washington). High seasonal snowfall totals occurred in the northern half of the Northeast in the winter of 2000/01 and across the entire domain during the winter of 2002/03.

4) CONTRIBUTION OF MODERATE AND HEAVY SNOW EVENTS TO SEASONAL SNOWFALL

An analysis of the snowfall records shows that a significant portion of the seasonal snowfall record in the five largest cities in the Northeast is composed of moderate [greater than 4 in. (10 cm)] and heavy snowfall events [greater than 10 in. (25 cm)] (Fig. 2-10). The monthly distributions of moderate and heavy snow events from 1950 through the turn of the century are also summarized in Fig. 2-11. The monthly and 50-yr totals are presented for Boston, New York, Philadelphia, Baltimore, and Washington, to provide additional background on the distribution and frequency of moderate and heavy snow events in this region.

During the 50-yr sampling period, the total number of moderate events that yielded 4-in. (10 cm) amounts or greater ranged from 65 at Washington to 95 in New York City and 158 in Boston. Based on these numbers, the cities of Boston, New York, Philadelphia, Baltimore, and Washington have averaged about 3, 2, 1.6, 1.6, and 1.2 10-cm snowfalls, respectively, per season during the last half of the 20th century. January and February account for the majority of these occurrences with slightly greater numbers of events in February in Washington, Baltimore, Philadelphia, and New York. The months of December, January, February, and March account for nearly all of the events for each city.

Over the last half of the 20th century, the total number of events that produced 10-in. (25 cm) or greater accumulations ranged from 11 at Washington to 14 at Baltimore, 13 at Philadelphia, 18 at New York City's Central Park, and 35 at Boston.[10] February displays the

---

[10] With additional storms from 2000 to 2003, the numbers stand at 13 at Washington, 16 at Baltimore, 14 at Philadelphia, 21 at New York, and 38 at Boston.

 # Snowfall Events > 4 in (10 cm)

A

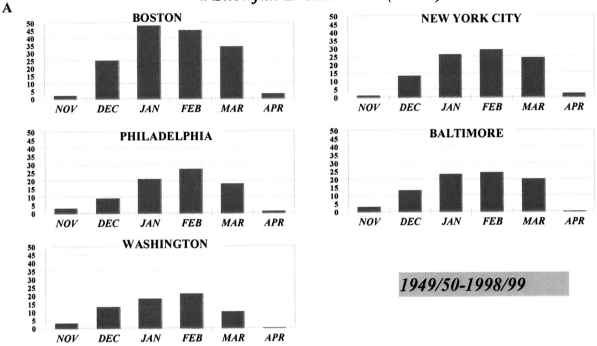

 # Snowfall Events >10 in (25 cm)

B

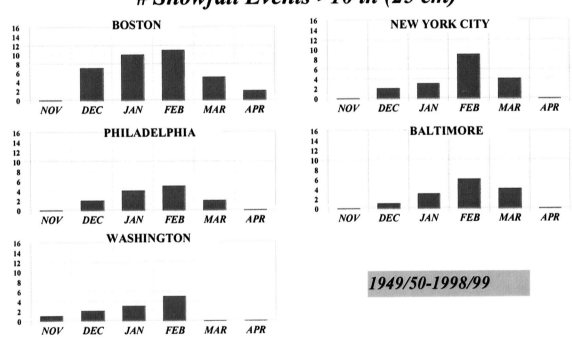

FIG. 2-11. (top) Monthly distribution of snowfall events greater than 4 in. (10 cm) (by city) for the winter seasons 1949/50–1998/99 in Boston, New York City, Philadelphia, Baltimore, and Washington. (bottom) Monthly distribution of snowfall events greater than 10 in. (25 cm) (by city) for the winter seasons 1949/50–1998/99 in Boston, New York City, Philadelphia, Baltimore, and Washington.

greatest frequency of heavy snow events in every city. The frequency of 10- and 25-cm snow accumulations exhibits only small variations between Washington, Baltimore, Philadelphia, and New York. Washington recorded marginally fewer events and New York slightly more than the other cities. In contrast, Boston had a 50%–100% greater incidence of 10-cm snows and two to three times the number of 25-cm events than the other cities. Such a substantial difference indicates that Boston has a significantly greater potential for heavy snowstorms than any of the other major cities of the Northeast urban corridor.

All snowfall events exceeding 10 in. (25 cm) for each of the major cities are listed in the appendix. Boston has recorded 72 events over a 130-yr period, an average of approximately of one storm every 1.5–2 seasons, followed by New York, with 53 events at Central Park over 133 yr, an average of about one storm every 2.5 seasons. Philadelphia, Baltimore, and Washington all report similar numbers of storms with 27, 32, and 27 events over 117, 118, and 118 yr, respectively, or approximately one storm every 4–5 seasons.

These large storms, however, are not distributed evenly. For example, Boston averaged one 10-in. (25 cm) storm every 1.5–2 yr. Yet, during the 10-yr period between February 1983 and March 1993, only one 10-in. storm occurred. In contrast, four such storms occurred in Boston during the winter of 1995/96. Meanwhile, Boston had recorded no 20-in. (50 cm) events in nearly 100 yr prior to 1969 but experienced five events in the past 34 yr. Two of those storms occurred during one 2-week period in early 1978.

There are other examples of snowstorm "clusters." In chapter 9, volume II, several of the historical examples, including the "great snow" of 1717 and the periods 28 December 1779–7 January 1780 and 4–10 December 1786, were instances of major snowstorms occurring in clusters. More recently, during the winter of 1957/58, significant snowstorms occurred throughout the winter. During the winter of 1960/61, three great snowstorms occurred across much of the urban Northeast, one on 11–12 December, the "Kennedy Inaugural Storm" of 19–20 January, and the final storm on 2–4 February 1961.

Sometimes, the storms occur in quick succession. During January 1966, three storms occurred in succession within a period of 1 week at the end of the month culminating in the "Blizzard of '66" that crippled a large area from Virginia to north-central New York (see chapter 10, volume II). In November 1968, three successive storms missed the urban corridor but dumped copious snow on interior sections of the Northeast. In January 1978, three storms in succession began with a significant ice storm on 13–14 January, an interior snowstorm on 16–17 January, and a major snowstorm on 19–21 January. In Washington, 10 yr could easily pass between consecutive 10-in. snowstorms (see the appendix). However, during the calendar year 1987, four

storms produced 10 in. or greater within the metropolitan Washington area.

Therefore, major snowstorms in the Northeast corridor may *average* one every 2–4 or 5 yr, but the historical record shows that these storms tend to occur in clusters. Many years may pass without an occurrence of a major storm. Yet, during some years, the storms seem to occur frequently, occasionally two or more in one season. These clusters of heavy snow events lend supporting evidence to the episodic nature of major Northeast snowstorms, as is discussed in section 2d.

5) PERIODIC CHARACTER OF SEASONAL SNOWFALL VARIABILITY

The highly variable nature of the seasonal snowfall in the Northeast United States' five major metropolitan areas described above begs a question. Do the snowfall records exhibit any periodic behavior? To answer that question, a spectrum analysis is applied to the seasonal snowfall data to separate the variance of the data in a time series into contributions arising from oscillations occurring over different periods. Information about the periodicity of the data is obtained when autocorrelation values are calculated for various time intervals, or lag times. For a given lag time, starting at an arbitrary year, if all of the data values scattered through the time series separated by that interval tend to be either all near normal (the mean), all above normal, or all below normal, then the autocorrelation values for that interval will be high.

This calculation is carried out for many different intervals ranging from 0 to 40 yr These intervals were chosen for series of seasonal snowfall records that ranged generally between 80 and 130 yr. Only the intervals of 20 yr or less are shown for Boston, New York, Philadelphia, Baltimore, and Washington in Fig. 2-12. Each lag is applied starting from the first point in the time series continuing all the way to the latest point. To simplify the interpretation of results, the autocorrelation values are divided by the total variance of the time series so that a normalized spectrum is generated. Therefore, the height of each bar in a histogram depicting the spectrum is proportional to the percentage of the variance contributed by that particular period of oscillation.

Computing the smoothed spectrum in this way effectively eliminates the occurrence of peaks that are not representative of the behavior of the time series of which the data constitute a sample. This is because peaks (or valleys) in an unsmoothed spectrum are not significant unless surrounded by high (or low) values. Thus, the percentage is implied not by the height of the histogram column at the peak in the smoothed spectrum, but rather by the sum of it and the neighboring columns that rise to form the peak.

While the spectral analyses for the major metropolitan areas are far from conclusive since the lengths of the seasonal snowfall records are fairly short (only about

SPECTRAL ANALYSES OF SEASONAL SNOWFALL

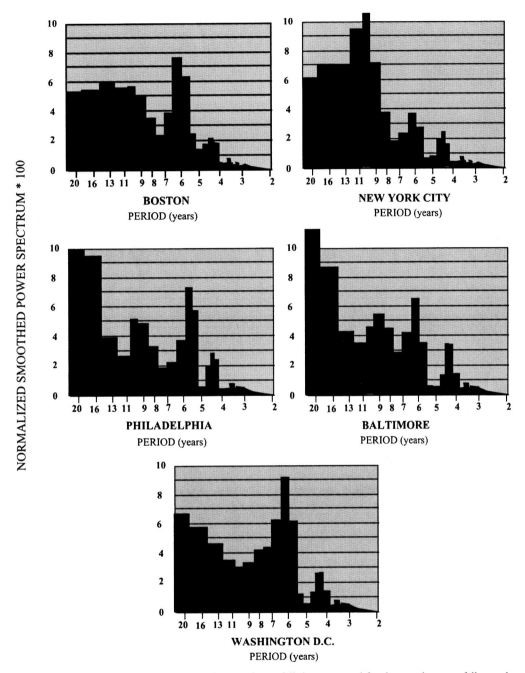

FIG. 2-12. Normalized spectral analyses of seasonal snowfall data computed for the complete snowfall records at Boston, New York, Philadelphia, Baltimore, and Washington. Autocorrelation values are shown only for intervals, or lag times, of 20 yr or less. Peaks in the spectrum represent percentages of the total variance over the time intervals they cover.

100–130 yr), and yield relatively weak signatures supporting a periodic behavior, the time series analyses of the five cities support the idea that seasonal snowfall appears to have nonrandom fluctuations on time scales of 5–7 and 8–12 yr. The 5–7-yr signal is most clearly apparent in all but New York City's time series (Fig. 2-12). Using Boston as an example, the 5–7-yr period, constituting less than 2% of the 127-yr record, contained approximately 20% of the variance in the data. This nonrandom behavior indicates that there may be climatic factors related to the evolution of the large-scale circulation regimes operating on these time scales that may influence these values (Livezey and Smith 1999). These controls will be examined in the next section.

b. Possible contributing factors to seasonal snowfall variations

Spectral analyses of time series of seasonal snowfall in the previous section revealed fluctuations in the seasonal snowfall records on the order of 5–7 and 8–12 yr. These nonrandom fluctuations indicate that "circulation anomalies" (Livezey and Smith 1999) related to the Southern Oscillation (SO or El Niño or ENSO) and the North Atlantic Oscillation (NAO), and may play a role in contributing toward seasonal snowfall variations.

1) THE SOUTHERN OSCILLATION

The Southern Oscillation (SO; Walker and Bliss 1932; Trenberth 1976, 1997) is a well-defined climatic perturbation that has potential effects on the weather pattern of the North American continent, a pattern that could contribute to snowstorms. The SO refers to sea level pressure anomalies that develop between the Indian Ocean–western Tropical Pacific and the east-central tropical Pacific (i.e., Bjerknes 1969; Halpert and Bell 1997; Bell and Halpert 1998). The sea level pressure anomalies are related to the abnormal warming and cooling of the equatorial Pacific waters, termed El Niño for warming and La Niña for cooling episodes. During El Niño episodes, an increase in horizontal pressure variations across the Pacific affects atmospheric circulation features such as the jet stream over the Pacific Ocean. During a warm El Niño episode, the jet stream over the Pacific is stronger than normal; while in a cool La Niña event, the jet stream is weaker than normal, affecting the subsequent development and movement of extratropical cyclones that influence the United States. The El Niño/La Niña represents a strongly coupled ocean–atmosphere interaction and is the dominant mode of climate variability in the global Tropics (Halpert and Bell 1997).

In a typical El Niño episode, the enhanced jet stream over the eastern Pacific tends to bring relatively warm air into the western and northern portions of North America while increased storminess is also noted across the southern United States. In a typical La Niña episode, storminess is inhibited over the North Pacific allowing cold air to develop over northwestern Canada that often spills southward into the northwestern and north-central United States. This circulation anomaly favors upper-level ridging, warming, and drying conditions over the southeastern United States (see Ropelewski and Halpert, 1986, 1987, 1989, 1996; Halpert and Ropelewski 1992). Shapiro et al. (2000) establish a link between El Niño and La Niña with the intensity and location of the North Pacific jet stream. They then relate the subsequent

downstream trough–ridge systems over the United States with a tendency toward negatively tilted troughs during El Niño and positively tilted troughs during La Niña. Therefore, the El Niño/La Niña signal may have a powerful effect on the tracks and intensities of storms that affect the United States.

An SO index has been kept since 1882 with no breaks in the record since 1932. One measure of the SO is made utilizing the mean sea surface temperature anomalies over the central Pacific (5°N–5°S, 120°–170°W; see Trenberth 1997), and is shown for the last half of the 20th century in Fig. 2-13. Significant El Niño episodes have occurred many times over the last half of the 20th century. Many of these episodes have been associated with increased storminess and above normal winter precipitation across the southern and eastern United States.

Some of these El Niño episodes have been associated with increased seasonal snowfall in the Northeast urban corridor as well and include the winters of 1957/58, 1963/64, 1965/66, 1968/69, 1977/78, and 1986/87. These associations would be quite compelling if not for some significant associations between strong El Niño episodes and significant periods of very low seasonal snowfall. Two pronounced El Niño seasons are especially prominent, the winters of 1972/73 and 1997/98, both of which were associated with some of the lowest seasonal snowfalls of the century (Table 2-1). It is interesting to note that during the nearly snow free winter of 1972/73, there was an active storm track over the eastern United States that resulted in very heavy precipitation in November and December, mostly in the form of rain. In January and February 1973, the storm track remained primarily over the southern United States and was associated with above normal snowfall in the southeastern United States, including one of the great snowstorms of all time in Georgia and South Carolina on 9–10 February 1973. During the winter of 1997/98, which experienced one of the strongest El Niño events of the 20th century, many storms battered the eastern United States, but snow only fell north of the Northeast urban corridor or in mountainous terrain. Most of these storms were associated with relatively mild air. Finally, another strong El Niño event in 1982/83 was associated with many storms in the eastern United States, resulting in a particularly snowy winter season for New England (see the seasonal snowfall peak in Fig. 2-4a), but produced primarily rain in the Northeast urban corridor. However, one storm termed the "Megalopolitan Storm" (Sanders and Bosart 1985a; Bosart and Sanders 1986; also see chapter 10.11.20, volume II) was one of the few events to occur while cold air affected the Northeast and produced record snowfall from northern Virginia to New England.

This suggests that the increased storminess attributed to El Niño may, on occasion, be associated with enhanced snowfall observed in the Northeast urban corridor, and is consistent with a composite study relating ENSO with regional snowfall distributions (Smith and

SST Anomalies NINO 3.4

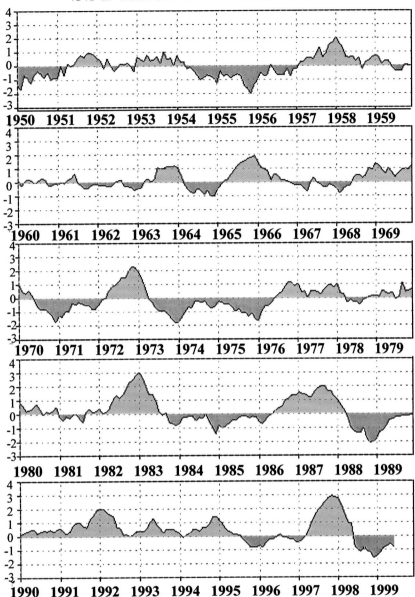

FIG. 2-13. A measure of the ENSO (or SO) sea surface temperature anomalies over the central Pacific (5°N–5°S, 120°–170°W) for the period 1950–99, derived from the NCEP/CPC.

O'Brien 2001). However, there may also be little if any snow during strong El Niño events, which are so warm that the precipitation falls mainly as rain in the Northeast urban corridor. Thus, El Niño episodes may be related with both excessive seasonal snowfall and minimal seasonal snowfall. With this in mind, it is important to recognize that strong El Niño conditions are not a prerequisite for snowy winters in the northeast United States. This is especially apparent since the two snowiest winters of the 20th century, 1995/96 and 1966/67, were not associated with El Niño conditions.

The strong La Niña episodes depicted in Fig. 2-13, such as 1954–56, 1964/65, 1971/72, 1974–76, 1984/85, 1988/89, and 1998/99 are mostly associated with relatively low seasonal snow totals. Given the tendency for upper-level ridging, warming, and drying over the southeastern United States during La Niña winters, it is not surprising to observe mostly seasonal snowfall minima (cf. with Fig. 2-3).

2) THE NORTH ATLANTIC OSCILLATION

According to Livezey and Smith (1999), three climatic "signals" strongly influence North American cli-

NAO vs Seasonal Snowfall Departures

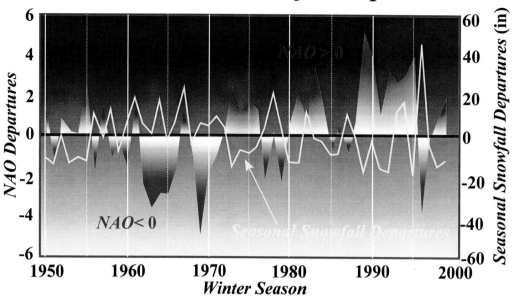

FIG. 2-14. Time series of the winter index of the NAO obtained by subtracting the mean pressure departures in Iceland from those over the Azores. Year on the *x* axis denotes the end of the winter season. Seasonal snowfall departures (thin line) from a 50-yr mean 24(1949/50–1998/99) for 18 selected sites in the Northeast urban corridor (see Fig. 2-2).

mate. One of these signals is the interannual variability of ENSO. A second signal that operates on interdecadal time scales (Hurrell 1995) is the North Atlantic Oscillation (NAO; see Walker and Bliss 1932; van Loon and Rogers 1978; Wallace and Gutzler 1981; Barnston and Livezey 1987). The NAO is a measure of pressure anomalies across the Atlantic Ocean, measured between the Azores and Iceland, that represent circulation changes over the Atlantic and surrounding land areas. The NAO is the dominant mode of variability in the surface circulation over the Atlantic (Hurrell 1995). The positive phase of the NAO is accompanied by lower pressures over the Arctic and higher pressures over much of the Atlantic, resulting in stronger than normal westerlies over the Atlantic into Europe with southerly flow common across the eastern United States, a pattern that would allow relatively mild conditions to occur (Hurrell 1995). The negative mode of the NAO is accompanied by an anomalously weak subtropical high and Icelandic low, with the Icelandic low displaced to near Newfoundland or eastern Canada with blocking patterns common aloft and polar anticyclones common over the eastern Atlantic and Europe. Winters are comparatively mild in Greenland and colder in the northeastern United States.

An analysis of the NAO during winter (December–March) shows that the positive and negative phases can persist for many winters. A time series of the "winter index" of the NAO, derived from analyses done at the National Centers for Environmental Prediction/Climate Prediction Center (NCEP/CPC), is shown in Fig. 2-14.

According to Hurrell (1995), the NAO index exhibited a downward trend from the 1940s through the 1970s, a period when European winters were warmer than usual, on average. From the 1970s through the 1990s, this trend reversed with strong positive values since 1980, with the winters of 1982/83, 1988/89, and 1989/90 marked by the highest values of the NAO index since 1864. In addition, according to Hurrell, the decadal variability in the NAO has become especially pronounced since 1950.

The seasonal snowfall for the 18 sites in the Northeast urban corridor discussed in section 2a and shown in Fig. 2-2 can also be described as variations about a mean value for the last half of the 20th century, exhibiting patterns of maxima and minima. These patterns are compared with the NAO shown in Fig. 2-14. What is intriguing about the NAO is that the seasonal snow patterns appear to correlate well with the index, where negative values of the NAO correlate with positive values of seasonal snowfall anomalies, and vice versa. For example, the late 1950s and 1960s show the greatest dip in the index, corresponding to the snowiest decade of the century for many Northeast locations. Meanwhile the 1970s show an upward trend, with a peak in the early 1970s, corresponding to one of least snowy winters on record, and then a minimum again in the late 1970s, corresponding to a period of snowy winters. Note the maximum in the early 1980s, a relative minimum in the mid-1980s, and a very strong maximum in the late 1980s and early 1990s. A dip in the NAO in 1995/96 correlates with the snowiest season of the 20th century.

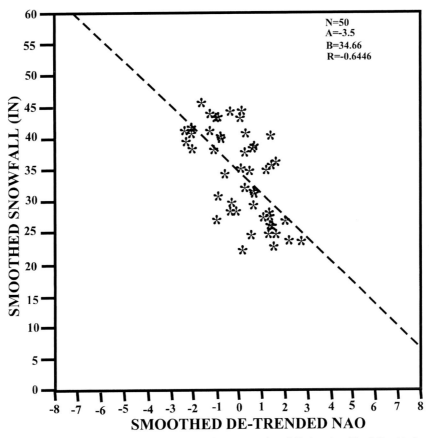

FIG. 2-15. Correlation of seasonal snowfall for a composite of 18 sites (see Fig. 2-3) with the detrended NAO (see text) for the 50-yr period 1949/50–1998/99. Here, N represents the number of cases, A and B are the linear slope and intercept values, and R is the computed correlation.

To examine the relationship between seasonal snowfall in the Northeast urban corridor and the North Atlantic Oscillation, correlation coefficients are computed for 18 sites shown in Fig. 2-2 during the winters between 1949/50 and 1998/99. A correlation coefficient of -0.57 is computed for the sample (not shown). Since an upward trend was detected in the NAO, especially since the end of the 1960s, likely reflecting the impact of warming during this period (see next section), the trend was removed from the computation and the subsequent correlation has a value of -0.64 (Fig. 2-15). This value reflects that a significant relationship exists between seasonal snowfall and the NAO between 1950 and the end of the 20th century.

When the 18 sites are subdivided into the northern, central, and southern areas shown in Fig. 2-2, reflecting the impact of the NAO on cities with seasonal snowfall averaging less than 20 in. (50 cm), from 20 to 40 in. (50 to 100 cm), and greater than 40 in. (100 cm), the correlations measured -0.55, -0.64, and -0.54, respectively. This north-to-south division of the relationship indicates that the mid-Atlantic region (as defined in Fig. 2-2) contributes the greatest to the total correlation between the NAO and snowfall observed in the Northeast

urban corridor. Therefore, the impact of the NAO on seasonal snowfall appears to be the greatest in the Northeast corridor experiencing 20–40 in. (50–100 cm) of snow, implying that the snow potential in the area including New York City, Philadelphia, Baltimore, Providence, and Harrisburg, Pennsylvania, is the most affected by conditions in the North Atlantic, as measured by the NAO index. Since the seasonal snowfall within this region is significantly influenced by the occurrence of moderate to heavy snowfalls, an important relationship between the NAO and the occurrence of significant snowstorms is indicated.

While the low-frequency variations in the NAO appear to correlate well to the seasonal snowfall record of the late 20th century, the relationship is not perfect. Some snowy winters are associated with negative seasonal values of the NAO, such as during the winters of 1957/58, 1963/64, 1968/69, 1977/78, and 1995/96. However, there are a significant number of exceptions, including the snowy winters of 1960/61, 1966/67, and 1993/94, which occurred during periods in which the winter index of the NAO was significantly positive. A comparison of monthly values of seasonal snowfall departures with monthly values of the NAO index also

TABLE 2-5. Dates of 30 Northeast snowstorms (see chapters 3 and 4) and 30 "near-miss" storms, including 15 "interior" snowstorms and 15 "moderate" snowstorms (see chapter 5). The NAO column shows the sign of the average daily values of the North Atlantic Oscillation during the storm dates (−, 0, and + represent negative values; <-1, near zero; −1 to +1, and positive values, $>+1$). Check marks represent the tendency of the NAO to become more positive during the storm period.

Heavy snow cases	NAO	√	Near-miss cases	NAO	√
18–19 Mar 1956	−		16–17 Feb 1952	−	
14–17 Feb 1958	−		16–17 Mar 1956	−	
18–21 Mar 1958	−		12–13 Mar 1959	+	
2–5 Mar 1960	−		14–15 Feb 1960	−	
11–13 Dec 1960	+		6–7 Mar 1962	−	√
18–21 Jan 1961	−		19–20 Feb 1964	−	√
2–5 Feb 1961	−		22–23 Jan 1966	−	√
11–14 Jan 1964	−		3–5 Mar 1971	−	
29–31 Jan 1966	−	√	25–27 Nov 1971	0	√
23–25 Dec 1966	−	√	16–18 Jan 1978	+	
5–7 Feb 1967	−		28–29 Mar 1984	−	√
8–10 Feb 1969	−		1–2 Jan 1987	−	
22–28 Feb 1969	−	√	10–12 Dec 1992	−	
25–28 Dec 1969	−		3–5 Jan 1994	+	
18–20 Feb 1972	+		2–4 Mar 1994	+	
19–21 Jan 1978	+		3–5 Dec 1957	0	√
5–7 Feb 1978	−		23–25 Dec 1961	−	√
17–19 Feb 1979	0	√	14–15 Feb 1962	0	
6–7 Apr 1982	−		22–23 Dec 1963	0	√
10–12 Feb 1983	−	√	16–17 Jan 1965	+	
21–23 Jan 1987	−		21–22 Mar 1967	+	
25–26 Jan 1987	−	√	31 Dec 1970–1 Jan 1971	−	√
22–23 Feb 1987	−		13–15 Jan 1982	0	
12–14 Mar 1993	0	√	8–9 Mar 1984	−	√
8–12 Feb 1994	+		7–8 Jan 1988	+	
2–4 Feb 1995	+		26–27 Dec 1990	+	
6–8 Jan 1996	−	√	19–21 Dec 1995	−	√
31 Mar–1 Apr 1997	0		2–4 Feb 1996	−	√
24–26 Jan 2000	−	√	16–17 Feb 1996	0	
30–31 Dec 2000	−	√	14–15 Mar 1999	−	√

illustrated a number of similar discrepancies (not shown).

Daily values of the NAO[11] were also examined to assess relationships with significant Northeast snowstorms. In chapters 3 and 4, 30 significant Northeast snowstorms are presented. In chapter 5, 30 "near miss" snowstorms are also presented. The daily values of the NAO were examined for these 60 cyclones during the period 1948–2000. Table 2-5 provides a description of the sign of the daily NAO values during each storm event. If the sign of the NAO changed value during a storm event, the event was assigned a neutral or 0 value.

Of the 30 significant snowstorm cases (Table 2-5), 22 occurred during periods when the NAO was clearly in its negative phase. Five cases occurred during periods when the NAO was clearly positive. Of these 22 cases, 10 occurred during a period when the negative phase of the NAO was weakening or becoming positive, similar to the signals shown for the three examples during January 2000, December 2000, and March 2001 (Fig. 2-16).

Of the 30 near-miss cases (Table 2-5; also see chapter 5), 16 or slightly more than half of the cases occurred

during distinct periods during which the NAO was in a negative mode, while 9 occurred during a positive mode. However, 14 of 30 occurred during periods when the negative phase of the NAO was either weakening or becoming positive.

The daily NAO values indicate how important the nature of the circulation pattern over the North Atlantic can be to Northeast snowstorms. During the very mild winter season of 1999/2000, both January and February exhibited a positive value of the NAO. However, when the daily NAO records are applied, it becomes clear that the period between 12 and 29 January was characterized by the negative phase of the NAO (Fig. 2-16). During this period, cold weather and significant snow affected the eastern United States, including a major snowstorm on 24–25 January (Fig. 2-16, top). During the following winter, two periods during which the NAO was distinctly negative occurred from mid-December 2000 through early January 2001, and then again during late February through early March 2001. During these periods, major Northeast snowstorms occurred on 30 December 2000 and 5–7 March 2001 (Fig. 2-16, bottom). These periods of negative NAO were marked by an upper ridge in the North Atlantic, sometimes referred to as the Greenland block (see chapter 4), and a deep

---

[11] Daily values of the NAO from 1948 to the present were provided by Gary Bates of the CIRES Climate Diagnostics Center.

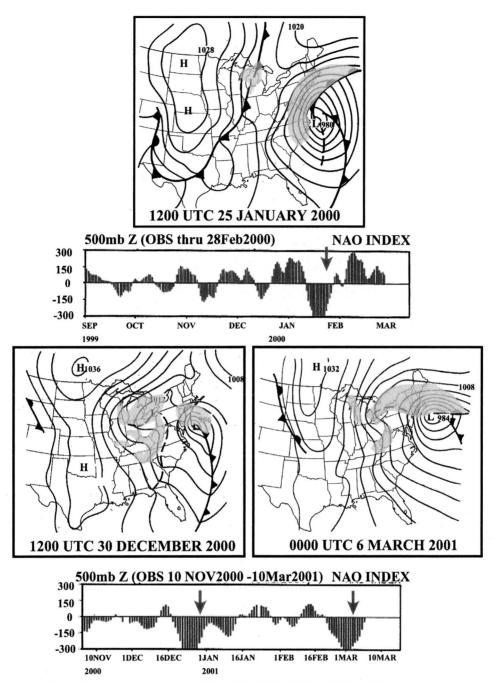

FIG. 2-16. Surface weather maps for the 25 Jan 2000, 30 Dec 2000, and 6 Mar 2001 snowstorms contrasted with daily values of the NAO over a 6-month period during which these snowstorms occurred. Red arrows point to the dates of the surface maps.

trough or cutoff low was observed in extreme eastern Canada.

Another common signature in the daily NAO index for the three cases noted above is that each of these three events occurred during a transitional period, after the NAO reached its nadir and was trending less negative (or positive). This suggests that the weather patterns occurring during the negative phase of the NAO

appear to establish conditions conducive to the formation of Northeast snowstorms, especially by locking in colder air throughout the Northeast. The transition period for the NAO also suggests that those conditions may "relax" just prior to or during the storm. If blocking patterns were to remain in the North Atlantic as a cyclone developed along the East Coast, it is possible that the storm could not evolve into a well-developed

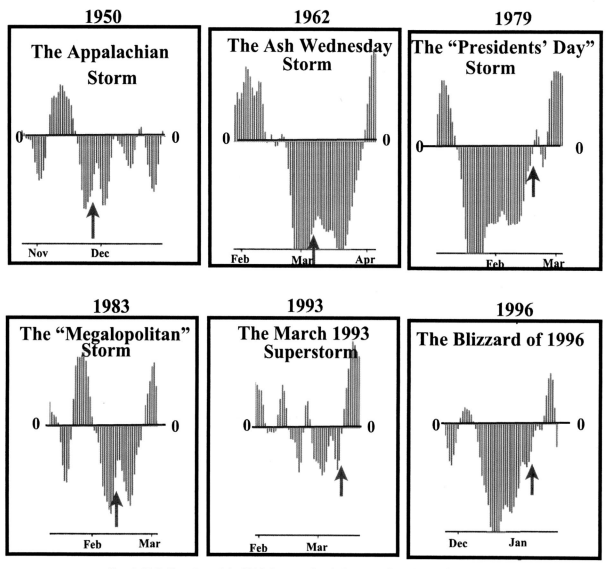

FIG. 2-17. Daily values of the NAO for several periods surrounding six major snowstorms.

cyclone and move up the coast. Rather, the larger-scale circulation might dominate and suppress the development and track of the storms farther south and east, so as not to affect the Northeast. While the monthly trends in the NAO correlate with cold northeastern winters and enhanced snowfall, the daily trends indicate that as a negative NAO pattern weakens or shifts, a storm system nearing the southeast U.S. coast has a better chance of moving up the east coast as a major snowstorm, rather than eastward over the Atlantic Ocean. While these conditions can contribute to an increasing likelihood of snowstorms in the Northeast urban corridor, they should not be considered a sufficient condition for their occurrence.

Examples of six major snowstorm events described later in this monograph that occurred during periods in which the storms developed within a period clearly char-

acterized by a positive rebound from the negative phase of the NAO (Fig. 2-17) provide supporting evidence to the examples shown earlier (Fig. 2-16) that there appears to be a significant relationship between the development of several significant Northeast winter storms and a sign change in the short-term fluctuations of the NAO.

This examination of the relationship between the NAO and Northeast snow events indicates that more than half of a sample of significant Northeast snowstorms appear to develop during periods in which daily values of the NAO are distinctly negative. It also appears that a significant number of these storms also occur during periods in which the negative phase of the NAO is weakening (daily values increasing toward zero or become positive). This offers some tantalizing associations that indicate that the onset of a period of negative NAO values, whether they occur for days, weeks, or months, may be indicative of a cold period for the Northeast and an increased potential for significant Northeast snowstorms.

3) GLOBAL WARMING

Another climatic signal that influences North American climate is warming that has been occurring since the 1960s (Livezey and Smith 1999; Lau and Weng 1999). The subject of global warming and the issue of natural variation versus anthropogenic contributions (i.e., Santer et al. 1996; Karl 1996; Houghton et al. 2001) has become one of the great scientific and political debates as the 21st century begins. Irrespective of its causes, the phenomenon of global warming is associated with rising global surface temperatures and is noted here to complete a discussion of seasonal snowfall.

Table 2-3 offers some possible insights into the possible influence of a warming trend on seasonal snowfall. For the cities of Boston, New York, Philadelphia, Baltimore, and Washington, seasonal snowfall appears to have been at a maximum in the late 19th century. For the last three decades of the 20th century, the means for New York, Philadelphia, Baltimore, and Washington were 7.0 (18), 1.8 (4), 4.2 (10), and 4.8 in. (12 cm), respectively, less than the averages for the first 50 yr of the century. While the decade of the 1960s included some of the snowiest periods of the century, there has been an overall decrease in seasonal snowfall from New York City southward since that decade that is consistent with a period of warming.

It is interesting to note, however, that the decade of the 1990s and the first three winter seasons of the 21st century brought greater variability in seasonal snowfall that is not reflected in the mean snowfall. Five seasons exhibited snowfall greatly exceeding all means, including 1992/93, 1993/94, 1995/96, 2000/01, and 2002/03. Meanwhile, several seasons exhibited much reduced snowfall to near-record levels, including 1990/91, 1991/92, 1994/95, 1997/98, 1998/99, and 2001/02. In addition, the three highest-ranked snowfalls using the Northeast Snowfall Impact Scale (NESIS; Kocin and Uccellini 2004; also see chapter 8), representing the most widespread snowfalls of 75 heavy snow events since the late 19th century, have occurred *since 1990*.

Whether this increased variability is merely a statistical blip or has some relationship to increased warming and associated extreme climate variability represents an interesting topic, but one that is beyond the scope of this monograph.

4) SUMMARIZING THE RELATIONSHIPS BETWEEN CLIMATE PATTERNS TO THE SNOWFALL DISTRIBUTION

There are no simple relationships between major climate anomalies such as the Southern Oscillation and the North Atlantic Oscillation, and also global warming,

to the observed variations in seasonal snowfall over the past 130 yr. These anomalies do appear to influence the distribution of seasonal snowfall. It appears that the occurrence or nonoccurrence of just a few storms can significantly influence whether the seasonal snowfall is above or below normal.

In some years characterized by the strong positive phase of the SO or an El Niño, there are more cyclones and more snowstorms. However, in some strong El Niños, there are also more cyclones but very mild conditions and little significant snowfall. The influence of El Niño appears to lead to increased storminess, but that only occasionally translates into more snow for the Northeast urban corridor. The influence of the positive phase of the SO, or La Niña, appears to be related to less, rather than more, seasonal snowfall.

The relationship between El Niño, La Niña, and mean snowfall for the entire United States over the period 1948–93 is illustrated in Fig. 2-18, which shows that during seven particularly strong El Niño years (compared to 19 years characterized by neutral El Niño conditions), snowfall has tended to be greater over the southwestern United States and diminished particularly over the northern Rockies and over the Ohio Valley. There has been a tendency for slightly more snow over the middle Atlantic region and Maine. The lack of a strong signal for much of the Northeast urban corridor corresponds with the tendency of El Niño winters to be associated with both significant above and below normal snowfall, as was discussed in the previous section. The relationship between eight strong La Niña winters and 19 years characterized by neither La Niña nor El Niño conditions, indicates a strong relationship with reduced snowfall for the Northeast urban corridor (Fig. 2-18).

In some years characterized by the negative phase of the NAO, a pattern that increases the likelihood of cold air in the northeastern United States, seasonal snowfall is commensurately higher, especially with regard to an increase in moderate and heavy snow events. However, some notable enhanced snowfall seasons are associated with positive monthly and seasonal values of the NAO. When the low-frequency behavior of the NAO is examined over interdecadal rather than interannual periods, there appears to be more of a relationship between the negative phase of the NAO and seasonal snowfall.

During the period 1948–93, an examination of snowfall over 15 years characterized by neutral NAO conditions versus 14 seasons with a significant positive versus negative phase of the NAO (Fig. 2-18) confirms a significant relationship between the negative phase of the NAO and snowfall across the Northeast urban corridor. Little relationship is shown between the positive phase of the NAO and seasonal snowfall.

c. The episodic character of seasonal snowfall

While there appears to be some periodic behavior in the seasonal snowfall records, the fluctuations in sea-

SNOWFALL ANOMALIES

El Nino and La Nina years   NAO - + and - years

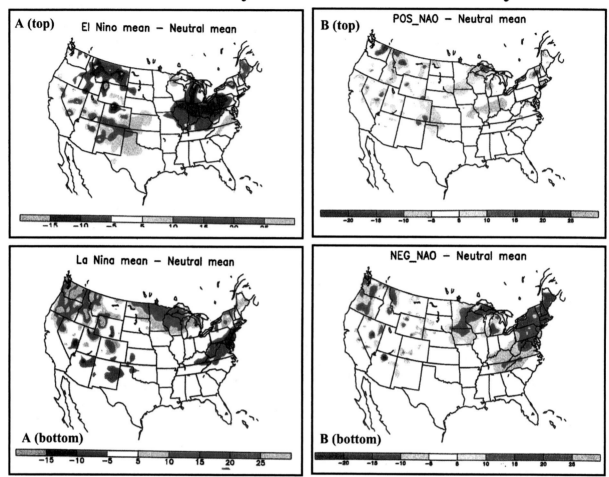

FIG. 2-18. (a) Yearly snowfall departures (in.) for El Niño and La Niña years compared to neutral El Niño conditions. Departures are computed for seven of the strongest El Niño years and eight of the strongest La Niña years, compared to 19 neutral years during the period 1948–93. (b) Yearly snowfall departures (in.) for both positive and negative phases of the NAO compared to neutral conditions. Departures are computed for 14 years characterized by + conditions and 14 years characterized by—conditions, compared to 15 neutral years during the period 1948–93. (Figure adapted from one provided by J. Janowiak, NCEP/CPC.)

sonal snowfall in all five major metropolitan areas appear to be closely linked to changes in the contribution of the moderate and heavy snowfall events. An assessment of the proportion of seasonal snowfall due to moderate and heavy snowfall events (greater than 10 and 25 cm, respectively) is summarized in Fig. 2-19.

During the last half of the 20th century, moderate and heavy events (snowstorms exceeding 10 cm) contributed to between 55% and 65% of the total snowfall for that period. Snowfall events exceeding 25 cm contributed between 18% and 24% of the total snowfall. When the 10 snowiest seasons for each of the five major cities are examined, moderate and heavy events account for an even larger fraction of the total snowfall. In these 10 seasons, moderate to heavy events account for between 70% and 86% of the total snowfall while the major events (>25 cm) account for between 38% and 46% of the total snowfall.

These numbers show that the moderate to heavy snowstorm events account for more than half to nearly two-thirds of the seasonal snowfall record in the Northeast urban corridor during the last half of the 20th century. During the 10 snowiest seasons, the increase in seasonal snowfall is accompanied by an increase in the moderate to heavy events, accounting for three-quarters or greater of the total snowfall. During these

% Snowfall due to mod and heavy events
1949/50-1998/99

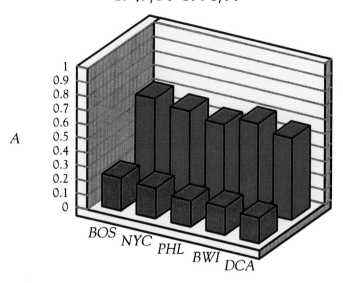

% Snowfall due to mod and heavy events
10 Snowiest winters (1950-2000)

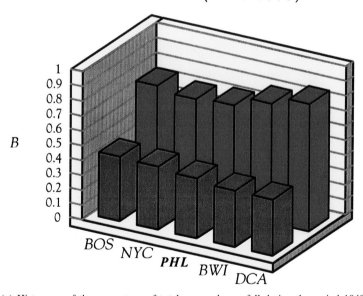

FIG. 2-19. (a) Histogram of the percentage of total seasonal snowfall during the period 1949/50–1998/99 due to moderate (>10 cm, blue) and heavy snow events (>25 cm, red). (b) Histogram of the percentage of the 10 snowiest seasons during the period 1949/50–1998/99 due to moderate (>10 cm, blue) and heavy snow events (>25 cm, red).

snowy seasons, the largest events, with snowfall greater than 25 cm, account for more than a third of the snow that falls. In the specific case of New York City, heavy snow events (greater than 25 cm) accounted for nearly half the total snowfall. The results illustrate the importance of these big, although rare, events in influencing the seasonal snowfall characteristics and related trends.

These results indicate that variability in seasonal snowfall appears to be more often "episodic" than pe-

TABLE 2-6. Dates of snowstorms responsible for accumulations exceeding 10 in. (25 cm) over at least two regions, and number of regions as defined in Fig. 2-9 (1949/50–1998/99). Italicized dates indicate the events composing the 30-case analyses.

1950–59 Date	No. of regions	1960–69 Date	No. of regions	1970–79 Date	No. of regions	1980–89 Date	No. of regions	1990–99 Date	No. of regions	2000– Date	No. of regions
13–14 Feb 1950	4	19–20 Feb 1960	2	22–24 Dec 1970	2	1–2 Mar 1980	2	30 Jan 1990	3	*24–25 Jan 2000*	6
5 Jan 1952	2	*2–5 Mar 1960*	18	3–4 Mar 1971	4	6–7 Dec 1981	3	11–13 Dec 1992	6	*30–31 Dec 2000*	7
17–18 Feb 1952	9	*10–13 Dec 1960*	15	20–21 Mar 1971	3	13–15 Jan 1982	2	16–17 Feb 1993	2	5–6 Feb 2001	10
8–9 Jan 1953	5	*18–20 Jan 1961*	11	25–26 Nov 1971	6	27–28 Feb 1982	2	*12–14 Mar 1993*	16	4–6 Mar 2001	10
6–7 Nov 1953	5	*2–5 Feb 1961*	14	*18–20 Feb 1972*	12	*5–7 Apr 1982*	11	6–9 Jan 1994	5	25–26 Dec 2002	8
16–17 Mar 1956	7	14–15 Feb 1962	7	15–16 Mar 1972	4	15–16 Jan 1983	5	17–18 Jan 1994	7	2–3 Jan 2003	8
19–20 Mar 1956	10	6–7 Mar 1962	3	29–30 Jan 1973	4	7–8 Feb 1983	7	*8–12 Feb 1994*	10	15–18 Feb 2003	17
3–4 Dec 1957	3	23–24 Jan 1963	2	9–11 Jan 1974	5	*10–12 Feb 1983*	17	1–4 Mar 1994	6	4–6 Dec 2003	15
7–8 Jan 1958	5	25–26 Jan 1963	2	5–6 Feb 1975	2	10–11 Jan 1984	2	*3–5 Feb 1995*	9		
14–16 Jan 1958	2	*11–14 Jan 1964*	14	20–22 Dec 1975	5	13–14 Mar 1984	6	19–20 Dec 1995	4		
14–17 Feb 1958	17	16 Feb 1964	3	7–8 Jan 1977	4	28–29 Mar 1984	3	2–3 Jan 1996	7		
14–16 Mar 1958	6	19–20 Feb 1964	3	10–11 Jan 1977	2	24–26 Jan 1986	2	*6–8 Jan 1996*	17		
18–21 Mar 1958	10	10–11 Mar 1964	2	6–7 Dec 1977	4	19 Dec 1986	3	12–13 Jan 1996	2		
12–13 Mar 1959	6	22–23 Jan 1966	5	13–14 Jan 1978	3	1–2 Jan 1987	5	2–3 Feb 1996	4		
29–31 Dec 1959	2	26–27 Jan 1966	3	*19–21 Jan 1978*	13	10–11 Jan 1987	3	7–9 Mar 1996	9		
		29–31 Jan 1966	10	*5–7 Feb 1978*	17	*22–23 Jan 1987*	13	9–10 Apr 1996	4		
		25–26 Feb 1966	9	24–25 Dec 1978	3	25–26 Jan 1987	7	5–6 Dec 1996	2		
		23–25 Dec 1966	9	25–26 Jan 1979	3	16–17 Feb 1987	2	7–8 Dec 1996	3		
		5–7 Feb 1967	11	*18–20 Feb 1979*	9	22–23 Feb 1987	5	*31 Mar–1 Apr 1997*	9		
		6–7 Mar 1967	5			11–12 Nov 1987	2	14–16 Nov 1997	3		
		15–16 Mar 1967	4			7–8 Jan 1988	2	23–24 Dec 1997	2		
		22–23 Mar 1967	2			25–26 Jan 1988	3	29–30 Dec 1997	2		
		27–29 Dec 1967	6			12–13 Feb 1988	6	13–14 Mar 1999	2		
		9–10 Feb 1969	11								
		22–28 Feb 1969	6								
		1–2 Mar 1969	2								
		22–23 Dec 1969	2								
		25–28 Dec 1969	11								

riodic in nature. In other words, winters in the Northeast urban corridor are characterized by prolonged snowfall droughts separated by sporadic episodes of significant seasonal snowfall. Furthermore, these episodes of significant snowfall are related almost entirely to changes in the frequency of moderate and heavy snowstorms. In the mean, these moderate and heavy snowstorms are relatively infrequent, averaging only about one to four events per season between Washington and Boston. But these numbers increase significantly during snowy winter seasons.

In addition, the occurrences of heavy snow events (>25 cm) are typically very rare, sometimes not occurring for several years at a time. However, during the 10 snowiest winters, the number of these events increases, with as many as four during a season in Boston to two to three from Washington to New York City. While an increase in moderate events can be related to a more active storm track or the presence of colder-than-normal conditions, the relatively rare heavy snow events that focus snowfall along the coastal plain warrant an examination of conditions that are conducive to their formation.

Table 2-6 contains a list of 115 "heavy snowstorm" events during the latter half of the 20th century and beginning of the 21st century that produced snowfalls of 25 cm or more in at least 2 of the 21 sections shown in Fig. 2-5 that encompass the Northeast urban corridor. These storms can be considered a listing of the most

widespread and significant snowstorms of the last half of the 20th century. While all of these events were responsible for significant areas of heavy snow across portions of the Northeast urban corridor, there are many variations in snowfall distribution from case to case. Some of these cases produced the heaviest snows primarily inland away from the coast and the centers of greatest population, or farther northward, usually from north of Boston to Maine. These storms are discussed in chapter 5. During these events, much of the highly urbanized sections from Virginia to extreme southern New England experienced a significant changeover from snow to rain, freezing rain, or ice pellets. These near-miss cases will be used to illustrate the sometimes subtle differences that may distinguish the crippling heavy-snow producers in the urban corridor from the near misses that are quickly forgotten by most the Northeast's urban inhabitants.

Thirty cases in which the snowfall is heaviest within the Northeast urban corridor (each italicized in Table 2-6) will form the bases of Chapters 3, 4, and 10 (volume II), where the general characteristics and structure of the most paralyzing snowstorms within the most heavily populated portions of the Northeast urban corridor are examined in depth. In the next two chapters, an examination of the surface and upper-level conditions that contribute to and are associated with some of the most significant snowstorms provides insight into those relatively rare conditions that can impact many millions of lives.

Chapter 3

SYNOPTIC DESCRIPTIONS OF MAJOR SNOWSTORMS: SNOWFALL AND SURFACE FEATURES

In this chapter, the distribution of heavy snowfall and surface features are summarized for 30 major snowstorms that affected the Northeast urban corridor between 1950 and 2000. Estimates of the areal coverage for snowfall amounts exceeding 10 (25), 20 (50), and 30 in. (75 cm) and a normalized population distribution (using the 1999 census) affected by heavy snowfall are provided in Fig. 3-1 and Table 3-1. These estimates are generated using a Geographic Information System (GIS) software package developed by Environmental Systems Research Institute, Inc. (ESRI), applied to the snowfall distributions for all 30 cases. The numbers contained in these estimates not only convey a comparative measure of the snowfall coverage of each storm but can also be viewed as a potential measure of the impact these storms had on the general population. Such measures are weighted toward the largest cities, precisely where the greatest impact on the largest numbers of people would be expected to occur. A measure of the areal coverage of the storms and its relation to the population lends itself to the formulation of an area- and population-weighted scale that provides a useful perspective on these storms' relative impact that can be communicated to the general public. A Northeast Snowstorm Impact Scale (NESIS; also see Kocin and Uccellini 2004), its formulation based on the principles above and application toward a quantitative rating of these storms and examples, is presented in chapter 8.

1. Snowfall distribution

In Fig. 3-2, the shaded regions for snowfall amounts exceeding 10, 20, and 30 in. (25, 50, and 75 cm) depict the typical southwest–northeast-aligned synoptic-scale banding that accompanies major snow events in the Northeast urban corridor. These bands of snowfall amounts greater than 10 in. (25 cm) average more than 500 km in length and from 200 to 500 km in width. The mean areal coverage for all 30 storms is 91×10^3 mi^2 (1.4×10^5 km^2), and on average over 35 million people were affected by each storm (Table 3-1), normalized to 1999 census figures. The largest coverage of 10-in. (25 cm) snowfall is observed with the March 1993 "Superstorm," which is likely the most widespread snowstorm east of the Mississippi River on record (Ko-

cin et al. 1995), with the heaviest snows extending southward to Alabama and westward into the Ohio Valley. Over 100 million Americans experienced accumulating snow from this storm and approximately 60 million from Virginia to Maine experienced accumulations of 10 in. (25 cm) or greater. [Other widespread storms include those of 14–17 February 1958, 2–5 March 1960, 3–5 February 1961, 11–14 January 1964, 19–21 January 1978, 11–12 February 1983, 21–23 January 1987, and 6–8 January 1996 (Table 3-1). Of the cases presented here, the smallest areal distributions are shown by the cases of 18–20 March 1956, 25–27 January 1987, 22–24 February 1987, and 31 March–1 April 1997.] Within the 13-state region composing the northeast United States from Virginia to Maine, 11 of the 30 storms produced accumulations greater than 10 in. (25 cm) over areas presently inhabited by more than 40 million individuals (Table 3-1). [The largest of these storms include the March 1993 Superstorm, the January 1996 "Blizzard of '96," the February 1983 "Megalopolitan" storm (50 million people affected), the two winter storms of January and February 1978 (48 and 43 million, respectively), and the snowstorms of February 1958, (50 million), March 1960 (50 million), and February 1961 (48 million).]

Snowfalls of 20 in. (50 cm) or greater were observed in every storm except that on 25–27 January 1987 [the most widespread 20-in. (50 cm) snowfalls were produced by the storms of 22–28 February 1969, 5–7 February 1978, 10–12 February 1983, 13-14 March 1993, and 6–8 January 1996 (Table 3-1)], with the storms of 1993 and 1996 covering the largest areas of the 30-case sample (Fig. 3-2b). During the storm of January 1996, an estimated 30 million people experienced accumulations of 20 in. (50 cm) or greater.

Snowfall accumulations exceeding 30 in. (75 cm) were analyzed in 12 of the 30 cases with the largest areal distributions exhibited by the storms of 22–28 February 1969, March 1993, and January 1996 (Table 3-1). The greatest number of people affected by 30-in. (75 cm) amounts occurred again during the January 1996 blizzard (Fig. 3-2b; Table 3-1) as the area of heaviest snowfall extended from eastern West Virginia northeastward to northern New Jersey, affecting an estimated 5 million people. In some cases, snowfall greater than

Population (in millions) affected by Northeast Snowstorms

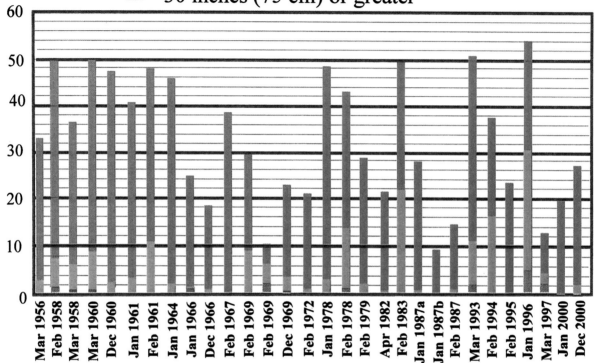

FIG. 3-1. Estimated population (in millions) affected by snowfall accumulations exceeding 10 (25), 20 (50), and 30 in. (75 cm) during 30 Northeast snowstorms. Population estimates are based on 1999 census figures.

30 in. (75 cm) occurred in a few isolated locations, while others produced significant areas, with localized totals of 40–50 in. (100–125 cm) or greater.

Notable examples of storms yielding significant areas of snowfall accumulations of 30 in. (75 cm) or more include the February 1958 "Blizzard of 1958," which left up to 3 ft (90 cm) of snow and greater on northeastern Pennsylvania and the Catskill Mountains in New York. The wet snowstorm of 19–21 March 1958 produced greater than 30 in. of snow (75 cm) across portions of northern Maryland and southeastern Pennsylvania, with a maximum of 50 in. (125 cm) at Morgantown, Pennsylvania, a record snowfall for the state. The snowstorm of 2–5 February 1961 produced many areas of 20 in. or greater, on top of unmelted snow remaining from earlier storms, and produced up to 40 in. (100 cm) at Cortland, New York. The January 1966 "Blizzard of '66" combined with "lake effect" snowfall to produce 101 in. (257 cm) in 3 days at Oswego, New York, and over 42 in. (107 cm) at Syracuse, New York. The "one hundred hour" snowstorm of late February 1969 dumped 2–3 ft of snow (60–90 cm) across eastern Mas-

sachusetts, up to 56 in. (142 cm) at Long Falls Dam, Maine (a state record), and 77 in. (198 cm) and 98 in. (248 cm), respectively, at Pinkham Notch and Mount Washington, New Hampshire (both state records). Nine years later, the great snowstorm of February 1978 produced 30 in. or more across eastern Massachusetts and northern Rhode Island, with 38 in. (97 cm) at Woonsocket, Rhode Island (the official state record), and unofficial measurements of up to 55 in. (139 cm). The Megalopolitan Snowstorm of February 1983 yielded 30-in. (75 cm) amounts across northern Virginia and western Maryland.

The 1990s brought two of the most widespread, heaviest snowstorms of the 20th century with the 12–14 March 1993 Superstorm and the 6–8 January Blizzard of 1996. Only a few months prior to the March 1993 Superstorm, an intense cyclone on 11–12 December 1992 produced 30–40-in. (75–100 cm) accumulations in the highlands of western Maryland and Pennsylvania and up to 4 ft (120 cm) in western Massachusetts (see chapter 5). The 1993 Superstorm produced its heaviest snowfall amounts along the spine of the Ap-

Table 3-1. Estimated area ($\times 10^3$ mi^2) and population (in millions, 1999 census), affected by snowfall accumulations of 10 in. (25 cm) or greater during 30 Northeast snowstorms (Fig. 3-2). In addition, estimated areas and populations are also listed for amounts exceeding 20 (50) and 30 in. (75 cm).

Date	10-in. area	10-in. pop	20-in. area	20-in. pop	30-in. area	30-in. pop
18–19 Mar 1956	28.6	32.8	2.6	2.6		
14–17 Feb 1958	126.0	53.8	20.2	6.0	3.4	0.8
18–21 Mar 1958	62.1	40.7	13.8	7.5	3.5	0.7
2–5 Mar 1960	133.7	53.9	7.6	8.5		
11–13 Dec 1960	74.5	48.0	0.6	2.5		
18–21 Jan 1961	62.3	43.0	5.7	2.9		
2–5 Feb 1961	112.2	50.3	19.4	8.7	1.4	0.2
11–14 Jan 1964	110.3	48.8	10.3	1.5		
29–31 Jan 1966	122.5	23.8	12.3	2.4	1.5	0.5
23–25 Dec 1966	83.4	18.1	9.9	1.4		
5–7 Feb 1967	50.9	44.8				
8–10 Feb 1969	66.4	31.2	11.6	9.6		
22–28 Feb 1969	48.4	10.3	40.8	8.2	24.2	4.2
25–28 Dec 1969	131.4	25.0	37.6	4.0		
18–20 Feb 1972	140.9	24.5	13.5	1.4		
19–21 Jan 1978	161.6	50.9	8.3	3.2		
5–7 Feb 1978	120.5	47.6	30.7	16.0	0.9	1.2
17–19 Feb 1979	56.9	31.5	4.3	3.0		
6–7 Apr 1982	76.8	22.5	21	0.6		
10–12 Feb 1983	111.1	51.4	33.7	25.7	0.9	0.2
21–23 Jan 1987	132.8	34.9	2.0	0.1		
25–26 Jan 1987	38.0	11.5				
22–23 Feb 1987	28.3	16.6	0.3	0.1		
12–14 Mar 1993	212.6	59.9	119.2	16.3	11.7	1.8
8–12 Feb 1994	55.0	39.0	4.4	13.4		
2–4 Feb 1995	98.0	29.9				
6–8 Jan 1996	137.9	56.6	88.7	39.7	15.5	4.7
31 Mar–1 Apr 1997	32.0	13.0	13.1	7.0	3.1	2.2
24–26 Jan 2000	59.6	19.7				
30–31 Dec 2000	56.5	28.0	3.7	1.4		

palachians with 20-in. (50 cm) amounts common from the mountains of North Carolina northward to New York and 30–40-in. snowfalls (75–100 cm) scattered across West Virginia, western Maryland, Pennsylvania, New York, and New England.[1] During the 6–8 January Blizzard of 1996, 30 in. (75 cm) or more were measured across a large region of northern Virginia, eastern West Virginia, northern Maryland, and portions of southeastern Pennsylvania and northeastern New Jersey.[2] During the early spring snowstorm of 31 March–1 April 1997, 30-in. (90 cm) amounts were recorded in portions of the Catskill Mountains in New York and across portions of central and eastern Massachusetts and northern Rhode Island.[3]

In contrast to the large synoptic-scale characteristics of the bands of heavy snowfall, the distance between regions experiencing heavy snow and little or no snow across the band can be surprisingly small, especially along the northern boundary of the heavy snow area. For example, during the Blizzard of 1996, Albany, New York, remained at the northern edge of the snow, reporting only 1 in. (2.5 cm), while only 25 mi to its south, snowfall approached 20 in. (50 cm). Another example is the Presidents' Day Storm of February 1979, which brought snowfalls of up to 16 in. (40 cm) across northeastern New Jersey and New York City, while the northern suburbs of New York City measured only 4 in. (10 cm) or less, and much of Connecticut and southern New England received no snow at all.

Other factors, such as changeovers from snow to either ice pellets, freezing rain, rain, or a mixture of any of these elements, can have a significant impact of the distribution of total snowfall. For example, during the March 1993 Superstorm, snow depths of 10–12 in. (25–30 cm) were common in the major Northeast cities with greater totals farther west, but a changeover to ice pellets and freezing rain kept accumulations at half of what would have accumulated if the precipitation that fell during the storm remained as snow. In this event, the combined mix of snow and ice pellets produced a dense icy mass that was resistant to snow removal efforts since it contained the liquid equivalent of a 20–30-in. (50–75 cm) snowfall. As another example, the storm of 9–10 February 1969 produced snowfall accumulations of greater than 20 in. (50 cm) in the northwestern suburbs

---

[1] A new state record for Maryland was established at Grantsville with 47 in. (119 cm), and 40-in. (100 cm) amounts were reported in New York, Pennsylvania, Vermont, and West Virginia.

[2] New state records for storm snowfall were established for Virginia with 49 in. (125 cm) at Big Meadows, and for New Jersey with 35 in. (89 cm) at White House Station.

[3] This includes 33 in. (83 cm) at Worcester, Massachusetts, their largest snowstorm on record, and 25.4 in. (63 cm) at Boston, Massachusetts, its greatest 24-h snowfall of record.

(a)
TOTAL SNOWFALL

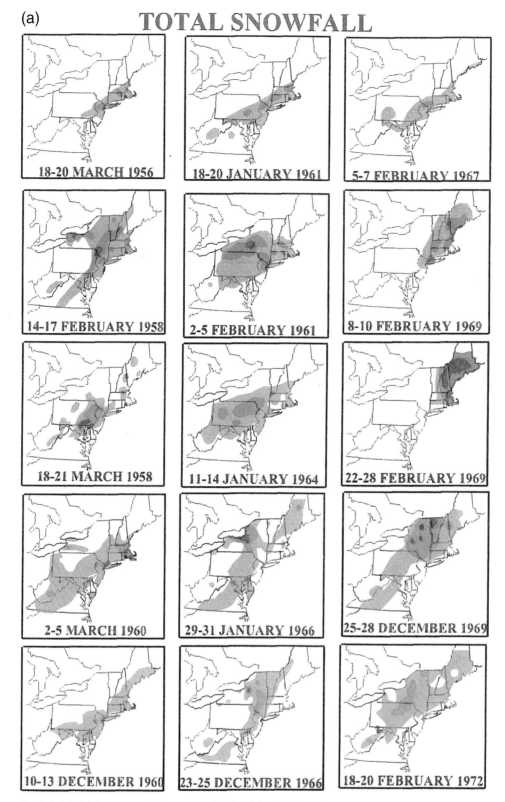

FIG. 3-2. (a), (b) Storm snowfall in excess of 10 in. (25 cm) (blue > 25 cm; green > 50 cm; red > 75 cm) for major snowstorms between 1950 and 2000.

(b)

TOTAL SNOWFALL

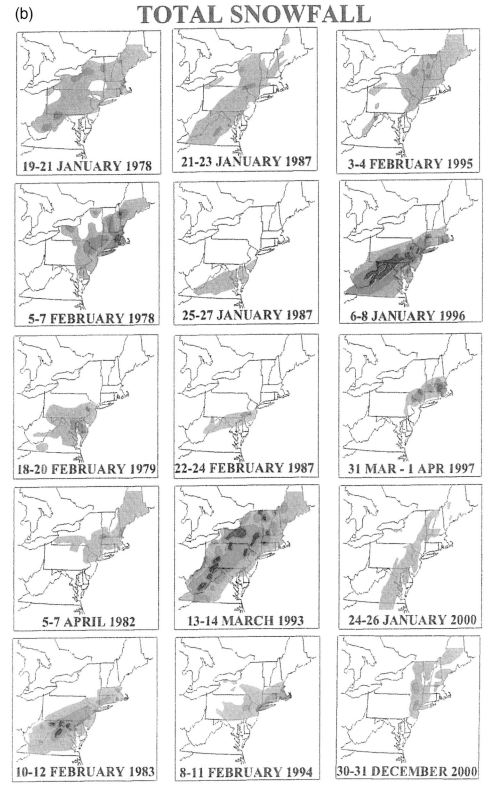

FIG. 3-2. (*Continued*)

of Boston, and 10–20 in. (25–50 cm) within the city limits, while insignificant amounts occurred in the southeastern suburbs, where the snow was mixed with rain and ice pellets. During other episodes, such as the storms of January and February 1961, the greatest snow amounts were measured in a narrow band along the northern fringe of the band of heavy snowfall, a factor that is discussed in chapter 6.

Elevation differences can also have a significant impact on the distribution of snowfall. During the storm of 19–21 March 1958, differences in elevation of only 100–200 m were the critical factor between mixed precipitation and snowfalls that exceeded 20 in. (50 cm) in parts of Pennsylvania and Maryland. Such extreme variability within a small area continues to present a tremendous challenge to the forecast community.

Every storm displayed local variations that are functions of the local topography and/or mesoscale processes. These local variations point to the difficulty in making precise forecasts of snowfall amounts and also indicate that a snowstorm scale based solely on areal coverage could limit the utility of such a scale, as is discussed in chapter 8. The mesoscale variations in snowfall will be addressed in chapter 6. The physical and dynamical processes that affect the spatial distribution and intensity of the snowfall associated with these storms are addressed in the following sections, as well as in chapter 7.

2. Cyclones

All 30 Northeast snowstorms identified in Table 3-1 are characterized by the development and propagation of well-defined cyclonic circulations along the east coast of the United States. Minimum sea level pressures for the cyclones associated with the 30-case sample range from 960 hPa (4 March 1960 and 13-14 March 1993) to 1009 hPa (11 February 1994), with an average minimum pressure of 978 hPa, indicating that many of these cyclones were quite intense. The initial development and subsequent movement of these low pressure systems appear to be influenced by the land–sea interface between the East Coast and the Atlantic Ocean and by the Appalachian Mountains. Once the cyclones move over the Atlantic Ocean, they tend to develop more rapidly. The rapid intensification of East Coast cyclones is addressed in section 2c of this chapter.

The coastline serves as a focal point for cyclogenesis since it represents the natural boundary between cold air trapped over the coastal plain or "dammed" east of the Appalachian Mountains and boundary layer air warmed and moistened by the Atlantic Ocean (Baker 1970; Bosart 1975; Richwein 1980; Forbes et al. 1987; Bell and Bosart 1988). The crucial role of cold-air damming and its relationship to coastal frontogenesis (Bosart 1975) and cyclogenesis are explored in section 3a of this chapter and in chapter 7.

The coastline also represents a discontinuity in surface roughness that influences low-level divergence and enhances cyclogenetic circulation patterns (Bosart 1975; Anthes and Keyser 1979; Ballentine 1980; Danard and Ellenton 1980). The shape of the coastline perhaps plays a role in the cyclogenetic process, since concave coastlines tend to promote cyclogenesis (Petterssen 1956; Godev 1971a,b; Bosart 1975; Ballentine 1980). The North Carolina and New England coasts exhibit a concave shape and are two areas prone to cyclogenesis and coastal frontogenesis (Bosart 1975).

The role of the Appalachians in the evolution of the East Coast cyclogenesis is harder to discern. Petterssen (1956, p. 267) showed that cyclogenesis occurs downstream of mountain chains and along coastlines, especially near the Rocky Mountains, the Alps, the Himalayas, and along the eastern coasts of North America and Asia. A mechanism termed "lee cyclogenesis" relates a tendency for cyclonic development on the leeward or downwind sides of mountain chains to enhancements in the relative vorticity when vortices are "stretched" as they cross mountain chains (i.e., see Holton 1979, 87–89; Smith 1979, 163–169). This mechanism appears to be most important in the lee of extensive and steep mountain chains, such as the Rocky Mountains (i.e., Newton 1956; Carlson 1961; Egger 1974; and others) and the Alps (i.e., Bleck 1977; Buzzi and Tibaldi 1978; Smith 1984; and others). The influence of the Appalachian Mountains on cyclogenesis appears to have a small but significant impact on the path of the storm (O'Handley and Bosart 1996), forcing it to take a path farther south and east of its initial track. In the absence of cold-air damming, the surface lows appear to develop farther south along the cyclone's cold front in the lee of the Appalachians. If cold-air damming is present, the lows tend to form farther east near the coastline (O'Handley and Bosart 1996). Processes related to the presence of the mountains are important factors in maintaining cold air over the coastal plain, contributing to the enhanced low-level baroclinicity at the coastal–ocean interface required for cyclogenesis (see section 3c).

a. Cyclone paths and redevelopment

The general characteristics of low pressure centers that develop along the East Coast have been described previously by Austin (1941), Miller (1946), and Mather et al. (1964). Miller examined 200 cases of varying intensity spread over a 10-yr period ending in 1939 and categorized East Coast storm development into two types: A and B. Type-A storms resemble the classical polar front wave cyclone studied by the Norwegian school (Bjerknes 1919; Bjerknes and Solberg 1922). The surface low, defined by the formation of at least one closed isobar, develops along a frontal boundary separating an outbreak of cold continental air from warmer maritime air. Type-B storms represent a more complex cyclonic development that Miller describes as a phenomenon unique to the east coast of the United States. Type-B systems feature the development of a secondary area of low pressure along the East Coast to the southeast of an occluding primary, or initial, low pressure center that moved to and filled over the Ohio Valley, west of the Appalachians. The secondary low pressure center forms along the primary cyclone's warm front, which separates a shallow wedge of cold air between the Appalachian Mountains and the Atlantic Ocean from warmer air over the ocean. While there continues to be a debate concerning a physical basis for the A and B classifications proposed by Miller, his study cites secondary sea level development as one of the most intriguing aspects of cyclogenesis along the east coast of the United States. The formation of a secondary low pressure center is of particular concern to forecasters, since their ability to predict the onset of cyclogenesis and the subsequent track of the surface low pressure center is a key factor in the forecast of when and where the heaviest snows will occur.

The cyclone paths for the 30 snowstorm cases and their relationship to the snowfall and 500-hPa vorticity maxima are depicted in Figs. 3-1a–e. A comparison of the cyclone tracks from all of the cases and their snowfall distributions (Figs. 3-3a–e) indicates that the axis of heaviest snow accumulations is usually found 100–300 km to the left of, and roughly parallel to, the paths of low pressure centers that track northeastward along the Atlantic coast. While snowfalls exceeding 10 in. (25 cm) occur typically well to the left of the cyclone track, the March 1993 Superstorm may be in a class of its own since more than 10 in. (25 cm) fell in areas where the center of the storm later passed. This is unusual because it is rare to receive heavy snow accumulations along the path of the surface low since warming associated with the cyclone's circulation will usually allow snow to change to rain before substantial accumulations occur.

Most of the cyclones moved along tracks that have some similarities to the type-A and -B storms described above. In the first scenario, a low pressure system develops over or near the Gulf of Mexico in association with a distinct 500-hPa vorticity maximum moving east-northeastward from the southwestern United States. The cyclone then tracks northeastward along the Atlantic Coast with snow falling approximately 200 km to the left of the storm track. The tracks of the surface cyclones associated with the January Blizzard of 1966 (Fig. 3-3b), 21–23 January 1978 (Fig. 3-3c), the March 1993 Superstorm (Fig. 3-3d), and the January Blizzard of 1996 (Fig. 3-3e) are the best examples of this scenario. These storms are similar to the type-A storms defined by Miller and represent classic "nor'easters" since they move northeastward along much of the Atlantic coast, affect the entire northeast United States, and are often accompanied by strong northeasterly winds.

In the second scenario that emerges from Figs. 3-3a–e, cyclones that develop initially over a variety of locations such as Canada, the central or western United States, or even over the Gulf of Mexico track toward the Appalachian Mountains. Once the cyclone reaches the Ohio Valley just west of the spine of the Appalachian Mountains, a new or "secondary" cyclone develops near or along the coastline as the 500-hPa vorticity maximum continues moving toward the East Coast.[4]

Some other cases have primary lows, which originate farther north, across southern Canada or the northern plains. These southeastward-propagating low pressure systems originating over Canada have been referred to by forecasters as "Alberta clippers" or simply "clippers." In some cases, these clippers are associated with amplifying upper-level troughs, which then initiate a secondary low off the East Coast that produces a major snowstorm [e.g., the March 1956 snowstorm (Fig. 3-3a) and the February 1978 blizzard (Fig. 3-3c)]. In each of these cases, the development of a secondary cyclone near or along the coastline is very similar to the Miller type-B cyclone.

Sixteen of the 30 cases developed either over the Gulf of Mexico or in one of the states bordering on the Gulf and then moved northeastward along the Atlantic coast and are grouped within a "Gulf of Mexico–Atlantic coast" category (Fig. 3-4). These systems passed south of the Appalachian Mountains and headed northeastward along the East Coast, near or offshore of the middle Atlantic and southern New England coasts. Many of these cases followed similar paths once off the middle Atlantic, New England, and Nova Scotia coastlines. A few cases passed inland across New England, and these cases were typically associated with a changeover to rain along the New England coast (also see chapter 5). Although the development of a secondary low pressure center was not observed in these 16 cases, 10 of these cases exhibited a "center jump" (see Figs. 3-3a–e; note the dashed portions of the storm tracks along the East Coast in the cases of February 1958, January 1966, December 1966, February 1967, January 1978, February 1983, the two storms of January 1987, February 1987, and January 1996). In these cases, the surface low pressure center propagated northeastward along the Southeast coast and appeared to suddenly redevelop farther to the northeast along the coast. This redevelopment occurs along the same path as that of the primary low center, in contrast to the separate tracks that mark "Atlantic coastal redevelopments" (Fig. 3-5). The sudden acceleration of the surface low that characterized center

[4] The cyclones associated with the December 1960 and March 1960 snowstorms (Fig. 3-3a), the February 1961 snowstorm (Fig. 3-3b), the "Lindsay" snowstorm of early February 1969 (Fig. 3-3b), and the April Blizzard of 1982 (Fig. 3-3d) are all examples of "secondary" developments along the East Coast while a primary low stalls and fills over the Ohio Valley, West Virginia, or western Pennsylvania, west of the mountains.

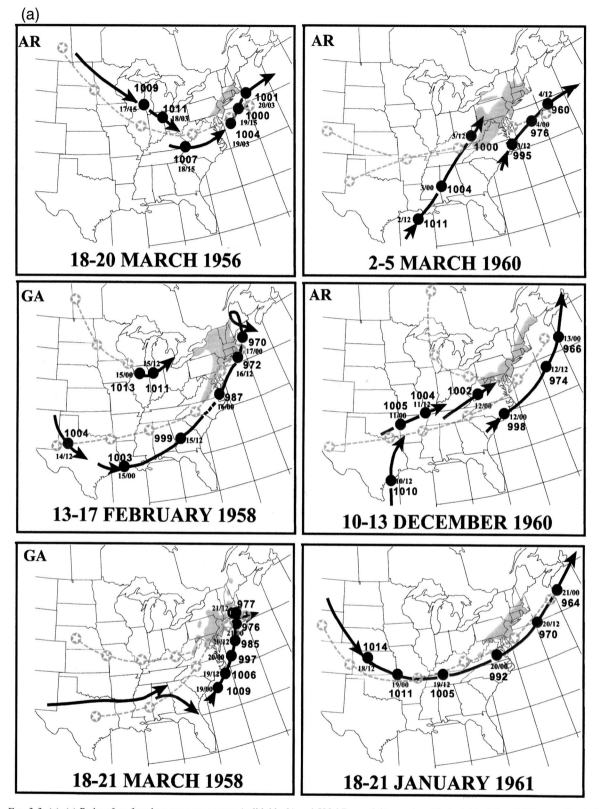

FIG. 3-3. (a)–(e) Paths of surface low pressure centers (solid, black) and 500-hPa vorticity maxima (dashed, blue) for all 30 cases. Symbols along path represent 12-hourly positions (solid circles for surface centers and starred circles for vorticity maxima) and numerical values are the sea level central pressures (hPa) of the cyclones at 12-hourly intervals (date/time included). Shading indicates snowfall exceeding 10 in. (25 cm) north of the Virginia–North Carolina border. AR and GA refer to Atlantic Coastal Redevelopment and Gulf of Mexico/Atlantic Coast categories, respectively, as discussed in text.

(b)

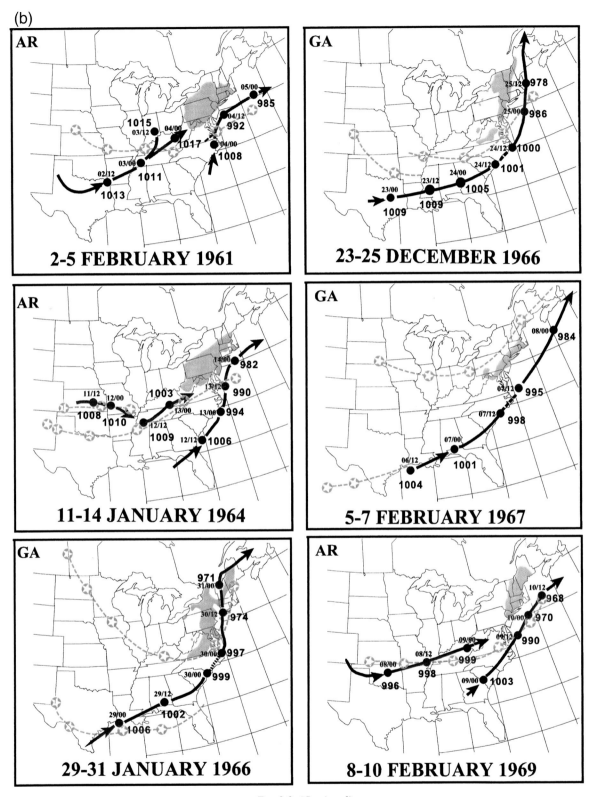

FIG. 3-3. (*Continued*)

(c)

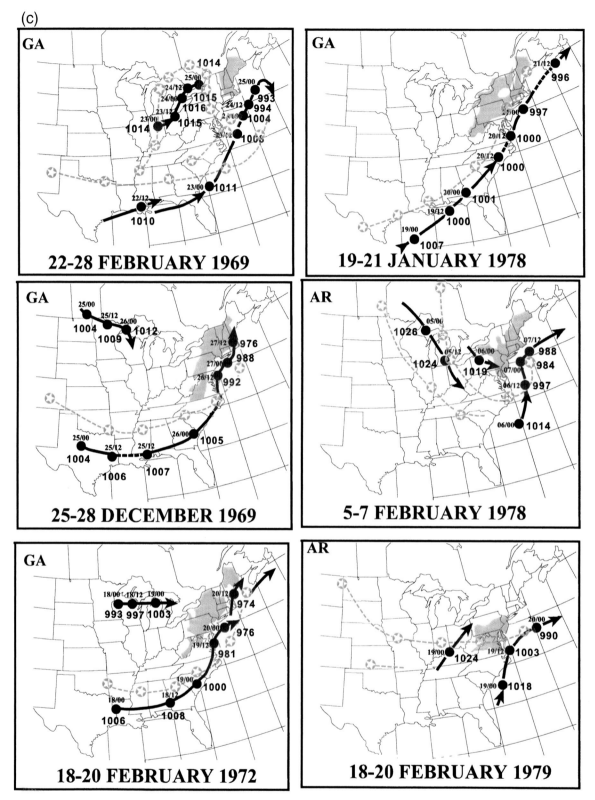

FIG. 3-3. (*Continued*)

(d)

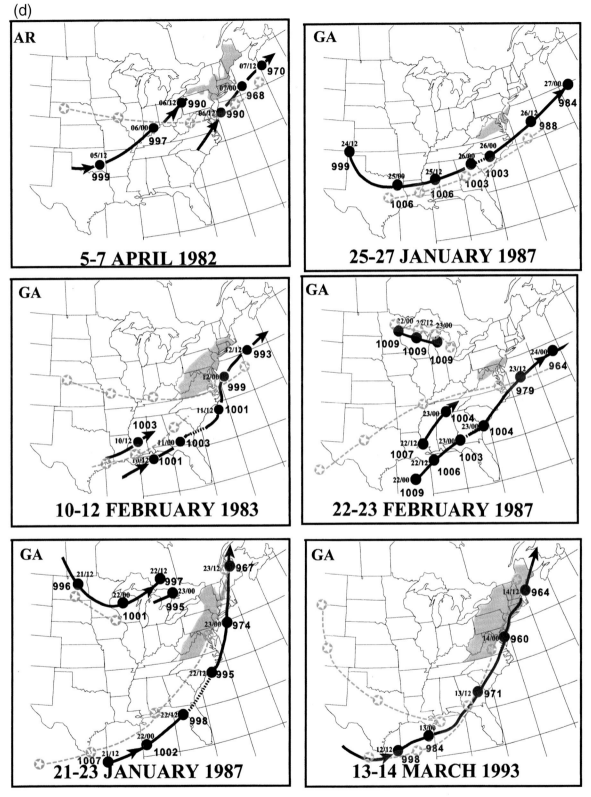

FIG. 3-3. (*Continued*)

(e)

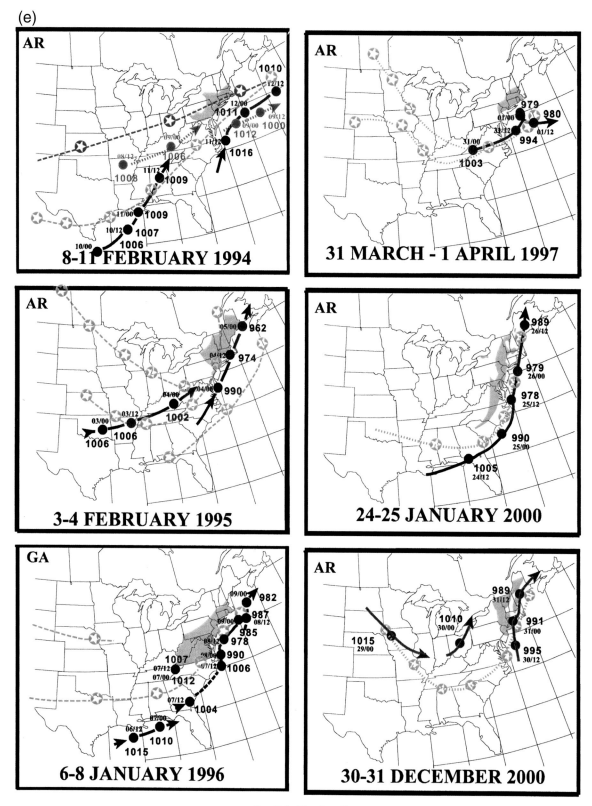

FIG. 3-3. (*Continued*)

GULF OF MEXICO-ATLANTIC COAST PATHS

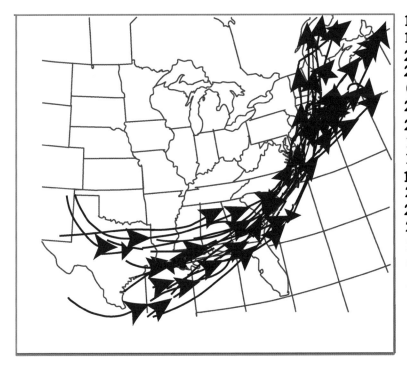

13-17 FEB 1958
18-21 MAR 1958
29-31 JAN 1966
23-25 DEC 1966
05-07 FEB 1967
22-28 FEB 1969
25-28 DEC 1969
18-20 FEB 1972
19-21 JAN 1978
10-12 FEB 1983
21-23 JAN 1987
25-27 JAN 1987
22-23 FEB 1987
13-14 MAR 1993
06-08 JAN 1996
24-26 JAN 2000

EXAMPLES

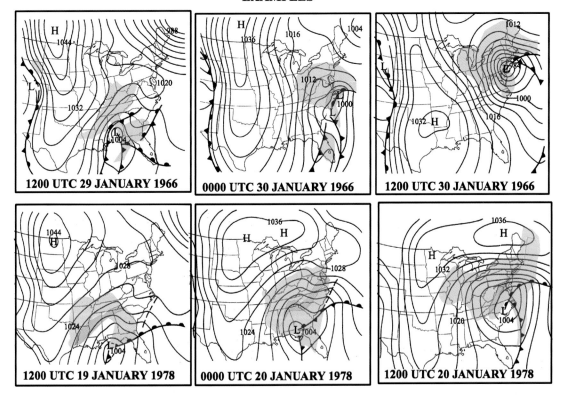

FIG. 3-4. Paths of all low pressure centers grouped according to Gulf of Mexico–Atlantic Coast category. Examples are shown beneath using 12-hourly surface analyses of the Jan 1966 and Jan 1978 snowstorms.

ATLANTIC COASTAL REDEVELOPMENT PATHS

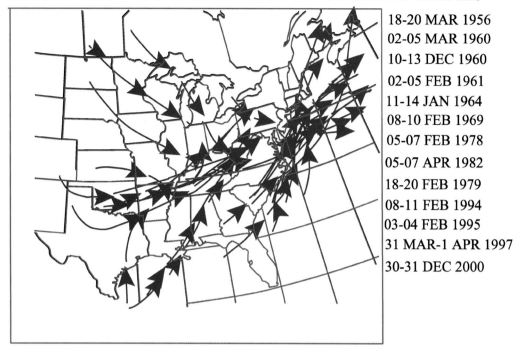

18-20 MAR 1956
02-05 MAR 1960
10-13 DEC 1960
02-05 FEB 1961
11-14 JAN 1964
08-10 FEB 1969
05-07 FEB 1978
05-07 APR 1982
18-20 FEB 1979
08-11 FEB 1994
03-04 FEB 1995
31 MAR-1 APR 1997
30-31 DEC 2000

EXAMPLES

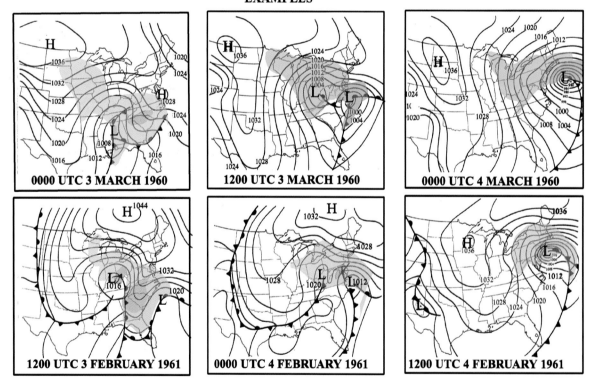

FIG. 3-5. Paths of all low pressure centers grouped according to Atlantic coastal redevelopment category. Examples are shown beneath using 12-hourly surface analyses of the Mar 1960 and Feb 1961 snowstorms.

jumps occurred almost exclusively along or near the Carolina coastline.

Thirteen cases fit into the Atlantic coastal redevelopment scenario (Fig. 3-5), in which a distinct second low pressure center forms to the southeast of a preexisting primary low (this scenario is well represented by the storms of March 1960, February 1961, January 1964, 8–10 February 1969, April 1982, and February 1995). The primary low pressure centers originated over diverse areas including the western Gulf of Mexico, the southern plains states, and the upper Midwest, with a locus of tracks oriented from the southern plains states to the upper Tennessee and Ohio Valleys (see Fig. 3-5). As each primary low pressure center moved toward the Appalachian Mountains, a secondary center developed over the Carolinas, southeastern Virginia, and the offshore Atlantic Ocean. Once the secondary low developed, the primary low then dissipated within 6–24 h. The secondary low pressure centers followed similar paths from eastern North Carolina northeastward to the waters offshore of the middle Atlantic states and southern New England, passing to the south of New England northeastward toward the offshore waters east of Nova Scotia. The one exception is the low pressure center associated with the January 1961 "Kennedy Inaugural Storm" (see Fig. 3-3a).[5] A study of a 7-yr sample of wintertime cyclones that crossed the Appalachians showed that although 30% of all cyclones did not conclusively exhibit redevelopment, "no cyclones appeared to cross the mountains completely intact" (O'Handley and Bosart 1996). They attributed this uncertainty to the limited spatial and temporal data resolution near the mountains.

In total, 23 of the 30 cases, including the 10 cases exhibiting "center jumps" and the 13 cases of Atlantic coastal developments clearly show a redevelopment and/or sudden acceleration of the surface low along the East Coast, especially along the Carolina coastline. The clustering of cyclone tracks near the coastline in systems that underwent secondary cyclogenesis or a center jump indicates that these storms are influenced by the underlying topography and coastal thermal contrasts. The Appalachian Mountains mark a transition between the areas where primary storm systems weaken and the region in which secondary storms develop and intensify farther east. Secondary cyclogenesis and center jumps also seem to be confined to the vicinity of the Carolina coast, lending support to the idea that the coastline and the land–ocean interface have a fundamental impact on

the dynamic and the thermodynamic processes that contribute to storm development.

b. Snowstorm duration and propagation rates

Storm duration is another factor that influences its impact, especially with respect to the total snowfall associated with each storm. Storm duration is shown for all 30 storms in Fig. 3-6 for all five major urban centers. The durations for snowfall, other frozen precipitation (freezing rain, ice pellets), and rain are included to help distinguish those storms that occurred as all snow from those storms that also reported a period of rain or other frozen precipitation for more than an hour. In general, about two-thirds of all cases occurred as all snow, while about one-third of the cases involved some combination of snow, rain, and ice pellets.

In the five major urban centers, 11 storms produced all of their precipitation in the form of snow (the storms of March 1956, February 1958, March 1960, January 1961, January 1964, February 1967, February 1978, February 1979, February 1983, 25–27 January 1987, and February 1987). For several of the other storms, the more southern cities of Washington and Baltimore received more hours of mixed precipitation than did their northern counterparts of New York and Boston. For example, mixed snow, rain, and/or ice pellets occurred for several hours during several storms (for the storms of December 1960, February 1961, February 1969, January 1978, April 1982, February 1994, January 1996, and 31 March– 1 April 1997), diminishing final accumulations. Conversely, precipitation fell mainly as snow in Washington, Baltimore, and Philadelphia but changed to rain farther north in New York and Boston in four cases (the storms of January 1966, December 1966, 21 January 1987, and February 1995), when the surface low propagated on a northward, rather than northeastward, path as it reached the Atlantic coast (see relevant cases in Fig. 3-3). In the last category of cases, snowfall changed to mixed precipitation or rain in all five cities during four storms (March 1958, December 1969, February 1972, and March 1993). Some of the factors for the changeover include the inland track of the surface low and, in the cases of March 1958 and April 1982, the late season. The physical processes that distinguish between snowstorms, rainstorms, and those storms containing mixed precipitation will be discussed in chapter 5.

The storm duration also provides an estimate of a mean snowfall rate for individual storms. For example, the February 1979 Presidents' Day Storm and the January Blizzard of 1996 both yielded approximately 20 in. (50 cm) of snow at Baltimore. However, the Presidents' Day Storm produced 20 in. (50 cm) of snow in only 19 h while the Blizzard of 1996 yielded 22.5 in. (57 cm) of snow over a period of 37 h. It is obvious that while slightly less snow fell in 1979, the snowfall rate during the Presidents' Day Storm was twice as great, on average, as that in the Blizzard of 1996. The

---

[5] It followed a path more typical of cyclones that undergo secondary redevelopment, but no secondary development was observed. It is possible that the rapid propagation of this system masked a redevelopment of the sea level center in southeastern Virginia while the primary low dissipated over the Tennessee–North Carolina mountains, as is inferred from 3-hourly (National Meteorological Center (NMC, now known as the National Centers for Environmental Prediction) analyses.

Snowfall, Precipitation type and Duration (hours)

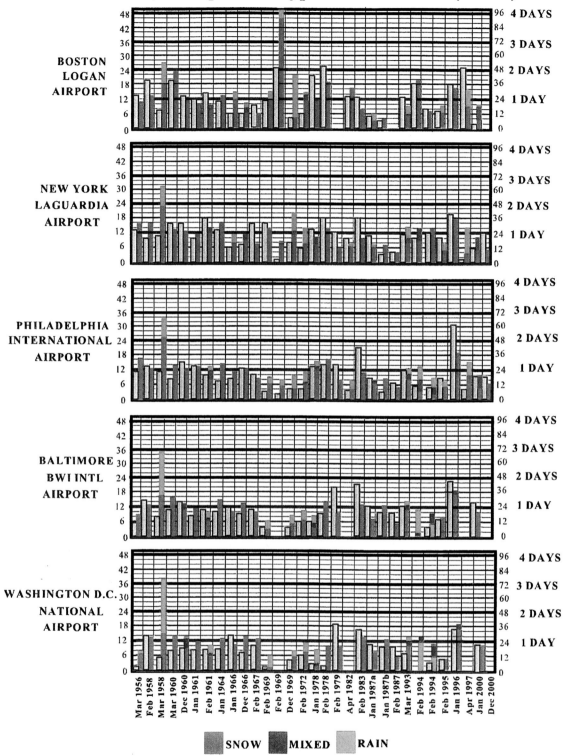

FIG. 3-6. Graph of total snowfall (gray, cm; scale on left) vs precipitation type and duration (h; scale on right) for 30 cases at five Northeast urban locations. Precipitation type is represented as snow (blue), ice pellets or mixed snow and ice pellets (violet), and rain or mixed rain, snow, and ice pellets (green).

length of each storm varied from around 10 h (for several storms that later changed to rain) to as much as 100 h during the late February 1969 snowstorm in Boston. The mean duration of the storm sample, calculated only for those cities in which greater than 6 in. of snow fell, increases from south to north from Washington to Boston with a mean of approximately 24 h at Washington to a mean of 31 h at Boston. The longer duration in the more northern locations is likely related to a well-known characteristic of the cyclones that decrease their forward motion as they intensify and "stall" near the New England coast (see below).

A comparison of the total snowfall (Fig. 3-2), storm duration (Fig. 3-6), and the paths of the surface low pressure centers (Fig. 3-3) indicates that differences in snowfall amounts and duration are associated with the considerable case-to-case variability in the propagation of the surface low pressure centers. In the examples cited above, the surface low pressure system associated with the January 1996 storm moved considerably slower than did the February 1979 Presidents' Day Storm. On one end of the spectrum are the rapidly moving cyclones, such as the surface low pressure center associated with the snowstorm of February 1967 (Fig. 3-3b), which moved from southern Alabama to a position south of Nova Scotia in a period of only 24 h at a mean rate of 28 m s^{-1}. Snowfall amounts along the axis of the heaviest snowfall were generally in the range of 8–12 in. (20–30 cm) since snow fell for a period of 15 h or less (Fig. 3-6).[6]

In contrast, the surface low pressure center associated with the snowstorm of March 1958 progressed northeastward at an unusually slow pace, averaging only 5 m s^{-1} over the 60-h period the storm was tracked (Fig. 3-3a). During this storm, snow fell continuously for 3 days at some locations, with accumulations of 30 to as much as 50 in. (75–125 cm) in Maryland and Pennsylvania. The 100-h snowstorm of late February 1969 was also associated with an exceptionally slow-moving cyclone that initially moved at a pace of 5–6 m s^{-1} and later wandered at a slower pace, resulting in the one of the heaviest snowstorms on record for eastern Massachusetts, New Hampshire, and Maine. The January Blizzard of 1996 was another slow-moving cyclone and was the most prolonged snowfall of the sample for Washington, Baltimore, and Philadelphia (Fig. 3-6), with snows falling for an average of 36 h.

In addition, most of the cyclones are marked by significant changes in speed during their evolution. Decelerations in the forward motion of the cyclones, sometimes termed "stalling" occurred most frequently in this

sample over the western Atlantic from east of the New Jersey coast to south and east of southern New England, and its most obvious effect was to prolong snowfall, especially in New England. Following the development of the New England snowstorm of February 1978, the surface low initially moved northward from off the Carolinas to a position east of the Maryland coast at a pace of approximately 13 m s^{-1} in the first 12 h after it developed, slowed to a mean pace of 8.5 m s^{-1} over a 12-h period as it neared Long Island, and then moved only at a pace of 4 m s^{-1} as it moved eastward south of New England in the 12-h period when much of southern New England received exceptionally heavy snowfall. In several storms (e.g., the blizzard of February 1958, the storm of March 1960, both February storms in 1969, January 1996, and that of 31 March–1 April 1997), the surface lows decelerated or stalled off the New Jersey or New England coasts, resulting in excessive snowfall amounts. In these slow-moving cyclones (the storms of February 1958, March 1958, February 1978, and March–April 1997), the surface low may even loop around itself. Many of these slower-moving storms occurred during late winter or early spring.

Although the number of heavy snow events diminishes as spring approaches, there may be a tendency for these storms to move more slowly and persist for a longer period as the global circulation patterns change. During winter, the midlatitude circulation is dominated by relatively fast-moving upper-level traveling shortwave troughs. As spring approaches, there is a transition to slower, higher-amplitude upper troughs and ridges that are more likely to cut off and form blocking patterns. Upper-level patterns are examined in chapter 4.

The most notorious example of a stalled cyclone that displays the temperatures of midwinter but the slower movement of a spring storm (which is often characterized by the establishment of large blocks) is the March 1888 "Blizzard of '88" (see volume II, chapter 9). This storm initially raced northeastward from the Carolina coast to southern New England, then remained nearly stationary close to the southern New England coast for 2 days, performing a slow counterclockwise loop south of Rhode Island. A protracted period of very heavy snow, strong winds, and extremely cold temperatures resulted. The deceleration occurs as a manifestation of the latter stages of cyclogenesis in which the surface low pressure center slows considerably as the upper-level low appears to cut off directly over the surface low before the entire system continues its eastward or northeastward track over the ocean. This factor had posed a considerable challenge to forecasters concerned with predicting an end to the snowfall and the final accumulation. Improved model forecasts appear to have reduced such forecast errors over the past 20 yr (see, e.g., Sanders 1987, 1992; Sanders and Auciello 1989; Grumm et al. 1992; Grumm 1993; Uccellini et al. 1995, 1999).

---

[6] Other relatively fast-moving storms include the low pressure system of 21–23 January 1987, which traveled from the central Gulf of Mexico to just south of Long Island in 24 h (Fig. 3-3b), averaging 27 m s^{-1}, and the Blizzard of January 1966, which moved from southern Alabama to New York City in 24 h (Fig. 3-3b), averaging nearly 20 m s^{-1}.

c. Cyclone intensification

It is a common perception that many Northeast snowstorms are associated with rapidly deepening cyclones. These storms are often marked by a distinct period during which the central pressure of the surface low decreases by 2–3 hPa h^{-1} with associated increases in the pressure gradients, surface winds, and precipitation rates. Moderate to heavy precipitation and snowfall are commonly observed along the Northeast coast during rapid surface intensification, resulting from the enhanced moisture and thermal advections that accompany explosive deepening. During the February 1979 Presidents' Day Storm, snowfall rates increased to 5–10 cm h^{-1} from Washington to New York City early on the morning of 19 February, coinciding with the onset of rapid surface deepening (Bosart 1981; Uccellini et al. 1984).

To examine deepening rates, the central sea level pressures of each storm were plotted at 3-h intervals for a 60-h period that includes storm initiation, development, and decay (Fig. 3-7). The period of deepening, marked by a decrease in central pressure of at least −1 hPa (per 3 h), generally occurred over a 24–48-h period for all but 3 of the 30 cases. The average deepening period was nearly 36 h with a mean decrease in central sea level pressure of 30 hPa. These values are consistent with Sanders's (1986a) observations of the duration and amount of deepening in a large sample of rapidly intensifying cyclones over the west-central North Atlantic Ocean and are also similar to 24-h pressure falls for a large sample of explosively deepening North Pacific cyclones (Gyakum et al. 1989). The maximum amount of deepening was 52 hPa (during the storms of 2–4 March 1960 and 18–21 January 1961), while five other cases deepened 40 hPa or greater.

Only four cases deepened 15 hPa or less during the 60-h period. Of the storms that deepened only marginally, the minimum amount of deepening of just 5 hPa was exhibited during the 11 February 1994 snow event. During the 19–20 January 1978 storm, sea level pressures decreased only 12 hPa, as central sea level pressures alternately rose and fell as the cyclone moved northeastward along the Atlantic coast. Most of the cases here contained a clearly defined and continuous period of intensification. Only a few storms (e.g., the storms of January 1964, December 1969, and January 1978; see Fig. 3-7) displayed multiple periods of deepening, separated by significant intervals of little or no change in central sea level pressure.

While most cyclones underwent a 36-h period of general deepening, periods of more rapid intensification occurred for 26 of the 30 cases. Rapid intensification is defined here as deepening rates exceeding −1 hPa h^{-1} over 12 h or greater (or −12 hPa per 12 h). Of these storms, 18 cases (with rapid deepening rates represented by the thicker pressure contours drawn in Fig. 3-7) exhibited periods of at least 6 h in which the cyclone's central pressure fell more than 10 hPa, the criterion used to select explosive cyclogenesis during the Experiment on Rapidly Intensifying Cyclones over the Atlantic (ERICA; Hadlock and Kreitzberg 1988). This rate of rapid intensification has been observed in individual events for periods of up to 27 h. Significant deepening rates were observed primarily along the coasts of the Carolinas, the middle Atlantic states, and east-northeastward over the Atlantic Ocean. The tracks of these storms are situated closer to the coastline than the locus of rapidly developing cyclone tracks described by Sanders [1986a; cf. his Fig. 2)] for a 4-yr sample of rapidly deepening cyclones over the western Atlantic Ocean. Conversely, the storms of March 1956, January 1978, February 1983, and February 1994 are the only ones that did not undergo at least a 12-h period in which sea level pressures fell 12 hPa or greater. Therefore, while rapid sea level development is a common characteristic of many of these storms, it is not a necessary condition for widespread heavy snowfall in the coastal sections of the Northeast.[7]

Among the 16 storms that exhibited either a center jump or no redevelopment, deepening rates exceeding −3 hPa (per 3 h) occurred in all cases, for periods of up to 27 h. Of the 13 events marked by secondary redevelopment, rapid deepening of the primary low pressure center was observed in only three cases and lasted no longer than 6 h. Meanwhile, 11 of the 13 cases exhibiting secondary low pressure development underwent a 12–24-h period of rapid deepening immediately along the coast. The only exceptions are the 18–20 March 1956 and February 1994 systems, in which rapid intensification was not observed with the development of the secondary low pressure center. As a rule, rapid deepening commenced with the formation of the secondary cyclone center. The primary low center either filled (its central pressure rose) or was absorbed within the expanding circulation of the secondary low within a period of 12–24 h following the commencement of secondary cyclogenesis. The sudden onset of a secondary cyclone and its subsequent, nearly immediate rapid deepening posed major problems for forecasters during most of the 20th century. It has only been during the final 10–15 yr of the 20th century that numerical simulations have been successful in predicting this type of behavior (Sanders 1987; Junker et al. 1989; Sanders and Auciello 1989; Mullen and Smith 1990; Sanders 1992; Grumm et al. 1992; Smith and Mullen 1993; Oravec and Grumm 1993; Grumm 1993; Uccellini et al. 1995, 1999) and contributed to the successful forecasts of these storms, as discussed in chapter 8.

---

[7] The February 2003 Presidents' Day Storm II is a dramatic case in point. Record snowfall fell in the Northeast while the central pressure of the surface low dropped only a few hectopascals.

d. Winds and pressure gradients

The impact of snowstorms within the Northeast urban corridor is also enhanced when the heaviest snow is accompanied by strong northeasterly winds, causing considerable blowing and drifting. Some Northeast snowstorms are particularly crippling because the heaviest snows and highest winds often occur at the same time. This is due to the effects of the strong temperature contrasts between the Atlantic Ocean and inland regions, the associated enhanced baroclinic processes, and reduced frictional effects, which impact the wind fields along the Northeast coast to a significant degree, particularly in advance of these cyclones.

Mean wind speeds were computed for all 30 cases (Fig. 3-8) in which 6 in. (15 cm) of snow or greater fell at any of the five major city airports (Boston, Logan; New York, La Guardia; Philadelphia, International; Baltimore, Baltimore–Washington International (BWI); and Washington, Reagan National). Wind speeds were averaged over the period from the hour in which snow began to the hour in which it ended. While it is recognized that these sites may or may not be representative of their particular urban region and are undoubtedly influenced greatly by their specific geography and instrumentation, these measurements provide at least a relative measure from one case to the next.

In general, Boston shows the highest average wind speeds of any of the five sites, due to the greater intensity of storms as they approach eastern New England and the location of Boston's Logan International Airport near the ocean, where wind speeds tend to be higher. Philadelphia, Baltimore, and Washington are located farther from the coastline than either New York or Boston, and, therefore, wind speeds tend to be less during these events. In Boston, mean wind speeds averaged 28 mi h^{-1} (13 m s^{-1}) for the entire set of storms.[8] The highest mean wind speed occurred during the storm of February 1978, which averaged 35 mi h^{-1} (16 m s^{-1}) with peak gusts reaching 79 mi h^{-1} (36 m s^{-1}). At New York's La Guardia Airport, the mean wind speed was 24.5 mi h^{-1} (11 m s^{-1}) and the highest mean wind speed was 36 mi h^{-1} (16 m s^{-1}) during the February 1961 snowstorm, the highest mean speed of any storm in the sample. During this event, wind gusts peaked at 83 mi h^{-1} (37 m s^{-1}). At Philadelphia, the mean wind speed was 19.5 mi h^{-1} (8.8 m s^{-1}) with the largest mean wind speed of 25 mi h^{-1} (11 m s^{-1}) occurring with the storm of January 1964. In Baltimore and Washington, the mean wind speeds were 15.5 (7) and 15.2 mi h^{-1} (7 m s^{-1}), respectively[9].

Rapidly falling pressures at the centers of these storms, combined with the decreasing distance between the propagating and developing cyclones and the persistent nearby anticyclones (whose high pressures are slow to erode over colder land surfaces), result in increased pressure gradients, especially north and east of the developing low pressure centers. During a 24-h period in which many of the cyclones underwent significant deepening (panels 3-5 of the surface analyses shown in chapter 10, volume II), the maximum gradients surrounding the low pressure centers increased by a factor of 2–10, with the gradients nearly tripling on average.[10] However, even in these cases with relatively weak gradients, gale force winds still occurred, especially along the coast of New England.

Several examples of cyclones that developed large sea level pressure gradients are shown in Fig. 3-9. The majority of storms developed the largest gradients in their direction of motion, northeast of the low pressure center, where strong northeasterly winds were usually observed. As these storms propagated northeastward along the coast and intensified, winds increased to gale force, or in some instances storm to hurricane force, along the immediate coastline.[11] Some of the more notorious storms also include the March 1993 Superstorm, whose vast circulation and dramatic low pressures brought gale to storm force winds along the entire length of the Atlantic coast. Another powerful example is the February 1978 snowstorm, which produced hurricane force wind gusts with heavy snow to widespread areas of southern and eastern New England and Long Island, New York. Both storms were responsible for crippling windblown heavy snowfall and severe coastal erosion. Although wind speed and direction are influenced by friction, isallobaric effects, and the flow of air relative to a moving storm system (i.e., Naistat and Young 1973), the large increases in pressure gradient associated with rapid deepening reliably indicate the high wind speeds that often accompany heavy snowfall. The large increases in maximum sea level pressure gradients are consistent with the tendency of these systems to deepen rapidly.

3. Anticyclones

In this section, the surface anticyclones that provide and maintain cold temperatures for the heavy snow events are described, together with a discussion of "cold-air damming" and coastal frontogenesis, two commonly observed features associated with the development of Northeast snowstorms.

---

[8] The windiest events for Boston were in March 1956, March 1958, February 1961, February 1978, and February 1983, when average wind speeds during the snowfall exceeded 30 mi h^{-1} (15 m s^{-1}).

[9] The largest mean speeds occurred with the storms of March 1993 [19.5 mi h^{-1} (8.8 m s^{-1})] and January 1996 [20.7 mi h^{-1} (9.3 m s^{-1})]. Peak gusts were measured in the 40–50 mi h^{-1} range (17–22 m s^{-1}).

[10] The weakest surface pressure gradients were associated with the cases of 22–28 February 1969, 19–21 January 1978, and 18–20 March 1956, which were also the cases in which sustained periods of rapid intensification did not occur.

[11] The February 1961 snowstorm and the New England snowstorm of February 1978 are probably two of the best examples of heavy snowfall accompanied by storm force northeasterly winds.

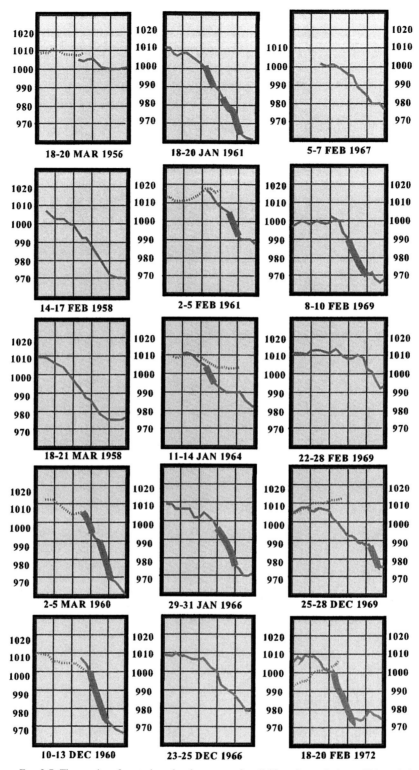

FIG. 3-7. Time series of central sea level pressures for all 30 cyclones during a 60-h period shown in the surface analyses in volume II, chapter 10. For cases of secondary cyclogenesis, the central pressures of the primary low are represented by dashed lines and those of the secondary low by solid lines. Thick solid lines represent periods during which the central sea level pressure fell by 10 hPa or more over 6 h.

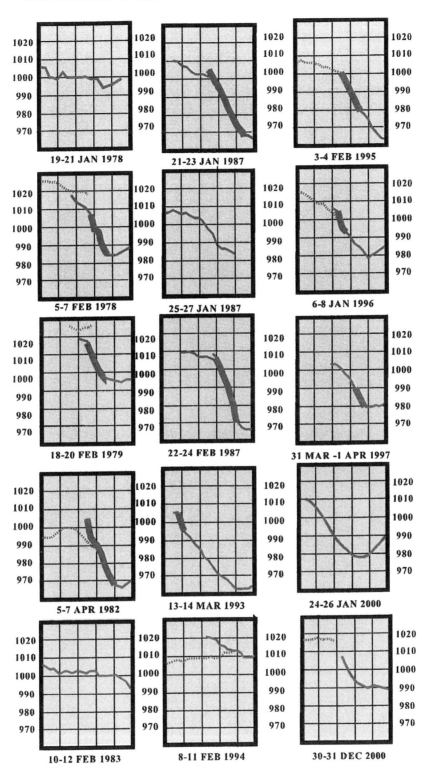

FIG. 3-7. (*Continued*)

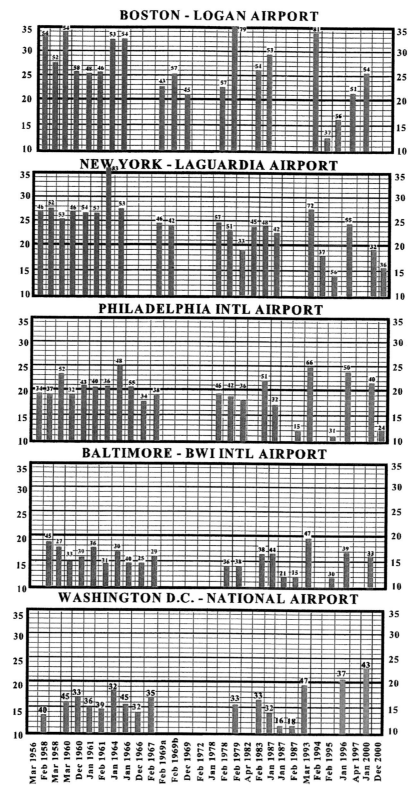

FIG. 3-8. Mean wind speeds (mi h⁻¹) at Boston, New York, Philadelphia, Baltimore, and Washington during snowstorms (mean speeds >20 mi h⁻¹, blue; mean speed < 20 mi h⁻¹, red). Mean winds were computed from hourly wind speed measurements beginning with the hour after the snow began and ending with the last hour in which snow was reported. Peak gusts (mi h⁻¹, top of bars) are also included. Computations were generated only for cases in which 6 in. (15 cm) or greater of snow fell.

Fig. 3-9. Examples of cyclones with extreme pressure gradients utilizing surface analyses derived from volume II, chapter 10 figures.

Major snowstorms along the Northeast coast occur with a wide range of temperatures, as shown in Fig. 3-10. As with the summaries describing storm-by-storm wind speed (Fig. 3-8), the range of temperatures and mean temperatures were computed for all 30 cases in which 6 in. (15 cm) of snow or greater fell at any of the five major city airports: Boston's Logan, New York's La Guardia, Philadelphia's International, Baltimore's BWI, and Washington's Reagan National. (Temperatures were averaged over the period from the hour in which snow began to the hour in which it ended.)

Some of the storms were accompanied by bitterly cold temperatures. At low temperatures of about 20°F (−6°C) or lower, snow tends to be powdery and drifts easily in the strong winds that often accompany these storms, creating dangerous traveling conditions. Historically, the "cold storm" of 1857 and the blizzards of 1888 and 1899 all produced significant accumulations accompanied by temperatures in the 0° to 10°F (−18° to −13°C) range. These represent some of the coldest conditions ever experienced in this region during a major snowfall.[12] Other major storms (including the storms of March 1960, January 1964, January 1966, February 1983, and January 1996) were accompanied by temperatures primarily in the teens to near 20°F (−13° to −7°C). The cities of Boston, New York, Philadelphia, Baltimore, and Washington have seen five, six, seven, six, and five storms, respectively, characterized by mean temperatures of 20°F or less (Fig. 3-10).

Conversely, the warmest storms were characterized by temperatures near or even slightly above freezing. At temperatures near freezing, snow clings to many surfaces, presenting a hazard for wires, flat roof structures, and shrubbery that are unable to withstand the load from heavy snow accumulations (the warmest storms of the sample occurred in March 1958, February 1987, and the Superstorm of March 1993). In some of these cases, snow changed to ice pellets, freezing rain, or rain during the storm period. It is interesting to note that during the storm of 22–23 February 1987, which occurred primarily at night, temperatures in the Washington area were near 50°F (10°C) the day prior to this nighttime snowstorm and during the daylight hours immediately following the major snowstorm. Most of the 10 in. (25 cm) of snow fell at night when the surface air temperature at Reagan National Airport failed to fall as low as the freezing level, remaining at 33°F (1°C) during the event.

Whether these storms were "wet" slushy snowstorms or fine, powdery blizzards, a source of cold air was required in each case to maintain temperatures low enough for the precipitation to fall as snow. In general, the cold air is associated with surface anticyclones,

which often originate over Canada. To identify common characteristics of anticyclones, centers of high pressure are plotted in Figs. 3-11a–e at 12-hourly intervals (this figure corresponds to the cyclone analyses in Figs. 3-3a–e and the surface analyses shown in chapter 10, volume II). The maximum central pressures of anticyclones occur within a range between 1020 and 1055 hPa for all cases. In many cases, well-defined high pressure cells were noted, with central pressures of 1040 hPa or greater; these events include the storms of March 1960 (1045 hPa), February 1961 (1047 hPa), January 1964 (1046 hPa), January 1978 (1046 hPa), February 1978 (1055 hPa), and February 1979 (1050 hPa). In some instances, the high pressure centers were ill defined with central pressures of 1030 hPa or less, including the storms of 8–10 February 1969 (1024 hPa), 21–23 January 1987 (1027 hPa), February 1987 (1025 hPa), and February 1995 (1021 hPa).

There are several different configurations that surface anticyclones take during these storms but a few basic patterns emerge from the variety of anticyclone paths prior to and during these intense snowstorms. In a number of instances, cold air was associated with anticyclone centers that typically followed paths from central Canada, either 1) as a primary high pressure center moved eastward across Ontario and Quebec, toward the Maritime Provinces, New York, or New England; or 2) as a secondary high pressure cell or ridge of high pressure extended eastward toward eastern Canada, New York, or New England from a primary high pressure center over south-central Canada or the north-central United States (Fig. 3-12).

Both sea level pressure anticyclone configurations shown in Fig. 3-12 produce a north- northeasterly flow of air along the coast that drives cold air southward toward the developing cyclone. The orientation of the anticyclone relative to the cyclone results in a low-level airflow that passes primarily over land, or for only a limited distance over the ocean before reaching the coastal sections of the northeastern United States. As a result, the cold air mass is not substantially modified by sensible and latent heating from the nearby ocean, allowing precipitation even in the coastal plain to fall as snow. The anticyclones also play a role in cyclogenesis, by maintaining or enhancing the low-level thermal fields and thermal advection patterns along the coastline, factors that contribute to cyclone development and enhanced deepening rates. Despite the frequency of cold-air outbreaks across the northeastern United States during the winter season and the passage of cyclonic weather systems toward the East Coast about every 3–5 days, they only rarely combine to produce heavy snowfall along the Northeast coast. The patterns of sea level pressure, wind, and temperature described here indicate that the position, orientation, and evolution of the anticyclone, relative to the coastline and the developing cyclone, are important factors for maintaining a cold low-

---

[12] More recently, the storms of December 1960, January 1961, February 1967, and the Presidents' Day Storm of February 1979 were accompanied by temperatures (°F) primarily in the single digits and teens (−15° to −7°C) throughout the Northeast urban corridor.

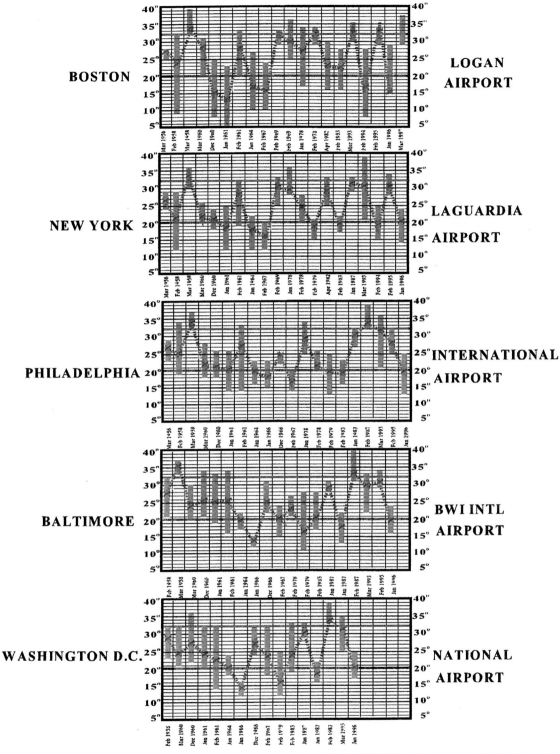

FIG. 3-10. Mean temperatures and temperature range (°F) at Boston, New York, Philadelphia, Baltimore, and Washington during snowstorms. Mean temperatures were computed from hourly temperatures beginning with the hour after the snow began and ending with the last hour in which snow was reported. Computations were generated only for cases in which 6 in. (15 cm) or greater of snow fell.

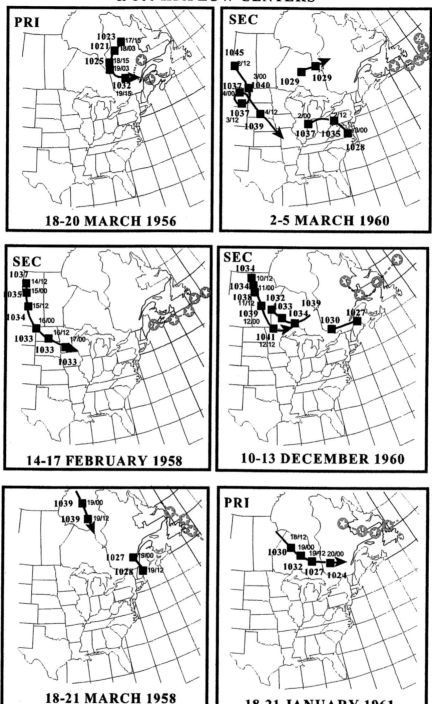

FIG. 3-11. (a)–(e) Paths of surface high pressure centers for all cases. Symbols along path represent 12-hourly positions and numerical values are the sea level central pressure (hPa) of the anticyclones. Positions of 500-hPa low centers associated with eastern Canadian upper troughs are also included. Starred symbols represent 12-hourly positions of the 500-hPa low center.

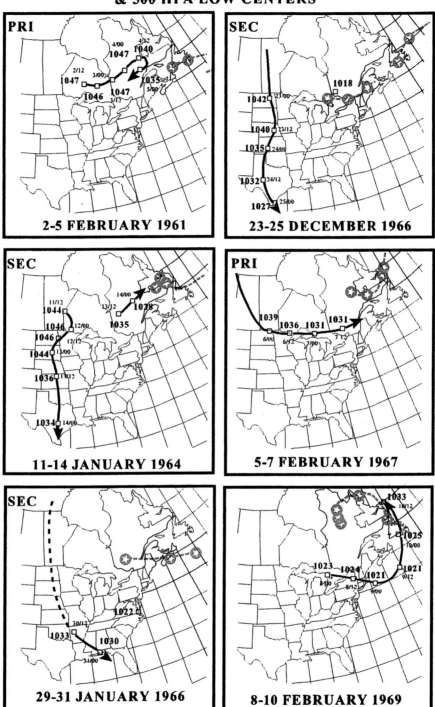

FIG. 3-11. (*Continued*)

level airflow either toward or along the coast that enables a significant snow pattern to develop throughout the Northeast urban corridor.

Heavy snowfall is less likely when the storm center passes nearby and especially to the north and west of a given location, since the cyclone's counterclockwise circulation draws in warmer air from the south and east. Heavy snow amounts also may not materialize when the

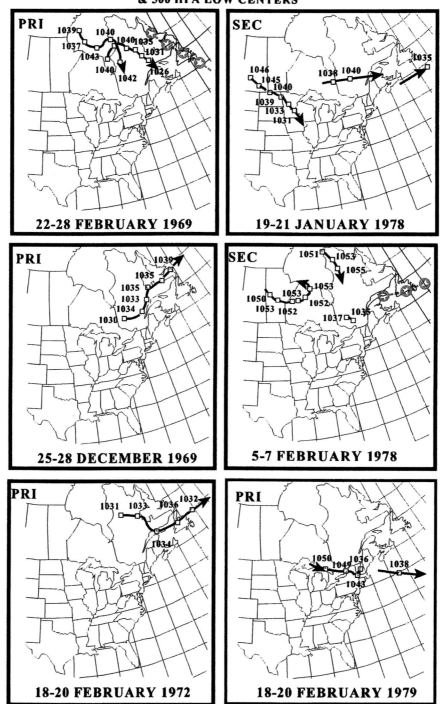

FIG. 3-11. (*Continued*)

anticyclone north of the region continues drifting eastward, producing a broad low-level flow of air passing over the ocean before reaching the coast. In these situations, precipitation often falls as rain along the immediate coast, with heavy snow more likely 100–500 km inland. Conditions that yield a changeover from snow to rain are explored in more detail in chapter 5.

In one category of anticyclone configurations typical to major snowstorms (Fig. 3-12), an anticyclone remains poised to the north-northwest of New England over the

(d) ANTICYCLONE TRACKS
& 500 HPA LOW CENTERS

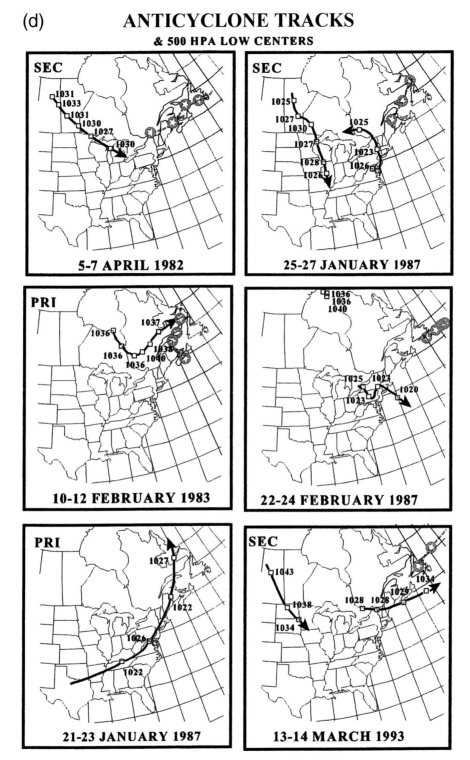

FIG. 3-11. (*Continued*)

provinces of Ontario and Quebec in Canada while the cyclone develops farther south along the East Coast. This scenario is represented by seven cases (March 1956, January 1961, February 1961, February 1967, February 1979, February 1983 and February 1994), whose composite paths are shown in Fig. 3-13. Two examples, February 1967 and February 1983, are shown at the bottom of Fig. 3-13. Four other cases also exhibit

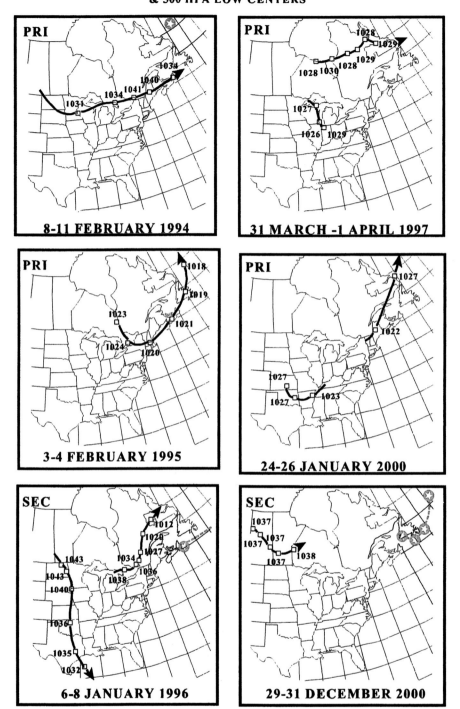

FIG. 3-11. (*Continued*)

cold-air sources in which cold air is provided by a solitary, primary anticyclone. These four cases (the storms of March 1956, January 1961, February 1961, February 1967, February 1979, February 1983, and February 1994) were associated with anticyclones that moved farther eastward than the previous examples and most were associated with a changeover from heavy snow to rain along the middle Atlantic and New England coasts (see chapter 5) and the paths of these anticyclones are shown in Fig. 3-13 in red.

TYPICAL ANTICYCLONE CONFIGURATIONS
FOR MAJOR NORTHEAST SNOWSTORMS

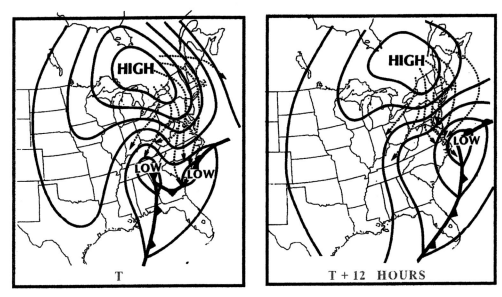

PRIMARY ANTICYCLONE OVER ONTARIO-QUEBEC

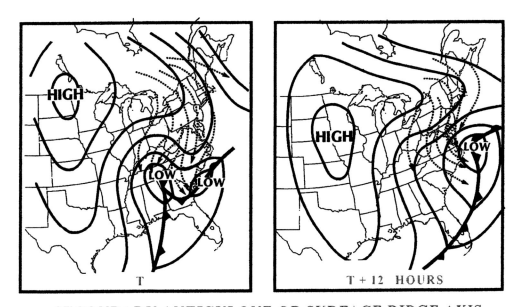

SECONDARY ANTICYLONE OR SURFACE RIDGE AXIS

FIG. 3-12. Schematic of two characteristic scenarios of anticyclone–cyclone couplets during major Northeast snowstorms. Schematic includes representative streamlines (dotted arrows) and sea level isobars (solid). Left- and right-hand columns depict surface analyses at 12-hourly intervals. Top panels show a well-defined anticyclone over Ontario and Quebec and bottom panel shows a high pressure cell over the plains states ridging eastward to the northeastern United States.

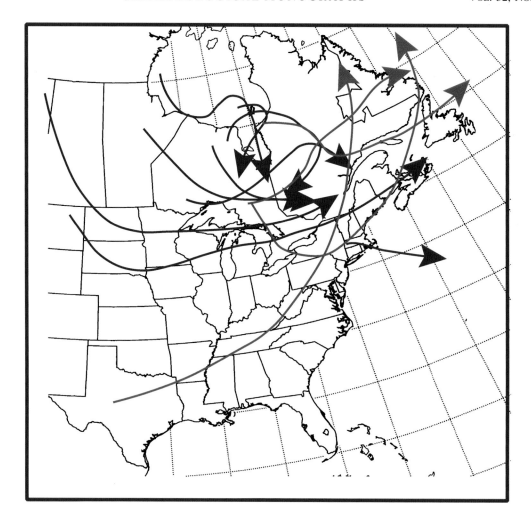

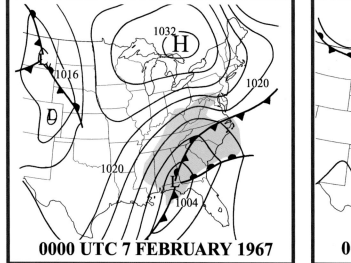

0000 UTC 7 FEBRUARY 1967

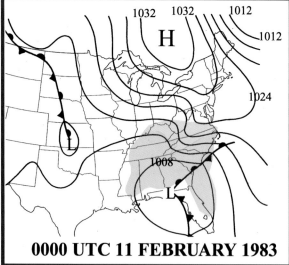

0000 UTC 11 FEBRUARY 1983

FIG. 3-13. Composite paths of all anticyclones with only one primary center over Ontario and Quebec. The cases of Feb 1967 and Feb 1983 are shown as examples.

In the second category of typical anticyclone configurations, a primary anticyclone remains nearly stationary; drifts southeastward across south-central Canada, particularly across the provinces of Saskatchewan and Manitoba; or passes southward over the north-central United States. While the primary anticyclone passes well to the west of the Northeast urban corridor, it is associated with either a surface ridge axis or a secondary high pressure center farther east over Quebec, Ontario, or the northeast United States (Fig. 3-12); examples of this pattern are demonstrated by the 11 cases of February 1958, March 1960, December 1960, January 1964, January 1966, December 1966, February 1978, April 1982, 25–27 January 1987, 6–8 January 1996, and 30–31 December 2000. The composite paths of these anticyclones are shown in Fig. 3-14. In two additional cases, the anticyclone initially is found over south-central Canada and drifts southeastward as a new anticyclone develops to the north and follows a path toward eastern Canada. This subcategory includes the storms of January 1978 and the Superstorm of March 1993 (paths are shown in red in Fig. 3-14), which also drifted east of Nova Scotia (these cases are similar to the storms of December 1969, February 1972, 21–22 January 1987, and February 1995), and was also associated with a changeover from snow to rain along the middle Atlantic and New England coasts.

An additional subcategory of anticyclones comprises those that appeared to have little impact on the Northeast snowstorms in that they were located too far north to have a significant influence (the three cases of March 1958, late February 1969, and 31 March–1 April 1997). The two cases of early February 1969 and February 1987 are enigmas in that heavy snow fell with little or no rain despite the retreat of the surface high offshore, a position typically unfavorable for heavy snow. Finally, the snowstorm of 24–25 January 2000 was indirectly associated with two separate anticyclones, one building eastward from the plains states, and one retreating northeastward across eastern Canada. This case does not fit neatly into any category.

a. Cold-air damming and coastal frontogenesis

Cold-air damming and coastal frontogenesis are linked to the underlying topography of the eastern United States and appear to have an important influence on major snow events along the northeast coast. Cold air at low levels of the atmosphere frequently becomes trapped or "dammed" between the Appalachian Mountains and the Atlantic coast when a cold surface anticyclone passes to the north of the middle Atlantic states and southern New England (see Baker 1970; Bosart et al. 1972; Richwien 1980; Ballentine 1980; Forbes et al. 1987; Stauffer and Warner 1987; Bell and Bosart 1988, 1989; Nielsen 1989; LaPenta and Seaman 1990, 1992; Nielsen and Neilley 1990; Riordan 1990; Doyle and Warner 1993a,b). The forecasting of damming events

by operational large-scale numerical models (Richwien 1980; Forbes et al. 1987) was problematic through the 1980s. With damming misrepresented in earlier operational and research numerical models, lower-tropospheric thermal fields would be poorly resolved, resulting in temperature forecasts that were too high and contributing to the difficulty in delineating frozen versus liquid precipitation along the coastal plain. Research efforts in the 1980s and 1990s paved the way for improvements in numerical model forecasts of cold-air damming, as well as coastal frontogenesis, principally related to improvements in the vertical and horizontal resolution of the atmosphere's planetary boundary layer and an improved representation of topography (e.g., Doyle and Warner 1993a,b).

Damming is characterized by a narrow, shallow wedge of cold air that persists east of the Appalachians in association with significantly ageostrophic flow near the earth's surface. Damming develops as cold air supplied by the anticyclone to the north is channeled southward along the eastern slopes of the Appalachians. The cold-air wedge is then maintained by a combination of frictional and isallobaric effects, evaporation, adiabatic ascent, and orographic channeling (Forbes et al. 1987; Bell and Bosart 1988). Damming is represented in the sea level pressure field by an inverted high pressure ridge between the mountains and the coast, which may extend as far south as Florida. The distinctive appearance of cold-air damming on the surface weather charts is especially evident in the examples shown in Figs. 3-13 and 3-14.

As a cold anticyclone passes across the northeast United States during winter, the eastern flank of the air mass moves over the Atlantic Ocean. A portion of the airstream connected with the anticyclone often flows from an easterly to northeasterly direction over the western Atlantic, while low-level winds over the coastal plain retain a more northerly component, as a result of cold-air damming. The deformation and convergence of these airstreams result in a zone of enhanced low-level baroclinicity near the coastline, since the air mass over the land remains cold, while air over the ocean is modified by sensible and latent heat fluxes within the ocean-influenced planetary boundary layer. These developments describe the process of "coastal frontogenesis" (Bosart et al. 1972).

In general, the coastal fronts that form during damming episodes are easily identified by an inverted sea level pressure trough near the coast, or just offshore, and are marked by a distinct temperature contrast and by changes in wind direction. These characteristics of coastal frontogenesis were observed in 25 of 30 cases; the five cases not exhibiting coastal frontogenesis are March 1956, January 1961, December 1966, February 1978, and March–April 1997. Sometimes the airflow in the cold wedge can be highly ageostrophic with north to northwesterly winds in the wedge. To the east of the coastal front, temperatures could be 20°F (11°C) higher

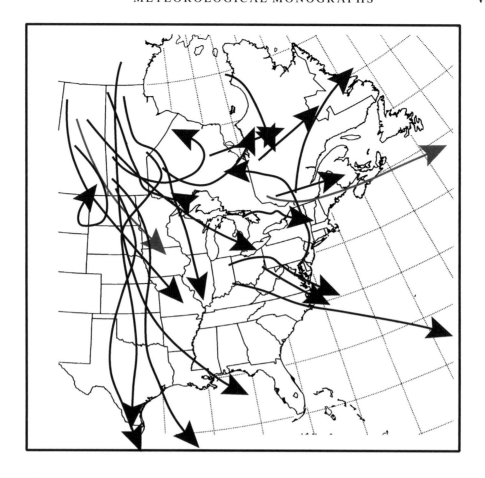

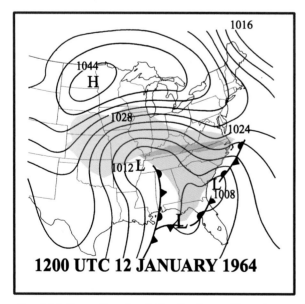

1200 UTC 12 JANUARY 1964

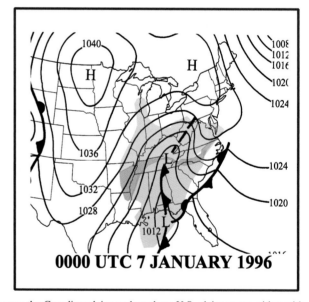

0000 UTC 7 JANUARY 1996

FIG. 3-14. Composite paths of all anticyclones with a primary path over the Canadian plains and northern U.S. plains states, either with a secondary high or surface ridge over Quebec and Ontario, New England, or the Northeast. The cases of Jan 1964 and Jan 1996 are shown as examples.

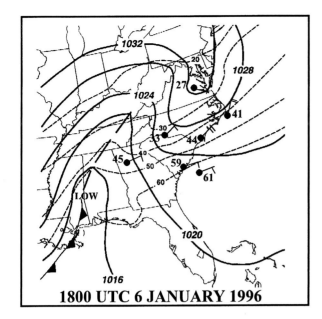

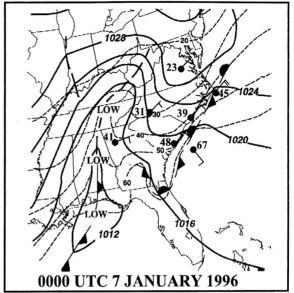

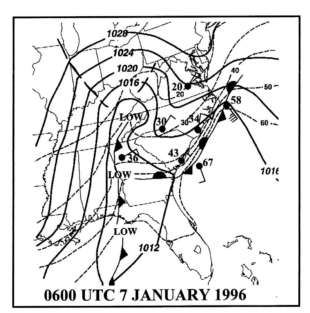

FIG. 3-15. Evolution of cold-air damming and coastal frontogenesis during the snowstorm of 6–7 Jan 1996, including isobars (hPa, solid), isotherms (°F, dashed), fronts, and selected observations with temperatures (°F) and wind barbs.

and winds will often blow from an east to southeasterly direction. Coastal frontogenesis can play an important role in the development of Northeast snowstorms since it provides a site where low-level convergence, baroclinicity, warm-air advections, and surface vorticity are maximized (Bosart 1975). Thus, the coastal front becomes a channel along which the surface cyclone rapidly develops and propagates. As noted by Keshishian and Bosart (1987), some cyclones remain weak and move along the coastal front, possibly as a series of low pressure systems, termed "zipper lows." Furthermore, the convergence associated with coastal frontogenesis

is probably responsible for enhancing the precipitation rate along the cold side of the frontal boundary, where the heaviest snowfall is often observed (Bosart et al. 1972; Bosart 1975; Marks and Austin 1979; Bosart 1981).

The January Blizzard of 1996 is used as an example of cold-air damming and coastal frontogenesis associated with major snowstorms (Fig. 3-15). An anticyclone center to the north of the Northeast urban corridor extends its broad circulation over the East Coast and adjacent Atlantic Ocean. As cold air within the eastern flank of the surface high is modified over the ocean, a

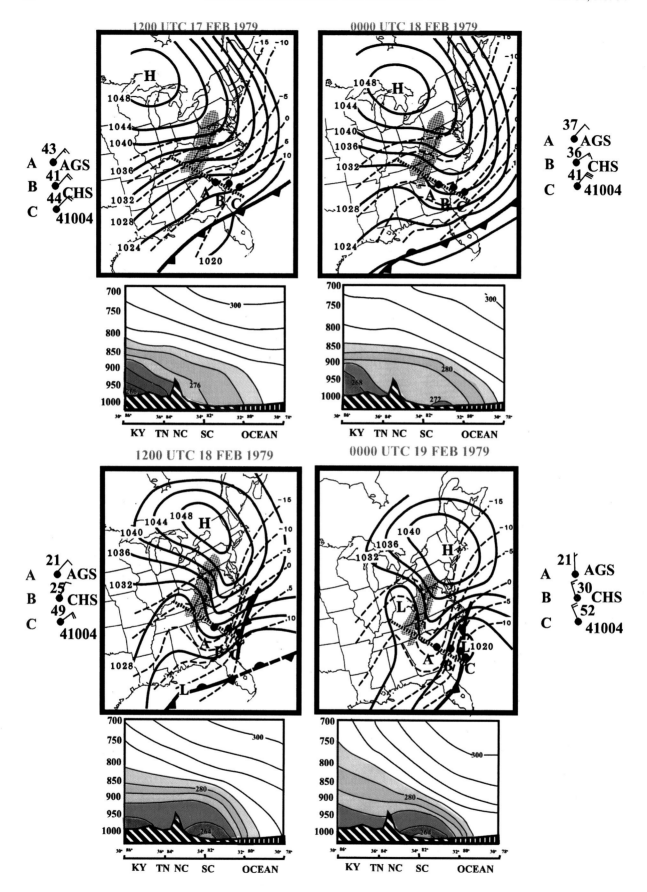

coastal trough develops, followed by coastal frontogenesis. The frontogenesis provides a favorable corridor within which cyclogenesis will occur and eventually track. When cold air is dammed against the Appalachians, low-level warm-air advection is suppressed over the coastal plain, thereby maintaining the thermal gradients and warm-air advection pattern along the coastline or farther offshore. As a result, when a cyclone and its associated upper-level forcing approach the Appalachians from the west, the low-level warm-air advection pattern often shifts to the east of the cold-air wedge and is focused along the coastline. This enhancement of the coastal baroclinic zone facilitates the formation of a secondary cyclone along the East Coast as the initial cyclone weakens in the Ohio Valley (see, e.g., Uccellini et al. 1987). Although quantifying the impact of damming is difficult and its presence is no guarantee that precipitation will fall as either snow, ice pellets, rain, or freezing rain, the inverted sea level pressure ridge and cold-air wedge were observed, to some degree, in all cases (possible exceptions include the storms of January 1961, 31 March–1 April 1997, and 24–26 January 2000).

A series of sea level pressure and surface temperature analyses and selected vertical cross sections were drawn for the 36-h period preceding the development of the February 1979 Presidents' Day Storm, to further illustrate the evolution of damming and coastal frontogenesis along the East Coast prior to a major snowstorm (Fig. 3-16). Between 1200 UTC 17 February and 0000 UTC 19 February 1979, a large anticyclone accompanied by record- or near-record-setting minimum surface temperatures drifted slowly eastward from the upper Great Lakes to New York. During this period, significant changes in the orientation of the sea level pressure field and surface isotherms indicated damming conditions. By 1200 UTC 18 February 1979, anticyclonically curved isobars and isotherms reflect the highest pressures and lowest temperatures located between the Appalachians and the coast. The coldest temperatures were nearly collocated with the ridge axis, as discussed by Baker (1970) and Richwien (1980). Meanwhile, the thermal gradient along the Southeast coast intensified, as temperatures decreased over the land and increased offshore, indicating the development of a coastal front.

Cold-air damming and coastal frontogenesis are also depicted by the vertical cross sections of potential temperature in Fig. 3-16. After 1200 UTC 17 February, decreasing potential temperatures near the earth's surface created the characteristic "cold dome" (i.e., see Fig. 14 in Bell and Bosart 1988), as the shallow, cold-air mass wedged between the mountains and coastline. The increasing horizontal potential temperature gradient along the coast identifies the coastal frontogenesis. A detailed description of the coastal frontogenesis associated with the Presidents' Day Storm can be found in Bosart (1981). A numerical-based diagnostic study of the evolution of the temperature and wind fields associated with the coastal front and the initial cyclogenesis can be found in Uccellini et al. (1987). The relationships between coastal frontogenesis and the development of mesoscale bands of heavy snow have been described by Marks and Austin (1979) and observed with a heavy snowband over Long Island, New York, in December 1988 (Roebber et al. 1994; Maglaras et al. 1995) and will be discussed in more detail in chapter 6.

To summarize, a majority of the 30 cases exhibit two primary patterns observed in the evolution of anticyclones over Canada that appear to allow cold air to remain within the Northeast urban corridor, enhancing the likelihood of snow versus rain. The interactions of the anticyclones, a source of cold air, with the developing cyclone play an important role in the subsequent evolution of the cyclone and associated snowstorm, especially with regard to the development of a strong baroclinic zone oriented near the coastline. Cold-air damming east of the Appalachian Mountains and coastal frontogenesis along or near the East Coast are two processes linked to the cold anticyclones and have a significant bearing on the development of Northeast snowstorms. These features provide an enhanced low-level baroclinic zone that shifts the cyclone path toward a track close to the Atlantic coast, or offshore.

In the next chapter, upper-level features such as trough–ridge configurations, confluence, and jet streaks will be linked to the evolution of the surface cyclones, anticyclones, and their roles in establishing conditions conducive for heavy snowfall in the Northeast.

←

FIG. 3-16. Example of a cold-air damming event and coastal frontogenesis prior to the February 1979 Presidents' Day snowstorm at (a) 1200 UTC 17 Feb, (b) 0000 UTC 18 Feb, (c) 1200 UTC 18 Feb, and (d) 0000 UTC 19 Feb 1979. Top panels show surface fronts, sea level isobars (solid, hPa), and surface isotherms (dashed, °C), and shading represents location of the Appalachian Mountains. Thick dotted line represents axis of cross sections shown at the bottom; A, B, and C represent Augusta, GA; Chaleston, SC; and NOAA buoy 41004, stations close to the axis of the cross section. Temperature and wind barbs are included. Bottom panels show cross sections of potential temperature (solid, °C; colored regions represent coldest potential temperatures).

Chapter 4

SYNOPTIC DESCRIPTIONS OF MAJOR SNOWSTORMS: UPPER-LEVEL FEATURES

In this chapter, a description of upper-level meteorological fields for 30 cases provides an overview of the geopotential height, wind, and temperature characteristics that support the development of Northeast snowstorms. Descriptions of 850-, 700-, 500-, 400-, and 250-hPa charts[1] focus on trough–ridge systems, thermal advections, and the horizontal and vertical orientation of jet streaks, which act to support the surface cyclone–anticyclone patterns discussed in chapter 3 that play such a critical role in the development of the snowstorms in the Northeast urban corridor. The chapter ends with a description of cloud features evident in satellite images.

1. Descriptions of upper-level features

The study of upper-level conditions required for surface cyclogenesis dominated meteorological research from the 1930s through the 1950s, concurrent with the introduction of techniques to monitor and measure atmospheric winds, temperature, and moisture above the earth's surface. These measurements, combined with concepts promoted by Sutcliffe (1939, 1947), Bjerknes and Holmboe (1944), Charney (1947), Sutcliffe and Forsdyke (1950), Bjerknes (1951), Palmén (1951), Petterssen (1955, 1956), and others, permitted weather forecasters to identify upper-atmospheric features and patterns that contribute to the development of surface cyclones. A detailed discussion is provided in chapter 7.

An outgrowth of these efforts was the recognition that a baroclinic environment marked by 1) upper-level troughs, 2) large thermal gradients in the middle and lower troposphere, 3) high wind speeds associated with the jet stream and embedded jet streaks, and 4) associated vertical and horizontal wind shears is essential

for the initiation and intensification of cyclones. Since all of the 30 snowstorms in this chapter were associated with significant cyclonic development, the following discussion on upper-level features will focus on characteristics of the upper-level troughs, ridges, and related vorticity and wind fields.

The 500-hPa geopotential height and vorticity fields and summaries of the 250- and 400-hPa wind fields are presented to describe the trough–ridge systems and jet streaks associated with the ensemble of 30 storms. The complex structure of the geopotential height and wind fields defies a simple description as noted by Palmén and Newton (1969, p. 273) in their review of numerous cyclone studies. Nevertheless, several generalizations will be offered for the 30 cases. The upper-level conditions that promote the characteristics of anticyclones that favor major snow events will also be explored in this chapter. Finally, a description of the 850-hPa temperature, wind, and moisture fields will illustrate the characteristics of lower-tropospheric thermal and moisture advections during cyclogenesis.

2. The large-scale upper-level environment prior to the development of Northeast snowstorms

Forecasters have long recognized that certain synoptic-scale patterns appear to favor the development of major snow events. These distinct upper-flow configurations provide a dynamical basis for examining the storms' physical mechanisms. In this section, upper-level signatures for 30 cases are examined at approximately 3, 2, and 1 days prior to heavy snow development along the Northeast coast. While there are many complex configurations that defy simple categorization, some generalizations can be made.

a. 72 h prior to heavy snow development

At 72 h prior to heavy snow development (Fig. 4-1), a number of major circulation features exist that provide an initial suggestion of conditions favorable for heavy snow. There appear to be at least two separate streams of air at upper levels that eventually produce conditions that result in heavy snowfall in the Northeast urban

[1] Upper-level analyses were derived from National Centers for Environmental Prediction (NCEP) analyses available on microfilm from the National Climatic Data Center (NCDC) in Asheville, North Carolina, and from datasets derived from the NCEP–National Center for Atmospheric Research (NCAR) reanalysis project (Kalnay et al. 1996). General Meteorological Package (GEMPAK; des Jardins et al. 1995) datasets were generated at the National Weather Service's (NWS) COMET program in Boulder, Colorado, and displayed on National Centers Advanced Weather Interactive Processing System (N-AWIPS) software developed at NCEP.

(a) 72 HOURS PRIOR TO HEAVY SNOW DEVELOPMENT

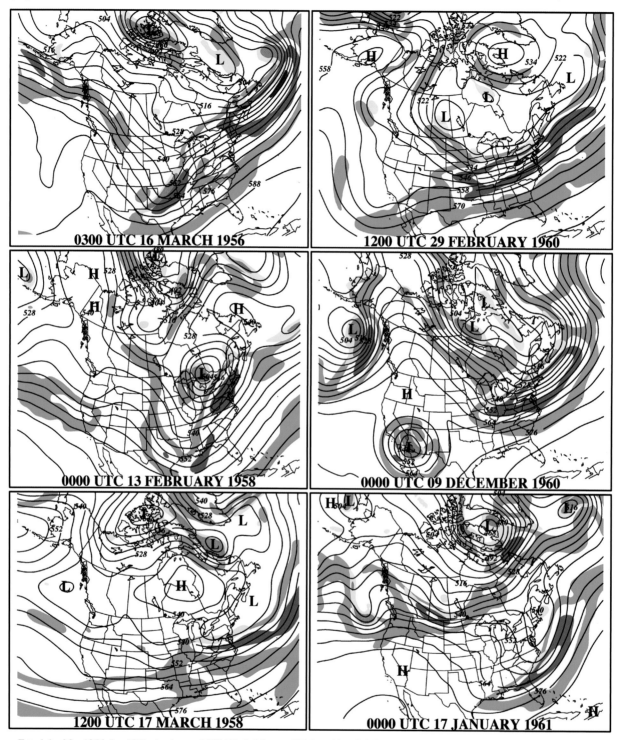

FIG. 4-1a. Mar 1956–Jan 1961. Analyses of 500-hPa heights, and high and low height centers, where yellow and orange shading represents absolute vorticity (greater than 16×10^{-5} s^{-1}, at 4×10^{-5} s^{-1} increments); and isotachs of maximum wind speeds observed at 400 hPa (alternating shading represents 5 m s^{-1} increments of wind speed starting with 40 m s^{-1}) 72 h prior to the development of heavy snow (times coincide with the analyses in Fig. 4-5).

(b) 72 HOURS PRIOR TO HEAVY SNOW DEVELOPMENT

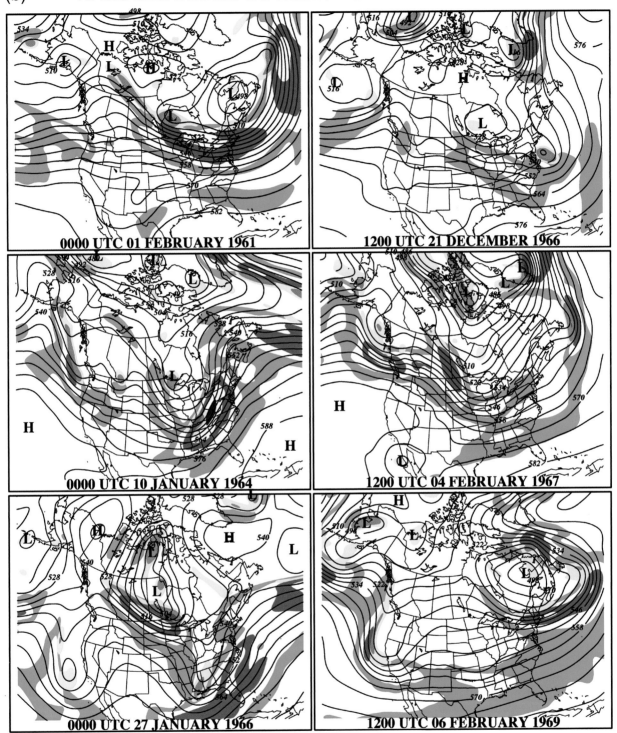

FIG. 4-1b. (*Continued*) As in Fig. 4-1a, but for Feb 1961–6 Feb 1969.

corridor. One stream of air is associated with cyclonic disturbances that are often responsible for the subsequent cyclonic development. The other stream of air may also be associated with a separate disturbance but

is more likely associated with features that allow cold air to penetrate and remain in the Northeast.

At 72 h, the upper disturbances that ultimately become associated with cyclonic development along the

(c)   72 HOURS PRIOR TO HEAVY SNOW DEVELOPMENT

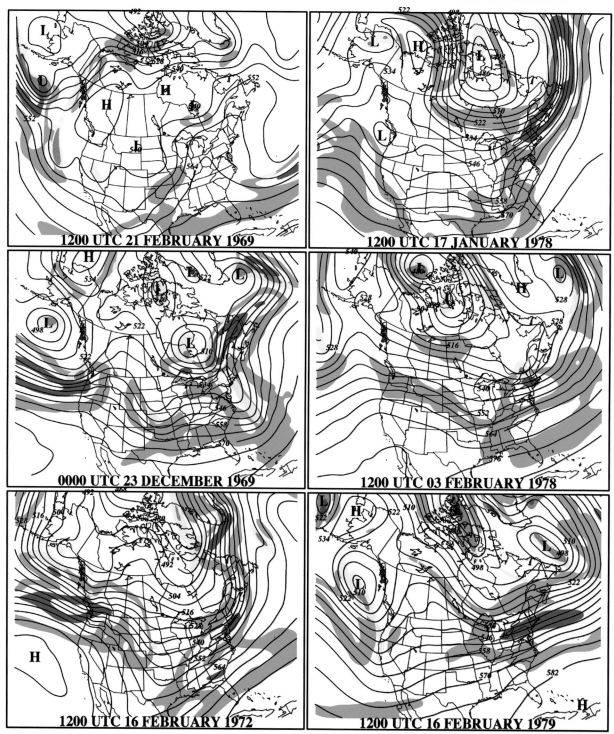

FIG. 4-1c. (*Continued*) As in Fig. 4-1a, but for 21 Feb 1969–Feb 1979.

East Coast may be divided into two major groups. The first group is accompanied by a very strong westerly jet along and to the west of the California coast that is suppressed farther south than is typical during the winter season. This strong zonal flow later breaks down with the amplification of a short-wave trough over the eastern United States. A short-wave trough embedded within this flow is represented by several cases (for 0000 UTC

(d)   72 HOURS PRIOR TO HEAVY SNOW DEVELOPMENT

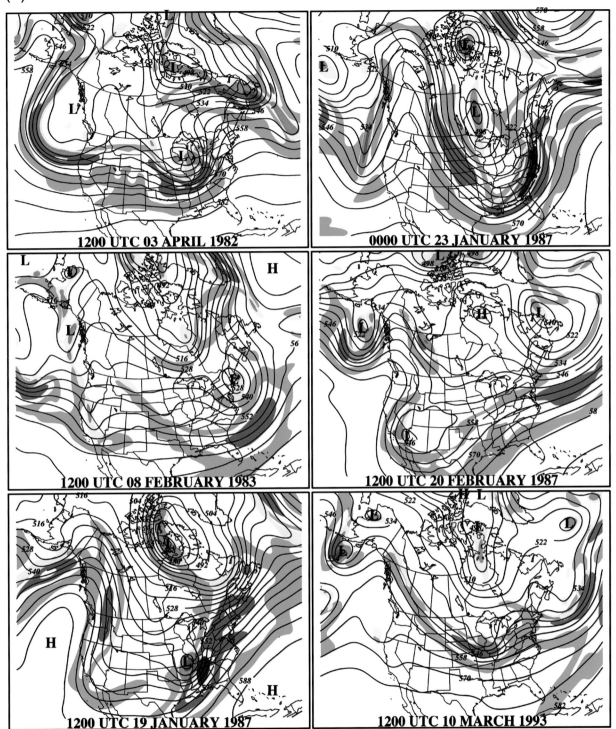

FIG. 4-1d. (*Continued*) As in Fig. 4-1a, but for Apr 1982–Mar 1993.

13 February 1958, 1200 UTC 17 March 1958, 1200 UTC 6 February 1969, and 1200 UTC 8 February 1983) illustrated in Fig. 4-1. Conditions during a strong ENSO (Halpert and Bell 1997) are often favored in this sce-nario and these four examples occurred during strong positive phases of the ENSO (see Fig. 2-19).

The second category of flow patterns conducive to subsequent East Coast cyclogenesis is an amplifying

(e) ## 72 HOURS PRIOR TO HEAVY SNOW DEVELOPMENT

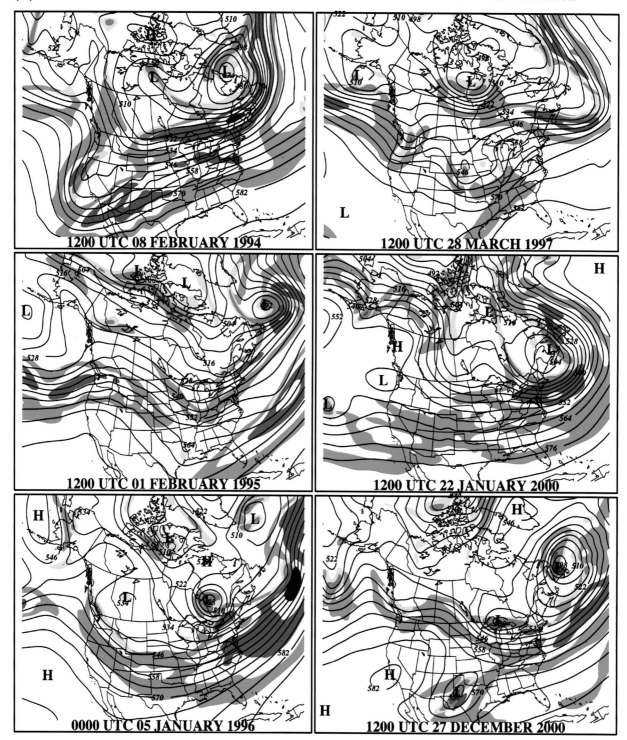

FIG. 4-1e. (*Continued*) As in Fig. 4-1a, but for Feb 1994–Dec 2000.

upper ridge typically near the Pacific coast and a subsequent deepening trough on the eastern flank of the upper ridge. In some of these cases, the short-wave trough or impulse that later spawns the cyclone may initially move within the mean upper-ridge position and as the impulse translates to the east-southeast, the upper ridge amplifies significantly. This scenario (for 0000 UTC 10 January 1964, 1200 UTC 19 January 1987, and 0000 UTC 5 January 1996) is also illustrated in Fig. 4-1. At 72 h, this amplifying ridge is often observed over the eastern Pacific Ocean in about half the cases. In some cases, the amplifying trough is composed of one primary impulse or short-wave trough while others are composed of a complex evolution of multiple impulses that merge or amplify (see section 3a of this chapter).

The other significant feature that precedes the development of these storms at 72 h is the existence of a trough or closed low aloft in eastern Canada that appears to support strong Canadian anticyclones that supply cold air to the northeast United States (see section 3b of this chapter). Many of the cases shown in Fig. 4-1 (0000 UTC 13 February 1958, 0000 UTC 9 December 1960, and 0000 UTC 1 February 1961) illustrate eastern Canadian troughs with closed centers and closed upper low–high couplets. While these features are not observed with every case 72 h prior the snow event, the other cases (March 1956, January 1961, February 1967, and February 1983) involve similar features that are progressing toward this trough configuration in eastern Canada, but at times closer to the heavy snow event.

b. 48 h prior to heavy snow development

Some very distinctive patterns become established 48 h prior to the development of heavy snow that signify the mechanisms that initiate cyclogenesis and provide a source of cold air to the Northeast (Fig. 4-2). A common signal at 48 h prior to East Coast cyclogenesis is a building upper ridge along the West Coast and an amplifying trough downstream over the Rocky Mountain or plains states. This regime is illustrated in Fig. 4-2 (for the cases at 0000 UTC 18 January 1961, 0000 UTC 2 February 1961, 0000 UTC 11 January 1964). In some cases (0300 UTC 17 March 1956, 1200 UTC 4 February 1978, and 1200 UTC 2 February 1995), a pronounced northwesterly flow on the east side of this ridge has yet to consolidate into a distinct trough 48 h prior to the heavy snowstorms. These features represent early stages of "downstream amplification" (Simmons and Hoskins 1979) in which the development of an upper-level trough is preceded by the amplification of a ridge immediately upstream.

Another important signal for subsequent cyclogenesis is the presence of a strong short-wave disturbance in the southwest United States. In some cases (0000 UTC 10 December 1960, 0000 UTC 28 January 1966, and 1200 UTC 20 February 1987), a nearly stationary upper low over the southwestern United States begins to move

eastward and is associated later with East Coast cyclogenesis. In other cases (1200 UTC 18 March 1958, 1200 UTC 7 February 1969, and 1200 UTC 9 February 1983), the observation of a nearly zonal jet stream over the southern United States with an embedded upper-level disturbance–trough reflects the trough observed off the California coast 24 h earlier.

Upper-level features that allow cold air to flow into the northeast United States are nearly always associated with an upper trough in eastern Canada. This trough, combined with a climatological ridge over the Atlantic Ocean, produce a confluent jet entrance region, an important factor that enhances the cold advection in the lowest layers (see chapter 7). The troughs in eastern Canada present several different appearances 48 h prior to heavy snow development along the East Coast. Most eastern Canadian troughs at this time (0000 UTC 10 December 1960, 0000 UTC 28 January 1966, and 0000 UTC 24 January 1987) are very large closed circulation systems; some cover large areas of northeastern Canada (Fig. 4-2), but others (0000 UTC 14 February 1958, 1200 UTC 1 March 1960, 1200 UTC 21 February 1987, and 0000 UTC 6 January 1996) are smaller but distinct closed lows (Fig. 4-2). Some systems (0000 UTC 18 January 1961, 1200 UTC 5 February 1967, 1200 UTC 4 April 1982, and 1200 UTC 11 March 1993) are open wave troughs that are amplifying and will later cut off and decelerate (Fig. 4-2), locking in a confluent flow pattern over southeastern Canada and the northeast United States.

c. 24 h prior to heavy snow development

By 24 h (Fig. 4-3), the major features are converging toward a heavy snow situation for the Northeast. Many of the upper-level troughs that will be responsible for East Coast cyclogenesis are amplifying and well defined, often located over the central United States, and are associated with distinct jet streaks propagating toward the East Coast. The distribution of upper vorticity maxima used to track the upper-level troughs are more consolidated than at 72 h, and several have begun to merge together (see section 3a below).

An upper trough over eastern Canada is now very pronounced in nearly every case, often defined by a distinct low center at 500 hPa and characterized by a confluent pattern of height contours and the related entrance region of a well-defined jet streak over the northeast United States or southeastern Canada. This pattern has a significant bearing on the evolution of surface anticyclones and their associated cold air (section 3b). The combination of two distinct trough–jet streak features acts to focus a vertical motion pattern and related heavy snowfall along the East Coast, a process that is explored in more detail in section 3d of this chapter.

In about half of the cases (especially true at 0000 UTC 15 February 1958, 0000 UTC 12 January 1964, 1200 UTC 10 February 1983, and 0000 UTC 25 January

(a) 48 HOURS PRIOR TO HEAVY SNOW DEVELOPMENT

FIG. 4-2a. Mar 1956–Jan 1961. Analyses of 500-hPa heights, and high and low height centers, where yellow and orange shading represents absolute vorticity (greater than 16×10^{-5} s^{-1}, at 4×10^{-5} s^{-1} increments); and isotachs of maximum wind speeds observed at 400 hPa (alternating shading represents 5 m s^{-1} increments of wind speed starting with 40 m s^{-1}) 48 h prior to the development of heavy snow (times coincide with the analyses in Fig. 4-5).

(b) 48 HOURS PRIOR TO HEAVY SNOW DEVELOPMENT

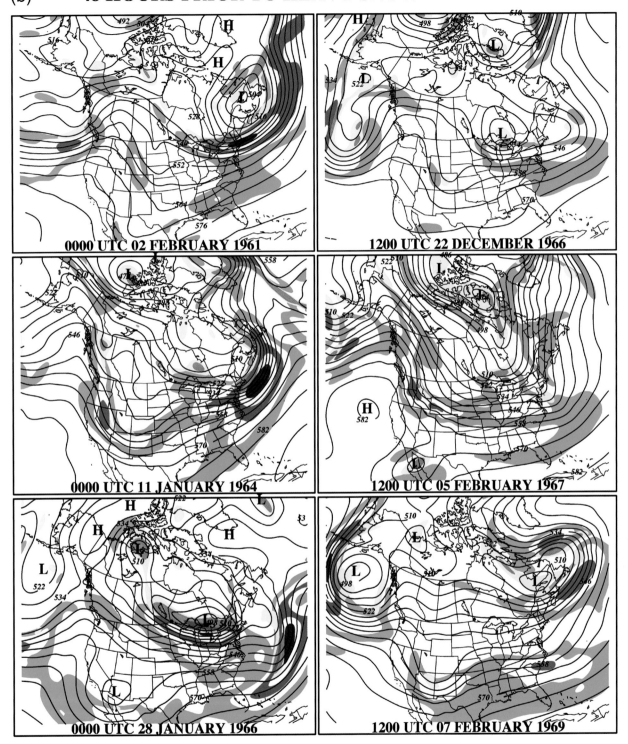

FIG. 4-2b. (*Continued*) As in Fig. 4-2a, but for Feb 1961–6 Feb 1969.

(c) **48 HOURS PRIOR TO HEAVY SNOW DEVELOPMENT**

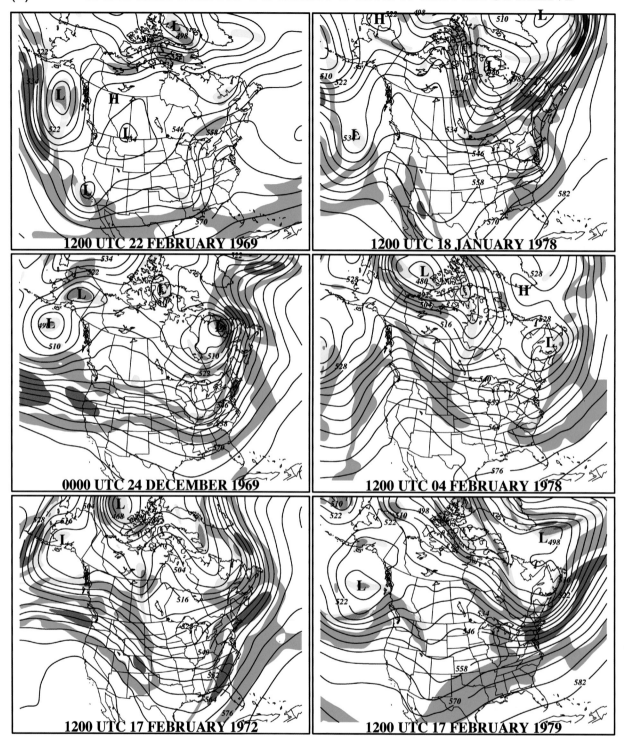

FIG. 4-2c. (*Continued*) As in Fig. 4-2a, but for 21 Feb 1969–Feb 1979.

(d) 48 HOURS PRIOR TO HEAVY SNOW DEVELOPMENT

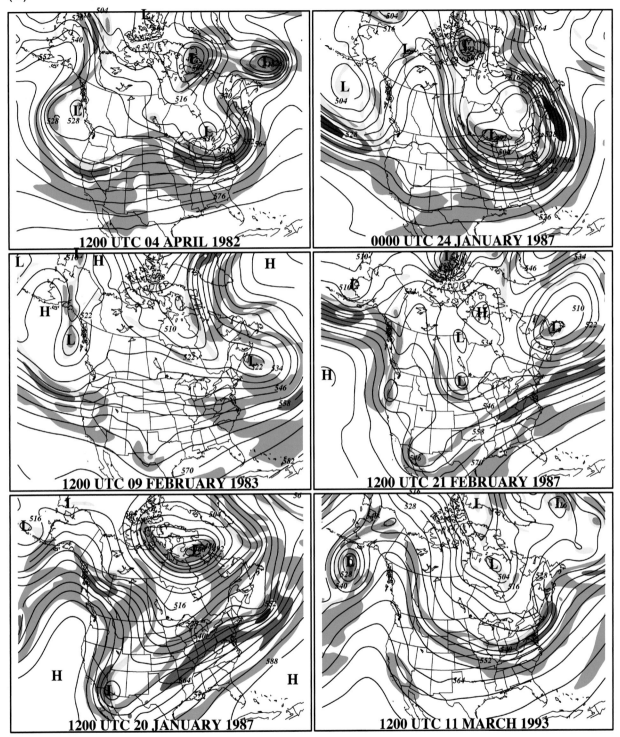

FIG. 4-2d. (*Continued*) As in Fig. 4-2a, but for Apr 1982–Mar 1993.

(e) **48 HOURS PRIOR TO HEAVY SNOW DEVELOPMENT**

Fɪɢ. 4-2e. (*Continued*) As in Fig. 4-2a, but for Feb 1994–Dec 2000.

1987), a blocking or cutoff ridge is observed downstream of the eastern Canadian trough. These features are also examined in more detail in section 3b. In summary, the 72-, 48-, and 24-h evolutions of upper-level features that predominate the precyclogenetic environment can be identified and illustrated as follows (Fig. 4-4):

1) The upper-level flow over eastern Canada is often dominated by a deep upper trough. The upper trough, often with the presence of a closed 500-hPa vortex, is typically located over Quebec, Newfoundland, or Nova Scotia in virtually every case with many cases serving as excellent examples. These upper troughs, combined with a ridge typically found over the Atlantic Ocean, establish a consistent pattern of confluence aloft over the northeastern United States or southeastern Canada. These patterns are evident in most, but not all cases; examples in which these features are less evident include 20–22 January 1987, 2–4 February 1995, and 31 March–1 April 1997.

2) An upper ridge over the western United States and the Canadian west coast is another common synoptic-scale feature observed prior to northeast snowstorms. This ridge typically extends from the west coast of the United States northward into western Canada. Only 3 of the 30 cases exhibit an upper trough, rather than a ridge, over the western United States or Canada. The amplification of the western ridge has been shown to be one of a number of possible dynamical "signals" related to the development of explosively developing cyclones in the western Atlantic Ocean (Lackmann et al. 1996). Sometimes, the upper ridge may be part of a circulation feature termed the split-flow regime by forecasters. The split flow refers to a jet stream pattern with a high-amplitude ridge over western Canada that is "split" from another active branch of the jet stream across the western United States, often allowing short-wave troughs to proceed eastward across the United States, south of the upper Canadian ridge. Sometimes, short-wave energy would be contained in both branches that could later merge to produce intense cyclogenesis (see section 3a). This flow regime could be conducive to Northeast snowstorms since two conditions prevail: 1) the upper Canadian ridge favors the development of cold surface anticyclones that track southeastward across central Canada toward the northern United States and 2) short-wave troughs embedded within either jet stream enhance the likelihood of cyclogenesis in the eastern and southeastern United States that later produce heavy snows in the northeast urban corridor.

3) Prior to the development of heavy snowfall, many cyclone-producing troughs are poorly defined only 1–2 days before the snow events and have not yet consolidated into a significant short-wave trough in the eastern United States. In half of the cases, the upper-level conditions are evolving toward, but are not yet conducive to, significant cyclogenesis 1–2 days prior to these major events. In the other half, cyclogenesis is already associated with significant upper-level short-wave troughs 1–2 days in advance of the heavy snow event. In the first set of cases, the upper trough(s) that later spawned the East Coast cyclones amplified, merged, or exhibited both characteristics during the following 36 h. In the second set, upper-level troughs were well-defined 36 h prior to the heavy snow events and are located primarily over the central and southern United States. Many of these troughs continue to amplify during the storm event and some involve trough mergers.

4) The presence of upper-level jet streaks is a common signature for all 30 snowstorms 24 h prior to the event. A polar jet is often found equatorward of the eastern Canadian trough with the axis of the jet located over the northeastern United States 12–48 h prior to the heavy snow event. The confluent entrance region of this polar jet is usually located over the anticyclone that supplies the cold air needed for significant snowfall. A significant subtropical jet streak is also found over the southeastern United States at this time in many of the cases. Jet streaks associated with the cyclone-producing troughs often display a more complex configuration during the early stage of the storm's evolution, but as the trough–jet systems approach the East Coast 24 h prior to heavy snowfall, the coastal cyclone appears to develop in the diffluent exit region of this jet streak. The roles of upper-level jet streaks in the development of Northeast snowstorms are examined in sections 3c and 3d.

5) In some instances, major snowstorms occur as a part of major changes to the larger-scale circulation pattern but their relationships are uncertain (i.e., Colucci 1985; also see Wallace and Gutzler 1981; Dole and Gordon 1983; Dole 1986, 1989). In particular, the snowstorm of 5–7 February 1978 occurred as part of a dramatic change in the circulation regime across North America (cf. the evolution in Figs. 4-1–4-3 and the 500-hPa analyses in volume II, chapter 10.17) that saw a blocking pattern emerge with an upper cutoff anticyclone over north-central Canada and a cutoff low over the northeastern United States and southeastern Canada. The blocking cyclone produced during this period remained a well-defined entity over eastern Canada for nearly the remainder of the month of February and suppressed the development of cyclonic systems across the eastern United States, resulting in a very dry period throughout the remainder of February 1978 (Colucci 1985). The January "Blizzard of '66" also occurred in a period of major upper-level circulation change (cf. Figs. 4-1–4-3 and the 500-hPa analyses in volume II, chapter 10.8).

(a) 24 HOURS PRIOR TO HEAVY SNOW DEVELOPMENT

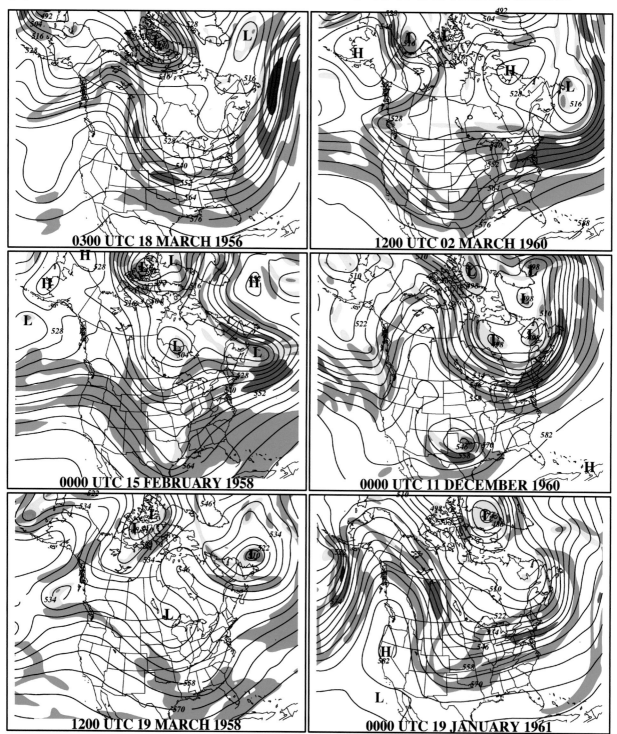

FIG. 4-3a. Mar 1956–Jan 1961. Analyses of 500-hPa heights, and high and low height centers, where yellow and orange shading represents absolute vorticity (greater than 16×10^{-5} s^{-1}, at 4×10^{-5} s^{-1} increments); and isotachs of maximum wind speeds observed at 400 hPa (alternating shading represents 5 m s^{-1} increments of wind speed starting with 40 m s^{-1}) 24 h prior to the development of heavy snow (times coincide with the analyses in Fig. 4-5).

(b) **24 HOURS PRIOR TO HEAVY SNOW DEVELOPMENT**

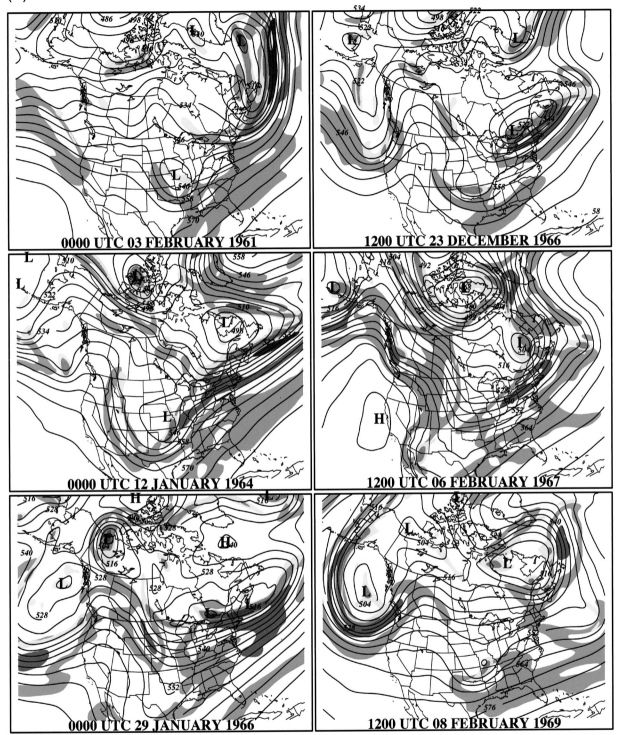

0000 UTC 03 FEBRUARY 1961

1200 UTC 23 DECEMBER 1966

0000 UTC 12 JANUARY 1964

1200 UTC 06 FEBRUARY 1967

0000 UTC 29 JANUARY 1966

1200 UTC 08 FEBRUARY 1969

FIG. 4-3b. (*Continued*) As in Fig. 4-3a, but for Feb 1961–8 Feb 1969.

(c) **24 HOURS PRIOR TO HEAVY SNOW DEVELOPMENT**

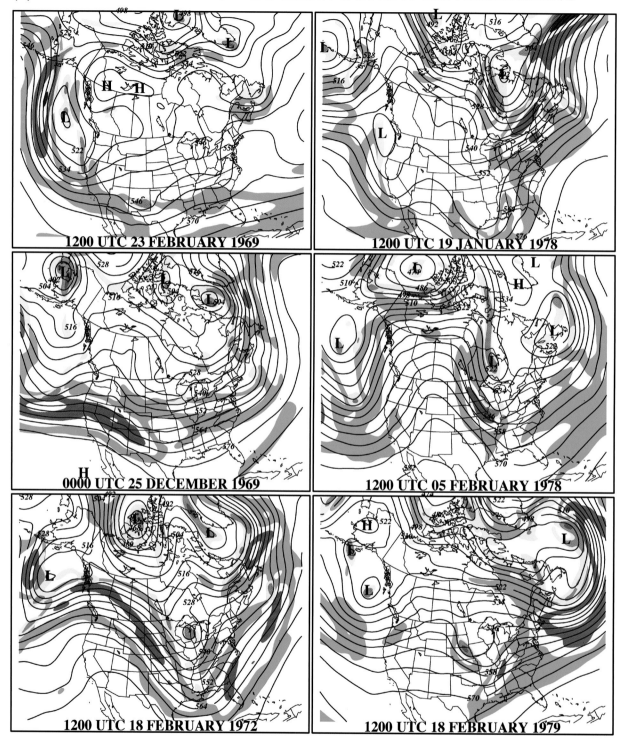

FIG. 4-3c. (*Continued*) As in Fig. 4-3a, but for 23 Feb 1969–Feb 1979.

(d) **24 HOURS PRIOR TO HEAVY SNOW DEVELOPMENT**

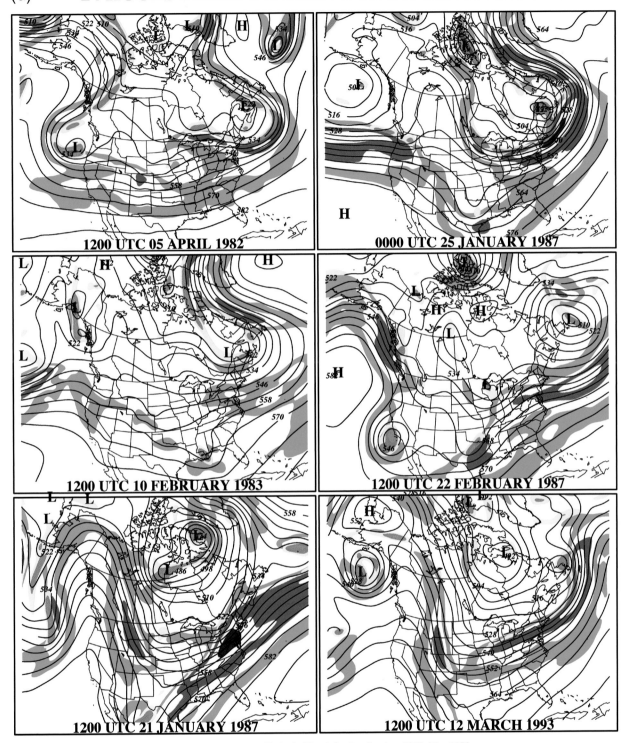

FIG. 4-3d. (*Continued*) As in Fig. 4-3a, but for Apr 1982–Mar 1993.

(e) **24 HOURS PRIOR TO HEAVY SNOW DEVELOPMENT**

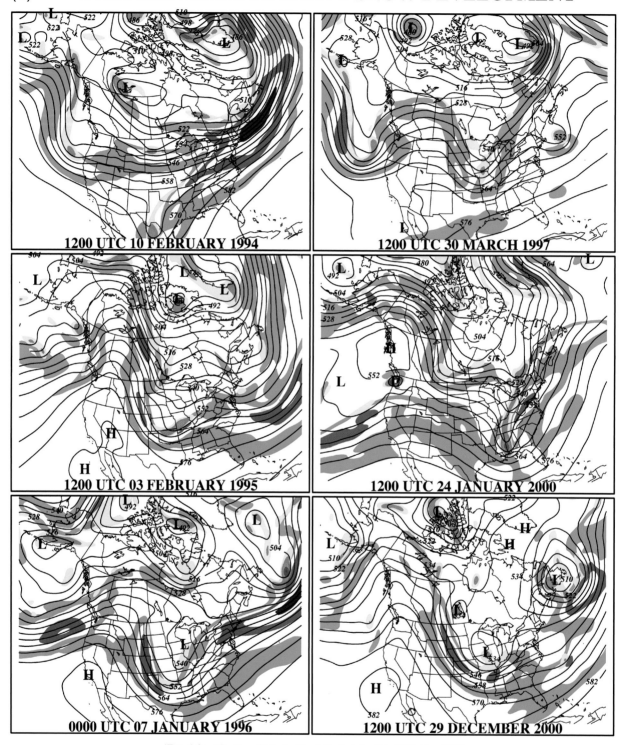

FIG. 4-3e. (*Continued*) As in Fig. 4-3a, but for Feb 1994–Dec 2000.

COMMON UPPER-LEVEL SIGNATURES
3,2 and 1 DAYS PRIOR TO NOTHEAST SNOWSTORMS

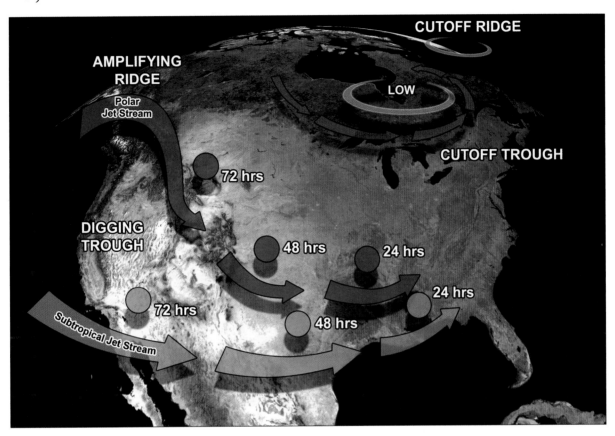

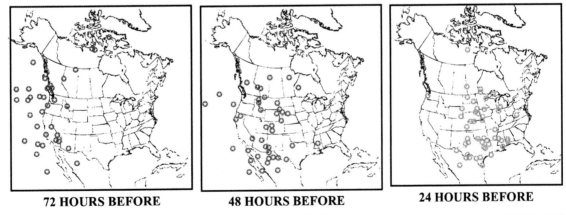

72 HOURS BEFORE   **48 HOURS BEFORE**   **24 HOURS BEFORE**

FIG. 4-4. Summary of upper-level signatures 3, 2, and 1 days prior to Northeast snowstorms, including summary of 500-hPa vorticity associated with Northeast snowstorms.

With the Presidents' Day Storm of February 1979, a major circulation change also occurred but not in the same manner as during either February 1978 or January 1966. This storm occurred during a transition from a persistent large-scale circulation pattern that brought ex-

tremely cold temperatures to the northeastern United States to one that became more zonal as the storm exited the East Coast, signaling the end of the cold regime. The North Atlantic Oscillation had been decidedly negative in much of February but became positive around

(a)

0300 UTC 19 MARCH 1956

0000 UTC 20 JANUARY 1961

0000 UTC 16 FEBRUARY 1958

0000 UTC 4 FEBRUARY 1961

1200 UTC 20 MARCH 1958

0000 UTC 13 JANUARY 1964

1200 UTC 3 MARCH 1960

0000 UTC 30 JANUARY 1966

0000 UTC 12 DECEMBER 1960

1200 UTC 24 DECEMBER 1966

FIG. 4-5a. Mar 1956–Dec 1966. Analyses of 500-hPa heights, where yellow and orange shading represents absolute vorticity (greater than 16×10^{-5} s^{-1}, at 4×10^{-5} s^{-1} increments), isotachs of maximum wind speeds observed at 400 hPa (alternating shading represents 5 m s^{-1} increments of wind speed starting with 40 m s^{-1}), and corresponding surface analyses during the development of heavy snow along the Northeast coast for all 30 cases.

(b)

1200 UTC 7 FEBRUARY 1967

1200 UTC 20 JANUARY 1978

1200 UTC 9 FEBRUARY 1969

1200 UTC 6 FEBRUARY 1978

1200 UTC 24 FEBRUARY 1969

1200 UTC 19 FEBRUARY 1979

0000 UTC 26 DECEMBER 1969

1200 UTC 6 APRIL 1982

1200 UTC 19 FEBRUARY 1972

1200 UTC 11 FEBRUARY 1983

FIG. 4-5b. As in Fig. 4-5a but for Feb 1967–Feb 1983.

(c)

1200 UTC 22 JANUARY 1987

1200 UTC 4 FEBRUARY 1995

0000 UTC 26 JANUARY 1987

0000 UTC 8 JANUARY 1996

1200 UTC 23 FEBRUARY 1987

1200 UTC 31 MARCH 1997

1200 UTC 13 MARCH 1993

1200 UTC 25 JANUARY 2000

1200 UTC 11 FEBRUARY 1994

1200 UTC 30 DECEMBER 2000

FIG. 4-5c. As in Fig. 4-5a, but for Jan 1987–Dec 2000.

the time of the storm (see chapter 2). A major strato-spheric warming also occurred during this period (Man-ney et al. 1994) but it is not known how the circulation changes associated with the warming are related to the evolution of the cyclone or the pattern shift that occurred during the period encompassing the cyclone. With the pattern change, much of the snow from the Presidents' Day Storm melted quickly. Historically, a similar cir-culation change may have occurred following the pas-sage of the February 1899 blizzard (see volume II, chap-ter 9) when a similar period of extreme cold eased fol-lowing the passage of the storm (Kocin et al. 1988).

3. Upper-level conditions during the development of Northeast snowstorms

The analyses in Fig. 4-5 highlight how the upper-level and corresponding surface features evolve at the time of heavy snowfall. In all cases, the upper troughs underwent major amplification during major East Coast cyclogenesis. Some of the more significant upper-trough amplifications are exhibited by the cases of March 1956, January 1966, February 1972, February 1978, March 1993, and January 1996 (cf. with Figs. 4-1–4-3).

The eastern Canadian trough, the most common fea-ture of the precyclogenetic period, continues to move eastward or northeastward for all cases, but a confluent pattern in the 500-hPa height contours and an associated upper jet streak remain upwind of the trough in southeast Canada. The jet streaks associated with the cyclone-producing trough in the southeastern United States ap-pear to become better defined with the exit region lo-cated over the surface cyclone. The evolution of these features; their relationships to the surface cyclones, an-ticyclones, and the development of heavy snowfall; the many variations of the patterns of the upper troughs and their associated jet streaks; and their magnitudes, tilts, and amplifications will all be examined in the following sections.

a. 500-hPa troughs

The evolutions of the upper-level cyclone-producing troughs associated with major Northeast snowstorms are described in the following manner. An examination of the 500-hPa surface shows that a majority of cases in-volved the transformation and amplification of an open-wave trough and a subsequent transformation into a closed circulation at 500 hPa, a process depicted sche-matically by Palmén and Newton (1969, p. 326). A total of 18 cases appear to develop in this manner with the scenario demonstrated by 4 snowstorms (March 1960, early February 1969, April 1982, and March 1993; see Fig. 4-6a). The 500-hPa trough evolves into a vortex as the rapidly developing sea level cyclone circulation ap-pears to grow upward while it intensifies. Following the formation of the upper low, the sea level and upper-level circulation centers become nearly vertically col-

located as the temperature gradients decrease, marking the decaying stage of the cyclone's evolution.

A second group of cases (February 1967, January 1978, February 1983, the two storms of January 1987, February 1994, and February 2003) is identified in which the upper-level short-wave trough remains an open-wave system of relatively constant amplitude and does not develop a closed circulation at 500 hPa, as illustrated in Fig. 4-6b. In general, these storms do not intensify to the degree exhibited by the previous cases that amplified from an open wave into closed upper-level circulations; however, the surface low for the Feb-ruary 1967 snowstorm and the 21–22 January 1987 snowstorm did undergo explosive development for a period of at least 12 h. In a few of the other cases (January 1978, February 1979, 23–25 January 1987, and February 1994), the upper-level short-wave trough ap-pears to remain nearly steady state or deamplifies with time. These storms illustrate that heavy snows can ac-company upper-level troughs that are not strengthening and may, in fact, be deamplifying with time.

A third group of cases (February 1961, January 1964, February 1978, and, to some degree, January 1996 and December 2000) featured a closed circulation in the middle and upper troposphere prior to sea level cyclo-genesis (Fig. 4-6c). Closed upper-level circulations, or "cutoff" lows, may form prior to, simultaneously with, or without surface cyclogenesis, and tend to drift slowly, as pointed out by Petterssen (1956, 244-245) and Pal-mén and Newton (1969, 274-283). The cutoff low re-flects a pool of cold tropospheric air that is detached from the primary region of cold air located at higher latitudes. While a cold pool may sometimes be the rem-nant of a previously occluded storm, the five cases in this sample were not associated with earlier major sur-face cyclones. In these cases, the surface lows developed in association with a closed cyclonic circulation aloft as it moved slowly eastward from the central United States to the east coast. Four of the five cases reflect a sample of storms marked by a relatively long duration of snowfall, indicative of their slow movement.

1) AMPLITUDE AND WAVELENGTH

The amplitude and wavelength of the upper troughs and their adjacent ridges, as well as their evolution, can be used as a measure of the magnitude of the upper-level divergence and the potential for sea level devel-opment. The increased amplitude of upper-level trough–ridge systems enhances vorticity gradients due to chang-es in curvature, and strengthens vorticity advection along the upper-level flow, which implies an increase of upper-level divergence (Palmén and Newton 1969, p. 325). Increasing amplitude is also indicative of wave growth, an essential component of baroclinic instability theories that describe the energy conversions that take place during cyclogenesis (Charney 1947; Eady 1949; Gall 1976; Farrell 1985). A decrease in the distance, or

(a) **"SELF-DEVELOPMENT" OPEN ⟶ CLOSED**

FIG. 4-6a. Representative examples of trough evolution at 500 hPa for major Northeast snowstorms. Illustration of "open wave" to "cutoff" system, using storms of Mar 1960, 8–10 Feb 1969, Apr 1982, and Mar 1993. Solid lines are geopotential height contours at 500 hPa, yellow and orange shading represents absolute vorticity (greater than 16×10^{-5} s^{-1}, at 4×10^{-5} s^{-1} increments), and isotachs of maximum wind speeds observed at 400 hPa (alternating shading represents 5 m s^{-1} increments of wind speed starting with 40 m s^{-1}).

EVOLUTION OF 500 hPa TROUGHS
OPEN-WAVE TROUGH

(b)

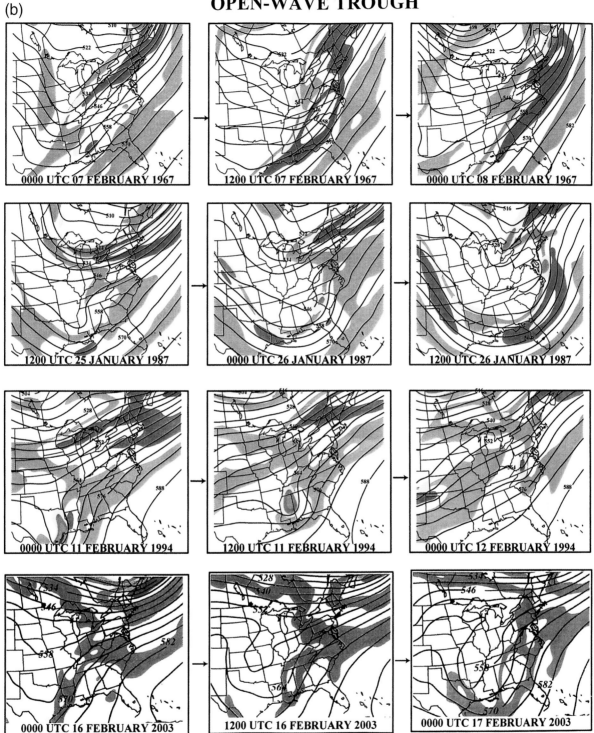

FIG. 4-6b. (*Continued*) Representative examples of trough evolution at 500 hPa for major Northeast snowstorms. Illustration of an open-wave system, using storms of Feb 1967, 25–26 Jan 1987, Feb 1994, and Feb 2003. Solid lines are geopotential height contours at 500 hPa, yellow and orange shading represents absolute vorticity (greater than 16×10^{-5} s^{-1}, at 4×10^{-5} s^{-1} increments), and isotachs of maximum wind speeds observed at 400 hPa (alternating shading represents 5 m s^{-1} increments of wind speed starting with 40 m s^{-1}).

(c)

FIG. 4-6c. (*Continued*) Representative examples of trough evolution at 500 hPa for major Northeast snowstorms. Illustration of cutoff lows, using storms of Feb 1961, Jan 1964, Feb 1978, and Jan 1996. Solid lines are geopotential height contours at 500 hPa, yellow and orange shading represents absolute vorticity (greater than 16×10^{-5} s^{-1}, at 4×10^{-5} s^{-1} increments), and isotachs of maximum wind speeds observed at 400 hPa (alternating shading represents 5 m s^{-1} increments of wind speed starting with 40 m s^{-1}).

half-wavelength, between a trough axis and the downstream ridge crest can also increase upper-level divergence. Changes in the amplitude and wavelength of upper-level trough–ridge systems were studied to determine whether developing cyclones act as "positive feedback" mechanisms for sustaining or even enhancing cyclogenesis, as described by Sutcliffe and Forsdyke (1950) and Palmén and Newton (1969, 324-326) and described in chapter 7.

To examine amplitude changes, the evolution of the 500-hPa trough associated with the main surface cyclone and its downstream ridge was examined.[2] As depicted in Figs. 4-7a–c, the 36-h changes in wavelength and amplitude are clearly apparent in most instances, while in a few cases, such changes could not easily be measured. Of 30 cases, 28 show evidence of increasing amplitude (largest increases shown by the blizzard of January 1966 and the storms of January 1961, December 1966, February 1978, 22 January 1987, the March 1993 Superstorm, and February 1995) while 2 cases (January 1978 and February 1994) show decreasing amplitudes while heavy snow was affecting the Northeast urban corridor; a mean increase of 7° latitude was noted for the 30-case sample, marked by a range of −4.5° (February 1994) to +21° (January 1966). These highly amplifying cases are all characterized by rapid sea level cyclonic development while the two cases showing a decrease in amplitude are accompanied by surface cyclones that only slowly intensify. (Small increases of 2° or less were noted for the cases of December 1960, February 1961, February 1979, and February 1983.)

Changes in the half-wavelength of the 500-hPa trough and downstream ridge are also estimated for 30 cases, using the contours selected for the analysis of amplitude changes. Decreases in half-wavelength are observed in 28 of the 30 cases (exceptions include the cases of 25 January 1987 and February 1994). The general observation that amplitudes are increasing during the period of shortening wavelength indicates that increases in vorticity advection and upper-level divergence are also characteristic of the cyclogenetic period along the coast. The consistent depiction of well-defined upper troughs and ridges with increasing amplitudes and shortening wavelengths provides a clear upper-level signature of these storms.

2) VORTICITY MAXIMA, DIFFLUENCE, AND TROUGH TILT

The increasing upper-level divergence and vorticity advection that accompany upper-level trough–ridge systems whose amplitudes (wavelengths) are increasing (decreasing) are associated with distinct 500-hPa vorticity maxima. The paths of these maxima are plotted for each case in Fig. 4-7.[3] A composite of all paths is shown in Fig. 4-8. In addition, one or two representative 500-hPa geopotential height contours are plotted at 24-h intervals to illustrate the movement and changing amplitude, wavelengths, and orientations of the upper troughs as they evolve during these major snow events.[4]

The vorticity maxima followed a variety of paths prior to the onset of East Coast cyclogenesis, but many are initially directed from northwest to southeast. These equatorward-propagating features indicate deepening or "digging" troughs, in which the amplitude measured between the trough and upstream ridge axes increases with time. An amplifying ridge is usually present upstream of a digging trough, as noted by Palmén and Newton (1969, p. 335), and illustrated by several cases (e.g., March 1956, January 1966, February 1978, and 31 March–1 April 1997; see Fig. 4-7).

Lackmann et al.'s (1996) composite study of the planetary- and synoptic-scale environment for explosive cyclogenesis over the western Atlantic Ocean found the presence of a stationary high-amplitude ridge over the West Coast with the cyclone-producing trough deepening east of the ridge over central North America. The amplification of the trough is a manifestation of a process termed downstream development (Simmons and Hoskins 1979; Orlanski and Chang 1993), in which the amplification of an upper-level trough and resultant cyclogenesis develops after the upper-level ridge amplification occurred upstream of the system. An especially auspicious example of this process is represented by the snowstorm of February 1978. The transformation of a relatively zonal upper flow occurred as an upper ridge amplified over the western United States and Canada, followed by the subsequent amplification of an upper trough along the east coast and the development of a very intense cyclone just off the east coast. The deepening troughs were marked by growing cyclonic vorticity due to increasing curvature and cyclonic wind shear associated with the propagation of an upper-level

---

[2] A 500-hPa geopotential height contour that exhibited the greatest amplitude (variation by latitude) between the trough axis and the downstream ridge was selected at two times. The first was selected at a time prior to heavy snow along the Atlantic coast while the second contour was selected 36 h later, a time during which heavy snow was affecting portions of the middle Atlantic states or New England. The first contour was selected at the time of the second panel in the six-panel analyses in chapter 10, volume II, while the second contour was the same contour, but selected from the fifth panel.

[3] Locations of the maximum absolute vorticity over the 60-h period covering each storm were obtained from computer-derived analyses generated by NCEP and available from NCDC for events since May 1961. For cases prior to May 1961, the positions of the vorticity maxima were derived from analyses performed on a 2° grid using a Barnes (1964) objective analysis scheme and modified for interactive analysis purposes by Koch et al. (1983). Locations were later verified with the NCEP reanalysis data (Kalnay et al. 1996).

[4] The starred symbols in Fig. 4-7 represent locations of maximum vorticity while the enlarged symbols correspond to the times when the height contours were selected.

500 hPA VORTICITY MAXIMA PATHS

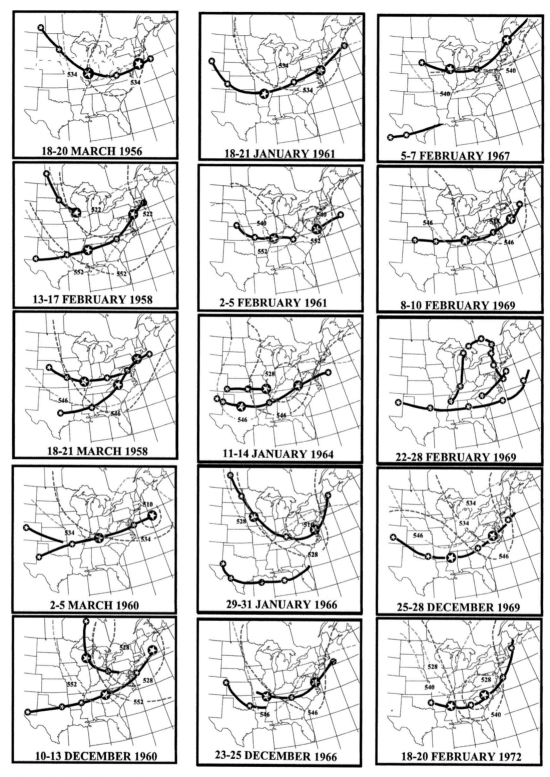

FIG. 4-7a. Mar 1956–Feb 1972. Paths of 500-hPa vorticity maxima for individual cases. Circled stars refer to 12-hourly positions. Dashed lines are representative 500-hPa geopotential height contours at 24-h intervals (light blue dashed is earlier contour; dark blue dashed is same contour 24 h later).

500 hPA VORTICITY MAXIMA PATHS

(b)

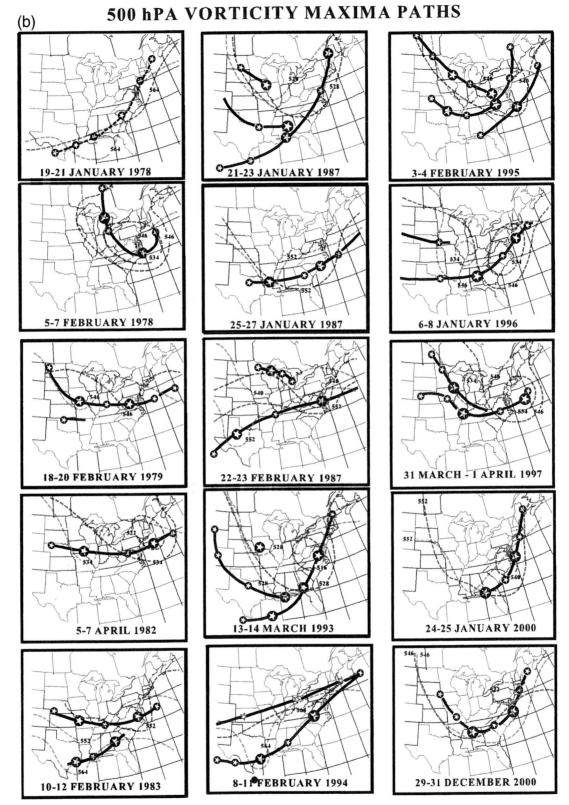

Fig. 4-7b. (*Continued*) As in Fig. 4-7a, but for Jan 1978–Dec 2000.

500 hPA VORTICITY MAXIMA PATHS
For 30 Northeast Snowstorms

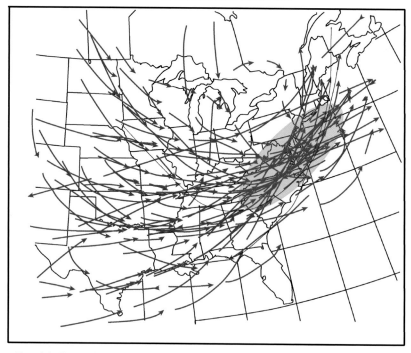

FIG. 4-8. Composite of 500-hPa vorticity maxima for all cases. Red area represents locus of paths over the middle Atlantic states when vorticity maxima are converging and moving east-northeastward over the Atlantic Ocean.

jet streak from the ridge toward the trough axis (Bjerknes 1951; Palmén and Newton 1969, p. 341).

Following the initial deepening or digging of the upper-level troughs in the central United States, the paths of the vorticity maxima typically curved to the east or northeast as they approached the Atlantic coast (see the clusters of paths indicated by the red area in Fig. 4-8). This scenario is similar to those described in a climatological study of heavy snowfall over the eastern United States by Goree and Younkin (1966) and in a composite study of rapidly intensifying cyclones over the west-central North Atlantic Ocean by Sanders (1986a). The heaviest snow often falls along or parallel to the west of the track of the 500-hPa vorticity maximum (see Fig. 3-3). With few exceptions, the centers of maximum cyclonic vorticity converge along a track off the coast between North Carolina and New Jersey to positions just south of the southern New England coast. Heavy snow seldom occurs over the coastal plain when the vorticity maximum associated with an amplifying upper-level trough curves to the northeast before reaching the East Coast (also see chapter 5).

One of the few cases in which the 500-hPa vorticity maximum passes west and north of the primary snowband across the Northeast urban corridor is the February 1967 storm (see Fig. 3-3). More typically, in cases where the vorticity maximum propagates northeastward into the Ohio Valley, the primary surface low pressure system usually maintains a well-defined circulation west of the Appalachians, advecting lower-tropospheric warm air over the coastal plain. Secondary cyclogenesis still occurs in some cases and cold air may remain in the lowest levels of the atmosphere east of the Appalachian Mountains due to cold-air damming. The warm-air advection aloft, however, often produces precipitation in the form of rain, freezing rain, or ice pellets, rather than heavy snow in the major metropolitan areas along the coast (see chapter 5). Therefore, the predicted path of the 500-hPa vorticity maximum is an important consideration for forecasters when trying to distinguish heavy snow events from other cases in which substantial snowfall is less likely.

The development of diffluence downwind of the trough and a change in the orientation of the trough axis from a positive (northeast–southwest) or neutral (north–south) tilt to a negative tilt (northwest–southeast) are also distinguishing characteristics of the upper-level trough evolution in all cases characterized by increasing amplitudes and decreasing wavelengths. Diffluence, defined in the *Glossary of Meteorology* (Glickman 2000) as the "rate at which adjacent flow diverges along an axis oriented normal to the flow at the point in question," has been recognized as the signature of increasing upper-level divergence immediately downstream of a trough axis (Scherhag 1937; Bjerknes 1954; Palmén and Newton 1969, 334-335). Diffluence becomes more pronounced as (i) the trough amplitude increases and the wavelength decreases, (ii) the trough axis becomes negatively tilted, and (iii) an upper-level jet streak propagates from the west-northwest of the trough axis toward the base of the trough. Bjerknes (1951) has also noted this process as an important factor contributing to surface cyclogenesis. In many of these cases, the evolution of the trough axis from a positive or neutral tilt to a negative tilt occurs as the trough–ridge amplitude increases and the wavelength decreases and an intense 500-hPa vortex forms as the surface cyclone undergoes explosive development.

3) TROUGH MERGERS

There is increasing evidence that intense cyclogenesis sometimes occurs as separate, distinct troughs appear to merge together (e.g., see Lai and Bosart 1988; Gaza and Bosart 1990; Dean and Bosart 1996; Hakim et al. 1995; Hakim et al. 1996; Huo et al. 1999a,b; Bosart 1999). With trough mergers, two or more troughs marked by separate vorticity maxima and distinct supporting baroclinic zones can interact to produce more intense cyclogenesis. The interaction of these systems enhances the upper-level vorticity advection and associated upper-level divergence as the merging of separate vorticity maxima into a more cohesive maximum produces a well-organized pattern of vorticity advection above the developing surface cyclone.

However, not all upper-trough interactions produce more intense cyclones since some interactions can result in the decay, or damping, of one or more upper-trough systems, impeding cyclogenesis. Dean and Bosart (1996) illustrate how one vorticity center may split into two or more centers, and termed this interaction a trough fracture. In other instances, two or more troughs may interact in a way to suppress or damp one or both of the troughs. For example, prior to a snowstorm on 22–23 February 1987 (see volume II, chapter 10.23), numerical models forecast a northern stream short-wave trough to progress eastward across the northern United States at a more rapid pace than a separate, southern stream short-wave system, damping the southern stream short-wave trough, and suppressing any subsequent cyclogenesis from the southern stream system. However, as the event drew closer and the northern stream short wave was resolved more accurately by the short-range forecast models, its movement was more southward and much slower than earlier forecasts, allowing the southern stream system to maintain its amplitude and intensify as it pivoted around the base of the northern stream trough. The result was an explosive cyclonic development that led to a burst of very heavy snow across Maryland, Pennsylvania, and New Jersey.

Theoretical and observational studies (Thorncroft et al. 1993; Shapiro et al. 2000) have also explored cyclone life cycles leading to different "types" of cyclone formation, the so-called anticyclonic (LC1) and cyclonic (LC2) modes, one that produces positively tilted troughs while the other leads to negatively tilted troughs. These studies appear to link background flow states of the atmosphere of specific climate regimes to upper-level trough interactions downstream that either promote mergers or fractures and provide a theoretical basis for distinguishing how the atmosphere's circulation may yield different "flavors" of cyclones.

Forecasters have long recognized the dilemma in determining whether separate troughs and their associated vorticity maxima will merge or remain separate, especially since numerical prediction models have not always been successful in simulating these interactions. An intense cyclone developed on 25 January 1978 over the Ohio Valley, and came to be known as the Cleveland "Superbomb" (Salmon and Smith 1980; Hakim et al. 1995, 1996). This storm represents one of the most dramatic examples of an intense cyclone that resulted from the interaction of two or more shortwave troughs, a merger that contributed to one of the most intense cyclones on record for the Ohio Valley and Great Lakes.

Trough mergers have also been responsible for the development of many East Coast cyclones and they play roles in the development of some of the snowstorms examined in this monograph. Trough mergers are suggested in varying degrees from 17 cases, which are those cases with multiple vorticity paths shown in Fig. 4-8; the more dramatic cases that evolved from an interaction of two or more distinct systems include December 1960, January 1966, February 1983, February 1987, March 1993, and February 1995. In these cases, there is evidence for the superpositioning of multiple vorticity centers that underwent various forms of amplification, consolidation, or both (i.e., see Lai and Bosart 1988).

A common occurrence in some Northeast snowstorms is that a short-wave trough propagating from the California coast across the southern United States will support the initial development of a surface cyclone and its associated precipitation shield. On occasion, a separate, significant (and possibly amplifying) trough located upstream of the southern stream short-wave trough will overtake the initial system and enhance the development of the surface cyclone and contribute to a sudden "jump" of the surface low farther northward in response to the impact of the upper-level forcing. New York City's "big snow" of December 1947 appears to have developed in such a manner (see chapter 9, volume II) in which a weak, southern stream short-wave trough merges with an amplifying northern stream short-wave system. At the time, this storm was poorly forecast possibly due to the difficulty in foreseeing the amplification of the northern stream trough and resulting trough merger off the Southeast coast.

The evolution of the Cleveland Superbomb and the March 1993 Superstorm (see volume II, chapter 10.24) are used as more recent examples of trough merging (Fig. 4-9) with Huo et al. (1999b) showing the important

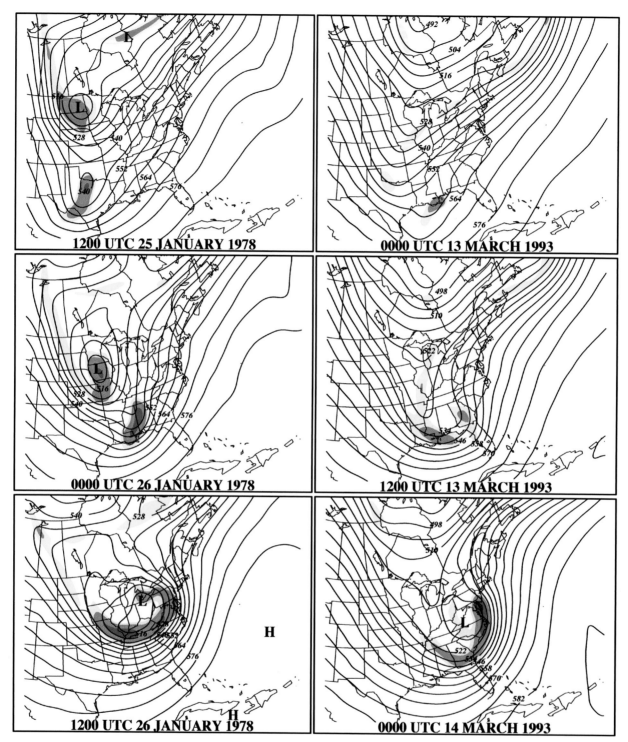

FIG. 4-9. Examples of trough "merging" at 500 hPa at 1200 UTC 26 Jan 1978 and 0000 UTC 14 Mar 1993. Solid lines are geopotential height contours, yellow and orange shading represents absolute vorticity (greater than 16×10^{-5} s^{-1}, at 4×10^{-5} s^{-1} increments).

role of the northern stream trough in the dramatic cyclogenesis during the March 1993 Superstorm (see chapter 10, volume II). Bosart et al. (1996) show that three separate trough features or anomalies played roles in the development of the March 1993 Superstorm. One

feature circumnavigated the globe before allowing cold air to plunge south and merged with two separate upper troughs or anomalies moving toward and across the Gulf coast (more details can be found in chapter 7 and chapter 10.24 in volume II).

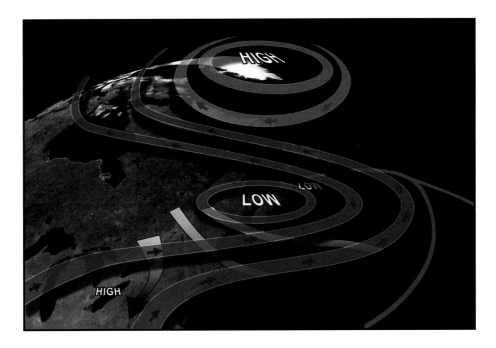

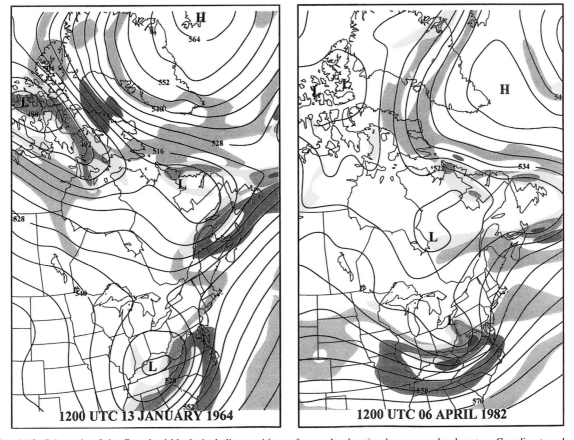

FIG. 4-12. Schematic of the Greenland block, including positions of upper-level anticyclone, upper-level eastern Canadian trough with confluent upper-level flow, and surface frontal and surface low and high pressure center positions. Also shown are examples of 500-hPa charts for cases exhibiting manifestations of the Greenland block. See Fig. 4-1 caption for details.

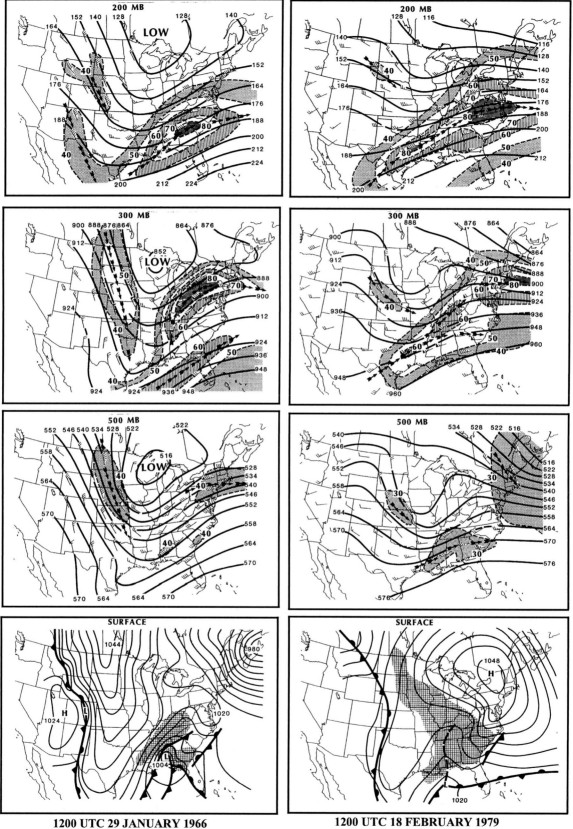

1200 UTC 29 JANUARY 1966   **1200 UTC 18 FEBRUARY 1979**

(Bjerknes 1951; Riehl et al. 1952; Newton 1956; Reiter 1963, 1969; Uccellini 1990).

Principal jet stream axes at 250 hPa are superimposed on the 12-hourly 500-hPa geopotential height analyses shown for 30 of the 32 cases in chapter 10, volume II. The upper-level wind analyses at 500, 400, and 250 hPa indicate that the sample of 30 storms is characterized by multiple and multilevel jet systems that cannot easily be described due to their complicated horizontal and vertical structure and evolution. Both polar and subtropical jet streaks are present concurrently in many cases. Polar jet streaks are described in the 250- and 400-hPa isobaric charts shown here, but may also be seen in the 200- and 500-hPa charts since they are associated with baroclinic zones that extend through a deep layer of the middle troposphere. Subtropical jet streaks are mainly observed at 250–200 hPa, since they are typically associated with baroclinic regions that are confined to the upper troposphere. Wind analyses in chapter 10, volume II, show that the polar jets attained typical speeds of 50–70 m s^{-1}, while maximum speeds in both polar and subtropical jet streaks approached 80–100 m s^{-1} in some cases.

The complex structure of jet streams and the jet streaks embedded within them is illustrated by the January 1966 and February 1979 cases (Fig. 4-13). At 1200 UTC 30 January 1966, the 200-, 300-, and 500-hPa charts reveal multiple jet streaks at various levels of the atmosphere, with the incipient surface cyclone located over the Gulf coast states (also see analyses in volume II, chapter 10.9). At 200 hPa, a jet streak extended from the western Gulf of Mexico through the southeastern United States with an 85 m s^{-1} wind speed maximum over Georgia. The concentration of maximum wind speeds in the upper troposphere and their relative equatorward location are characteristic of a subtropical jet streak. At 300 hPa, jet streaks were analyzed over the Ohio Valley and over Florida. The well-defined wind maximum over the Ohio Valley, with velocities exceeding 85 m s^{-1}, did not appear in any form on the 200-hPa analyses, although the lower extension of this jet system was evident in the 500-hPa analysis over the northeastern states, indicating that this was probably a polar jet. The relative wind speed maximum over Florida at 300 hPa may be a lower, sloped extension of the jet streak found over Georgia at the 200-hPa level. A separate upper-level jet was noted at 200, 300, and 500 hPa over the plains states on the upwind side of the amplifying trough over the central United States.

A second example of complex jet system structure is from the February 1979 Presidents' Day Storm. At 1200 UTC 18 February 1979, the 200- and 300-hPa charts show evidence of multiple jet streaks, at a time when heavy snow was developing across the southeastern United States (Fig. 4-13). At 200 hPa, the eastern half of the United States was located beneath an extensive belt of high winds, with an 85–90 m s^{-1} jet streak analyzed over North Carolina. There was also a small separate 40 m s^{-1} jet streak over western South Dakota. At 300 hPa, two jet streaks were evident within the belt of high wind speeds over the eastern United States. An 80 m s^{-1} jet streak was centered over southern New England, while a 65 m s^{-1} streak was located over the Tennessee Valley. The wind speed maximum over New England, which was not readily apparent at 200 hPa, represents a descending polar jet within the confluence zone upstream of a trough off the Canadian coast. The speed maximum over Tennessee was actually the lower extension of the ascending subtropical jet streak that had its maximum velocities at 200 hPa over North Carolina. The polar jet streak over the plains states was better defined at 300 hPa than at 200 hPa, and was also observed at 500 hPa. For more details of the jet structures in this case, see the analyses in volume II, chapter 10.18, and in Uccellini et al. (1984, 1985).

Large temporal changes in the strength and location of upper-level jets characterized many of the snowstorms. Large increases of wind speed, defined here as greater than 15 m s^{-1} over a period of 24 h or less, are observed in 19 of 30 cases both prior to and during cyclogenesis along the East Coast. These large wind speed increases are observed in an equal number of cases near or slightly upwind of the trough axis and along the downstream ridge crest, and two examples are presented in Fig. 4-14. One example in which significant increases in wind speed is noted near the base of the upper troughs is the 31 March 1997 case (Fig. 4-14, top). In this case, wind speeds at the 250–300-hPa level increased from 50 to over 70 m s^{-1} in 12 h, ending at 0000 UTC 31 March 31 1997. The increase occurred near the base of the upper trough as the upper trough amplified. In the cases in which upper-level wind speeds increase at the base of the trough, it does not appear that the increase occurs as an upper jet or wind maxima drops southeastward along the western side of the trough axis. Rather, as the upper shortwave increases in amplitude, a subtropical jet approaching the base of the trough appears to increase in speed as a response to the increasing pressure gradient related to the localized wave amplification. (Other cases that exhibit large in-

←

FIG. 4-13. Examples of multiple upper-level jets structure at (left) 1200 UTC 29 Jan 1966 and (right) 1200 UTC 18 Feb 1979. Arrows indicate jet axes. Top row: 200-hPa analyses, including geopotential height (solid, 140 = 1400 m) and isotachs (dashed, m s^{-1}, shading for alternating 10 m s^{-1} intervals above 40 m s^{-1}). Second row: 300-hPa analyses, including geopotential height (solid, 900 = 9000 m) and isotachs (dashed, m s^{-1}, shading for alternating 10 m s^{-1} intervals above 40 m s^{-1}). Third row: 500-hPa analyses, including geopotential height (540 = 5400 m) and isotachs (dashed, m s^{-1}). Bottom row: surface analyses, depicting fronts, sea level isobars (thin solid, hPa), and precipitation areas.

UPPER-LEVEL JET AMPLIFICATION
PRIOR TO MAJOR SNOWSTORMS

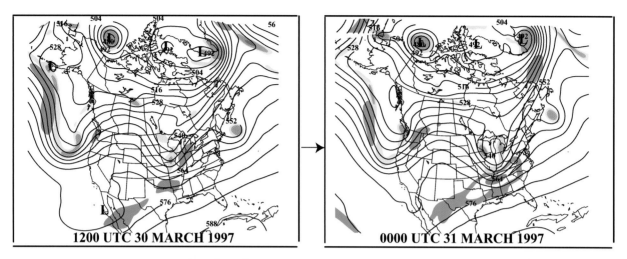

ALONG UPPER TROUGH AXIS

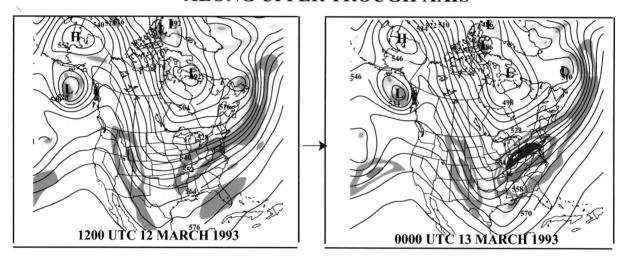

ALONG UPPER RIDGE CREST

Fig. 4-14. Examples of upper-level jet amplification prior to major snowstorms.

creases of wind speed at the base of the upper trough include February 1961, February 1967, and February 1983; see the upper-level wind analyses in chapter 10, volume II.)

While the speed increases are noted just upstream of the trough axis and along the downstream ridge, the greatest speed increases are often observed near the crests of the upper-level ridge downstream of the upper trough associated with the surface cyclone. The increases in wind speed prior to and during cyclogenesis are consistent with the increased geopotential height gradients and amplifying character of the upper-level trough–ridge systems observed with many of the 30 cases and demonstrate that in several of these cases the

wind velocities with the upper-level jet streaks are also undergoing significant changes near the upper-trough axis and within the downstream ridge.

These wind speed amplifications during cyclogenesis also reflect the upper-level impact of latent heat release (Chang et al. 1982; Keyser and Johnson 1986; Uccellini 1990; Dickinson et al. 1997) as significant precipitation development downstream of the upper-trough axis warms the atmosphere through latent heating, forcing the amplification of the upper jet to the north of the region (for more details, see chapter 7). The development of the January Blizzard of 1966 and the March 1993 Superstorm are both characterized by large wind speed increases in the downstream ridge crest. Prior to the Blizzard of 1966, the 300-hPa winds increased from 60 to 85 m s^{-1} in only 12 h over the Ohio Valley and middle Atlantic states as the surface low developed over the Gulf of Mexico. During the formation of the March 1993 Superstorm over the Gulf of Mexico, wind speeds at 300 hPa increased from 65 to over 90 m s^{-1} over the Ohio Valley, also in only 12 h (Fig. 4-14, bottom). Large increases in wind speed near the ridge crest prior to cyclogenesis were also shown by other cases (including January 1964, February 1979, and January 1996; see the upper-level wind analyses for these cases in chapter 10, volume II).

Despite the complexity in the upper-level jet structure near the trough, the upper-level confluence observed in southeastern Canada prior to many major Northeast snowstorms (see section 3b) was typically associated with the entrance region of one or more upper-level jet streaks (Fig. 4-10). Although some individual jet streaks were often difficult to define utilizing standard isobaric surfaces, there were consistent patterns of confluence and associated wind maxima, particularly over the northeastern United States and southeastern Canada. While the upper confluent signature was weak or absent in 5 of the 30 cases, the entrance region of the upper-level jet streak(s) was observed over the northeast United States or southeast Canada in all but 3 cases (the storms of 8–10 February 1969, February 1995, and 31 March–1 April 1997; see analyses in chapter 10, Volume II). The presence of an upper-level trough over eastern Canada, in combination with a mean climatological ridge over the western Atlantic Ocean may favor the appearance of confluence and an upper-jet entrance region across the northeastern United States and Canada, suggesting a favorable environment to maintain cold air in the northeast prior to many major snowstorms.

It is also well known that cyclones tend to develop within the exit region of upper-level jet streaks (see, e.g., Reiter 1969). While the jet exit regions were often located over the data-sparse Gulf of Mexico or the western Atlantic Ocean, the advent of the NCEP reanalysis data (Kalnay et al. 1996) allows a dynamically consistent diagnosis of winds that minimizes uncertainties regarding data-sparse areas and shows the jet streak exit region approaching the southeast coast during the cyclogenetic period. The mutual presence of both jet entrance and jet exit regions in the development of heavy snowfall will be explored next.

d. An orientation of jet streaks conducive to heavy snowfall

An orientation of upper-level jet streaks and their associated vertical ageostrophic circulations is an important signature of the cyclogenetic environment that appears to contribute to heavy snowfall in the Northeast, and is illustrated by three cases (Fig. 4-15) out of the eight used by Uccellini and Kocin (1987) to identify a lateral coupling of the ascent pattern related to these jet

systems. One jet is directed toward the base of the upper-level trough approaching the east coast and another is situated upstream of the trough crossing southeastern Canada or the northeastern United States. An upper-level jet streak approaching the base of the trough, with its diffluent exit region over the East Coast, is a common feature during the 12-h period when cyclogenesis was occurring along the coast. A jet streak embedded within the upper-level confluent zone across the northeastern United States or southeastern Canada was also a common feature during the same period. The cold surface anticyclone or ridgeline was often located beneath the region of confluent geopotential heights at 500 hPa that marked the entrance region of this jet streak.

Cross sections bisecting the entrance and exit regions of the two jet streaks reveal the presence of transverse vertical circulations that appear to be laterally coupled and link the jet streaks and their associated upper-level troughs to 1) the orientations of the surface cyclones and anticyclones, 2) the temperature advections, 3) moisture transports, and 4) vertical motions needed to produce heavy snowfall.

A kinematic methodology was developed by Keyser et al. (1989) to represent the three-dimensional ageostrophic circulations in terms of the contributions to the vertical motion by the cross-stream divergent ageostrophic flow associated with a jet streak from along-stream divergent ageostrophic flow associated with the upper trough–ridge pattern. Using Keyser et al.'s methodology, Loughe et al. (1995) found quantitative support for the interpretation that the "laterally coupled" jet streak circulation pattern makes a significant contribution to the ascent diagnosed for the February 1983 snowstorm. Furthermore, the cross-stream ageostrophic circulation pattern associated with the laterally coupled jets tends to focus the ascent pattern into a relatively narrow domain, coincident with the area of heaviest snow (Fig. 4-16).

In summary, Uccellini and Kocin's (1987) analyses of the February 1983 case and seven other cases and the resultant schematic representation of the lateral coupling of the transverse circulations around the separate jet streaks (Fig. 4-17) illustrates the following:

i) As an upper-level trough over the Ohio and Tennessee Valleys approaches the East Coast and a jet streak approaches the base of the trough and propagates toward the diffluent region downwind of the trough axis, a surface low pressure system develops. The cyclogenesis occurs downstream of the trough axis, within the exit region of the jet streak, where upper-level divergence (cyclonic vorticity advection) related to both the trough and the jet favors surface cyclonic development.

ii) A separate upper-level trough over southeastern Canada and a jet streak embedded in the confluence zone over New England are associated with a cold surface high pressure system, which is usually

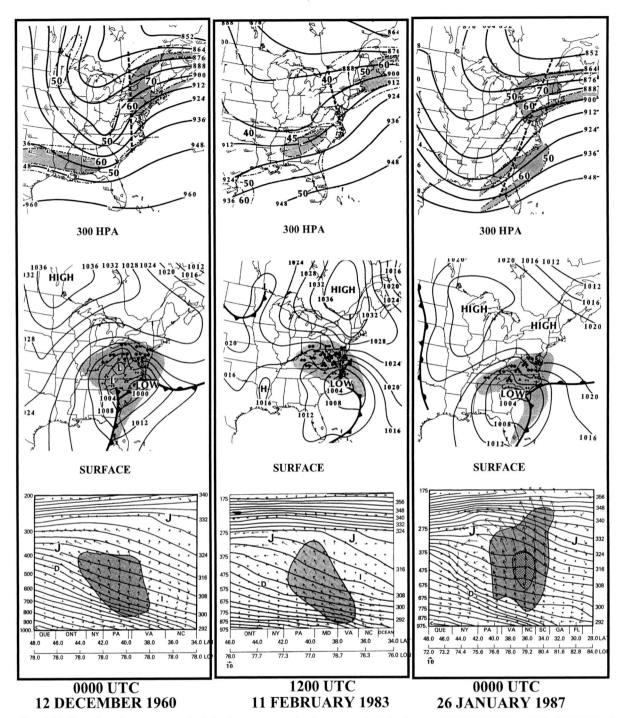

0000 UTC 12 DECEMBER 1960

1200 UTC 11 FEBRUARY 1983

0000 UTC 26 JANUARY 1987

FIG. 4-15. Examples of transverse vertical circulation patterns for three cases. (top) Analyses at 0000 UTC 12 Dec 1960, 1200 UTC 11 Feb 1983, and 0000 UTC 26 Jan 1987 include 300-hPa geopotential heights (528 = 5280 m), isotachs (dotted–dashed; 40 = 40 m s^{-1}), and each flag represents 25 m s^{-1}. Each barb denotes 5 m s^{-1} and each half-barb denotes 2.5 m s^{-1}. Thick dashed line represents axis of cross section shown in the bottom panels of the figure. (middle) Corresponding surface frontal and isobaric (solid, hPa) analysis with shading representing precipitation. Vector representation of vertical motions and ageostrophic wind components tangential to the plane of the cross section shown at the top of the figure, including isentropes (K) at 4-K increments. The horizontal vector components are scaled at the bottom of the figure (m s^{-1}). Shading represents ascent in excess of -5 μb s^{-1}. Positions of upper-level jet streaks are indicated by J; D and I denote centers of direct and indirect circulations, respectively. [Derived from Uccellini and Kocin (1987).]

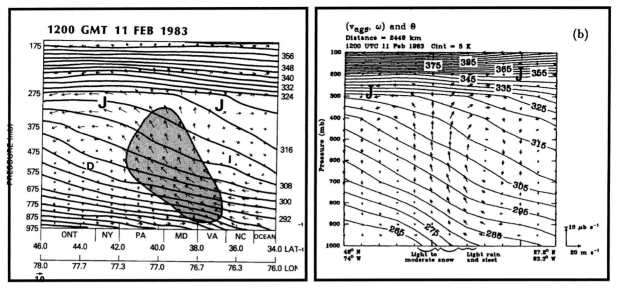

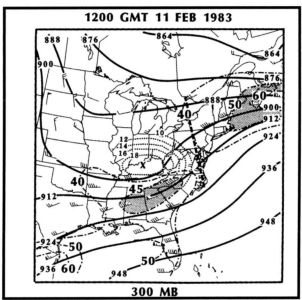

FIG. 4-16. (a) Vector representation of vertical motions and ageostrophic wind components tangential to the plane of the cross sections shown in (c) at 1200 UTC 11 Feb 1983, including isentropes (K) at 4-K increments. The horizontal vector components are scaled at the bottom of the figure (m s^{-1}). Shading represents ascent in excess of -5 μb s^{-1}. Positions of upper-level jet streaks are indicated by J; D and I denote centers of direct and indirect circulations, respectively (from Uccellini and Kocin 1987). (b) Cross section at 1200 UTC 11 Feb 1983 of transverse vertical circulation represented in terms of (v_{ags}, ω). Background fields are θ_E and θ (contour interval, 5 K), respectively (from Loughe et al. 1995). (c) The 300-hPa geopotential height and wind analysis (alternating shading represents 10 m s^{-1} increments of wind speed) at 1200 UTC 11 Feb 1983. Dashed line represents axis of cross section shown in (a); dotted line represents axis of cross section shown in (b).

found beneath the confluent entrance region of the jet streak.

iii) Indirect and direct transverse circulations that extend through the depth of the troposphere are located in the exit and entrance regions of the southern and northern jet streaks, respectively.

iv) The rising branches of the transverse vertical circulations associated with the two jet streaks appear to merge, contributing to a large region of ascent

along sloped isentropic surfaces, which produces clouds and precipitation between the diffluent exit region near the east coast and the confluent entrance region over the northeastern United States.

v) The advection of Canadian air southward in the lower branch of the direct circulation over the northeastern United States maintains the cold lower-tropospheric temperatures needed for snowfall along the east coast. In many cases, the lower

SCHEMATIC OF JET-RELATED CIRCULATION
PATTERNS DURING NORTHEAST SNOWSTORMS

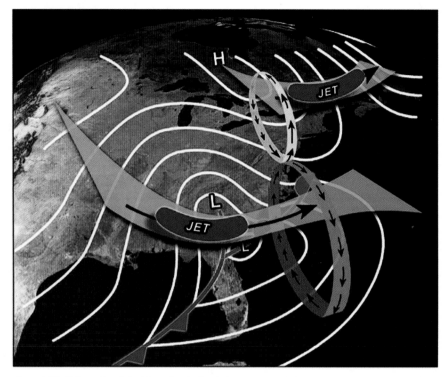

FIG. 4-17. Schematic of dual jet-related circulation patterns during Northeast snowstorms. Circulations are represented by pinwheels, jet streaks are embedded within confluent and diffluent regions, and solid lines are sea level isobars.

branch of this circulation pattern is enhanced by the ageostrophic flow associated with cold-air damming east of the Appalachian Mountains (see chapter 3).

vi) The northward advection of warm, moist air in the lower branch of the indirect circulation over the southeastern United States rises over the colder air north of the surface low. An easterly or southeasterly low-level jet streak (see the next section) typically develops within the lower branch of the indirect circulation beneath the diffluent exit region of the upper-level jet. This low-level flow enhances the moisture transport toward the region of heavy snowfall and focuses the low-level thermal advection patterns along the coast.

vii) The combination of the differential vertical motions in the middle troposphere and the interactions of the lower branches of the direct and indirect circulations appears to be highly frontogenetic [as computed by Sanders and Bosart (1985a) for the February 1983 storm], increasing the thermal gradients in the middle and lower troposphere during the initial cyclogenetic period. This finding is consistent with the increase in available potential en-

ergy observed by Palmén (1951, p. 618) during the early stages of cyclogenesis.

e. Lower-tropospheric jet streaks and warm-air advections

The northward advection of warm, moist air at low levels is an important component of the coupled jet pattern discussed in the previous section. The warm-air advection marks the lower branch of the indirect circulation in the exit region of the jet streak approaching the East Coast. The low-level advection of warm, moist air is usually associated with a pronounced isotach maximum or low-level jet streak (LLJ) that develops either prior to or during cyclogenesis (often near 850 hPa), and is directed toward the area of heaviest snowfall. With the onset of cyclogenesis along the East Coast, an LLJ often develops from a southerly to easterly direction near the coast (see the March 1960 example in Fig 4-18a). This LLJ is observed in all 30 cases, with maximum wind speeds typically ranging between 25 and 35 m s^{-1} at 850 hPa (Figs. 4-18a–c). The development of wind maxima near the 850-hPa level during East Coast storms may account for the secondary maximum of

LLJs frequently found along the North Carolina coast by Bonner (1965) in his climatological survey of low-level jet streaks in the United States; The primary maximum is located over the southern plains states; the primary maximum is located over the southern plains states.

The development of the LLJ is often initially observed over the southeastern United States with a south to southeasterly 850-hPa wind maxima located to the east-northeast of the developing surface low pressure center. In addition, the LLJ occurs beneath the exit region of the upper-level jet–trough systems approaching the East Coast (Figs. 4-18a–c) in a manner that is consistent with the isallobaric effects described by Naistat and Young (1973) and the jet streak coupling process described by Uccellini and Johnson (1979). Of the 30 cases examined, at least 18 exhibit the development of an LLJ along the Southeast or middle Atlantic coast within the entrance region of an upper-level jet streak over the northeastern United States and within the exit region of a separate jet streak approaching the east coast; this early onset is especially evident in the cases of February 1958, March 1960, January 1978, 22 January 1987, and March 1993; see Fig. 4-18). In many of the cases depicted in Fig. 4-18, the development of heavy snowfall along the middle Atlantic and Northeast coast was accompanied by the development of a southeasterly to easterly LLJ across the middle Atlantic states and New England with wind speeds usually exceeding 25 m s^{-1} and occasionally exceeding 35–40 m s^{-1}. Computations of low-level moisture transport associated with the development of an LLJ during the evolution of the February 1979 Presidents' Day Storm (Uccellini and Johnson 1979; Uccellini et al. 1984) show moisture transport values comparable to springtime convective environments.

Numerical simulations of the development of an LLJ with the February 1979 Presidents' Day Storm (Uccellini et al. 1987) showed that the development of this jet can occur over a period of only 2–4 h. High-resolution numerical simulations of cyclone events show that many LLJs can form over time periods of significantly less than the 12-h resolution of the operational rawinsonde network. Therefore, the examination of 30 cases utilizing the rawinsonde network was not sufficient to resolve the details of the development of the LLJ in every case. Nonetheless, the data were sufficient to observe a portion of the development of the LLJ in most cases. For example, during the development of the snowstorm of February 1978, 850-hPa wind speeds increased from 10 to 15 m s^{-1} at 1200 UTC 6 February as snow was beginning to develop along the mid-Atlantic coast to 35 to 40 m s^{-1} and greater over a large portion of the Northeast by 0000 UTC 7 February as heavy snow enveloped much of the Northeast (see chapter 10, volume II).

As shown by Uccellini et al. (1987), the development of a low-level jet streak prior to the development of the 1979 Presidents' Day Storm enhanced the lower-tropospheric thermal and moisture advections required for heavy snow development. It also was a crucial component of the initial cyclogenetic phase by enhancing the mass divergence in the low levels within a narrow region and focusing surface pressure falls immediately along the coast. A similar sequence of LLJ development with a significant ageostrophic component is also analyzed for the initial cyclogenesis for the February 1978 snowstorm (see volume II, chapter 10.17). Therefore, the LLJ serves a crucial role in enhancing and focusing the initial rapid development phase of the cyclone, in addition to transporting the moisture requirements for sustaining a large area of heavy snow.

f. Other lower-tropospheric characteristics

The development of cyclones depends on regions of enhanced baroclinicity that are often maximized in the lower troposphere, as well as their associated patterns of low-level warm-air advections. The development of low-level jet streaks within this lower-tropospheric flow provides a mechanism for enhanced thermal advections and serves as a mechanism of rapid lower-tropospheric mass divergence that can act to focus the pressure falls along the coast during the intense periods of cyclogenesis (Uccellini et al. 1987), a critical element for a cyclone's intensification.

In addition to the direct impact of the thermal gradients on cyclogenesis, many weather forecasters recognize that the demarcation between rain and snow is often found with 850-hPa temperatures near or slightly below 0°C. Other empirical approaches also stress critical values of 1000–850-, 850–700-, or 1000–500-hPa thicknesses (e.g., see Wagner 1957). The 850-hPa surface serves not only as an indicator of precipitation type, but is also representative of lower-tropospheric thermal and moisture advections. Thus, the 850-hPa surface is a valuable level for examining many of the factors that influence the precipitation form and intensity during snowstorms (Bailey 1960; Browne and Younkin 1970; Spiegler and Fisher 1971).

A comparison of the 850-hPa low tracks with the snowfall patterns in the 30 storms (Fig. 4-19) reveals that the heaviest snowfall occurred 50–300 km to the left of the center's path. This generally agrees with the results of Browne and Younkin (1970) and Spiegler and Fisher (1971). The paths of the 850-hPa low centers were located near or to the left of their sea level counterparts (cf. with Fig. 3-3), and generally followed east to northeasterly tracks along or off the middle Atlantic and southern New England coasts. Spiegler and Fisher stated that the 850-hPa low center is a better guide for forecasting snow amounts than is the sea level cyclone center, since its path appears to be more continuous. A review of the 850-hPa low tracks in the 30 cases, however, suggests that the movement of some of these centers may have been as discontinuous as the paths of the

(a)

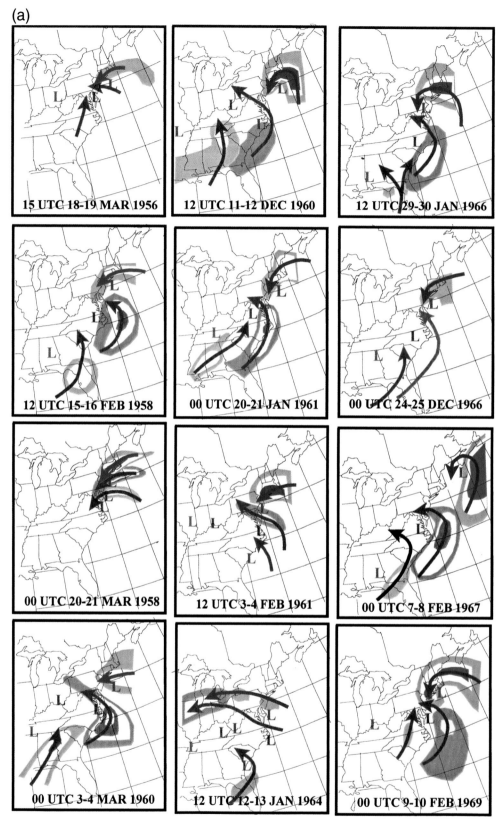

Fig. 4-18. (a)–(c) The 12-hourly evolution of the low-level jet at 850 hPa for a 24-h period during the development of heavy snowfall. Three colors represent three consecutive times at 12-h intervals. Arrows reflect axes of low-level jet and color gradations represent wind speeds > 20 m s⁻¹ (at 5 m s⁻¹ increments).

(b)

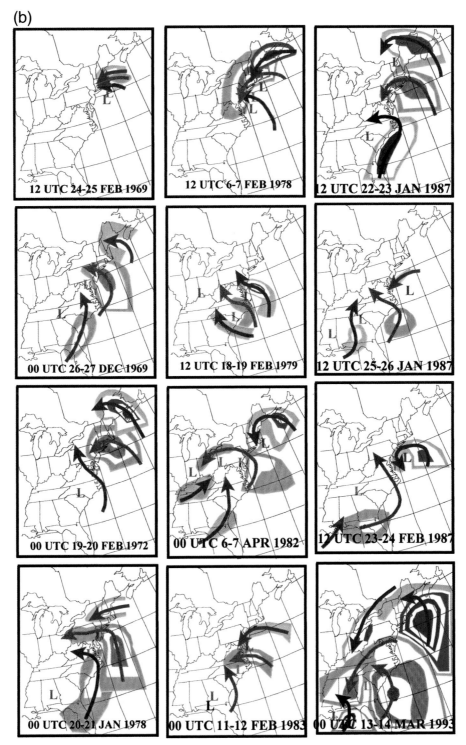

FIG. 4-18. (*Continued*)

surface low pressure centers that redeveloped along the coast. When the redeveloping surface low pressure systems intensified along the coast, the associated changes in the lower troposphere were displayed at the 850-hPa level as either a distinct secondary low center or as a separate maximum of geopotential height falls. It is difficult to properly monitor the secondary development at 850 hPa due to the 12-h interval between the operational radiosonde measurements, which cannot adequately resolve the rapidly evolving 850-hPa troughs

(c)

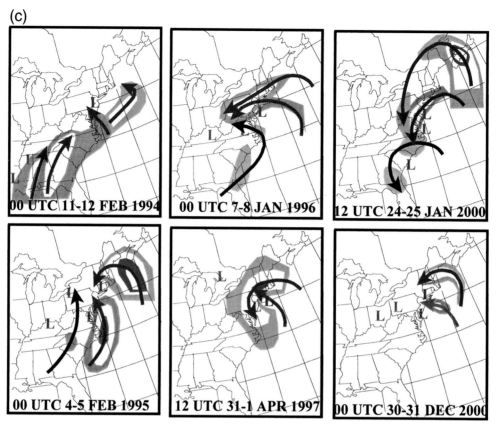

FIG. 4-18. (*Continued*)

during cyclogenesis along the coast (e.g., see Uccellini et al. 1987). The increasing availability of numerical model output at smaller time intervals allows an improved examination of this process.

The 850-hPa charts for the 30-case sample also revealed that prior to cyclogenesis, low-level northwesterly flow and cold-air advection usually covered the Northeast. Such advections were weak or absent in only six of the cases (including March 1958, both storms of February 1969, February 1987, February 1995, and 31 March–1 April 1997). The cold-air advection occurred to the rear of an intense 850-hPa low centered over southeastern Canada or the adjacent offshore waters. The northwesterly flow with strong low-level cold-air advections drives the 0°C isotherm south of the urban centers of the northeastern United States toward the Carolinas. The northwesterly flow and cold-air advections were associated with the surface anticyclones that influenced the northeastern United States (Fig. 3-11) that typically produce cold-air damming conditions to the lee of the Appalachians that also establish low-level coastal frontogenesis (see chapter 3). These lower-tropospheric features were also generally located beneath the regions of upper-level confluence within the entrance regions of the upper-level jet streaks.

Another characteristic of the 30-case sample is the evolution of a geopotential height minimum or low cen-

ter at 850 hPa along a preexisting isotherm gradient extending from the central United States to the east coast. The 850-hPa low center intensified during cyclogenesis. Meanwhile, the 850-hPa temperature gradients increased and evolved into an S-shaped pattern, as the advections of cold and warm air attending the 850-hPa circulation increased with time. Prior to rapid cyclogenesis, the 0°C isotherm progressed northward toward the middle Atlantic states and became nearly collocated with the 850-hPa low center. During cyclogenesis, the 0°C isotherm did not progress rapidly northward despite increasing south or southeasterly winds, implying that cooling resulting from strong ascent may have offset the effects of the thermal advection (Browne and Younkin 1970). The sea level cyclone was often located along the warm edge of the 850-hPa temperature gradient, either at or upwind of the apex of the isotherm ridge. As the cyclone continued to intensify and reach maximum intensity, the 850-hPa low frequently became nearly collocated with the sea level cyclone in an area where 850-hPa temperatures fell below 0°C.

4. Satellite signatures

Satellite-derived images of cloud patterns, especially the half-hourly high-resolution visible and infrared images provided by the Geostationary Operational Envi-

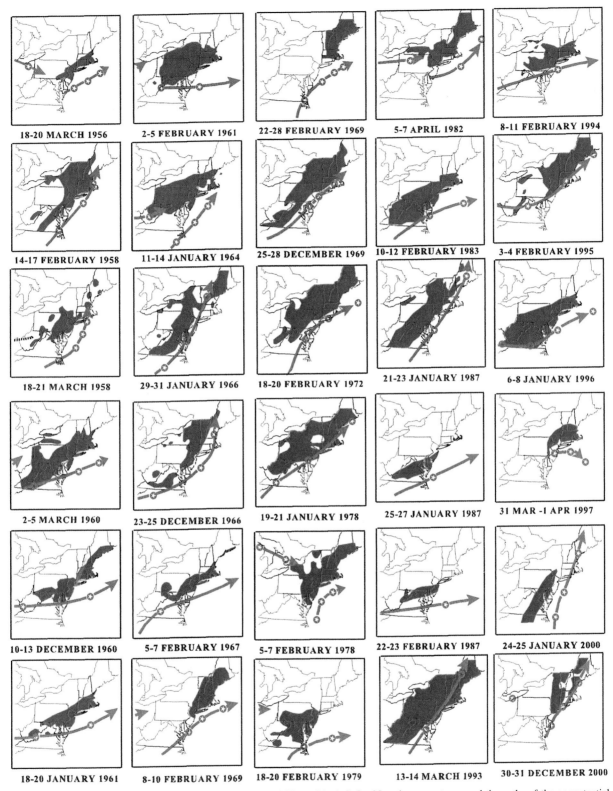

FIG. 4-19. A comparison between storm snowfall in excess of 25 cm (shaded) for 30 major snowstorms and the paths of the geopotential height minima at 850 hPa.

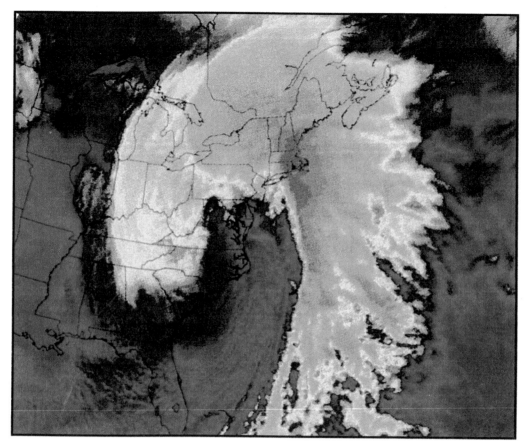

THE MARCH 1993 "SUPERSTORM"

Fig. 4-20. Infrared image of the Mar 1993 Superstorm, 13 Mar 1993.

ronmental Satellite (GOES) series, help in visualizing the evolution of cloud elements and the three-dimensional airflow associated with extratropical cyclones, which include heavy snowstorms along the Northeast urban corridor. Meteorologists who analyze satellite data have long recognized that the "comma cloud"-shaped pattern (i.e., Fig. 4-20) is a signature of these storms (e.g., see Boucher and Newcomb 1962; Leese 1962; Widger 1964; Weldon 1979; Carlson 1980; Browning 1990; Carlson 1991; Evans et al. 1994). Later studies (i.e., Browning 1990; Carlson 1991; Evans et al. 1994) have begun to categorize cloud patterns in satellite images to some of the dynamical patterns that give rise to cyclogenesis.

Browning (1990) describes three classes of cloud elements that define cyclone development that evolve into comma-shaped features in the satellite imagery: 1) comma cloud development without the presence of a "polar front" cloud band; 2) "instant occlusions," in which an organizing cloud cluster merges with a polar front cloud band to produce a comma-cloud-shaped structure; and 3) the frontal wave cyclone that evolves from only the polar front cloud band in which an upper-level disturbance overlies the cloud band. Evans et al. (1994)

extend the work of Browning and others by utilizing satellite imagery to categorize explosively developing cyclones into four distinct types: 1) emerging cloud head, 2) comma cloud, 3) left exit, and 4) instant occlusions. In their study, Evans et al. (1994) have examined four of the cases described here and ascribe the evolution of these cases to the categories of "emerging cloud head" or "left exit." The emerging cloud head is a category ascribed to the cases of January 1978 and February 1983 and is characterized by "two meridionally separated jet streams," a signature common to many of the storms examined. The emergence of the comma head occurs as the upper trough tends toward a diffluent orientation with a positive to negative tilt and the upper-level jet streak approaches the base of the upper-level trough and rapid surface deepening occurs within the exit region of the jet streak. The evolution of the cloud features associated with the March 1993 Superstorm reflects this type of cloud evolution (Fig. 4-21) with the letter A showing the development of the "comma head." This evolution appears to be similar to the frontal wave cyclone sequence described by Browning (1990).

"Left exit" is a category ascribed to the cases of

February 1978 and February 1979 in which there are two separate cloud features, one a "baroclinic leaf" cloud (Weldon 1979) poleward of a second, main polar front cloud band, roughly parallel to each other, that later merge (Evans et al. 1994). The evolution of the cloud features associated with the February 1979 Presidents' Day Storm illustrates this type of cloud evolution (Fig. 4-22) with the letters A and B representing the two merging cloud features. The surface low develops near the intersection of the merger of the southwestern edge of the leaf with the polar front cloud band, with the leaf taking on a comma-cloud shape. This category is similar to the "instant occlusion" signature described by Browning (1990) since it involves the merger of two separate cloud features but differs from the instant occlusion category in Evans et al. (1994) because the upstream cloud feature resembles a leaf cloud rather than a cluster of convective clouds. The difference between the left exit and instant occlusion categories may lie in the observation of whether the cloud merger occurs within a confluent synoptic pattern (instant occlusion) with rapid cyclogenesis occurring in the entrance region of a jet streak, or a diffluent pattern (exit region) with rapid cyclogenesis occurring within the exit region of a jet streak as was described for many of the cases.

An approach to describing midlatitude cyclones that relates the signatures apparent in satellite imagery to the relative airflow that interacts throughout cyclonic systems is described by Carlson (1980, 1991) and Browning (1990). This approach is based on trajectory analyses on constant potential temperature, or isentropic, surfaces in dry regions and on wet-bulb potential temperature surfaces in saturated regions (Eliassen and Kleinschmidt 1957; Green et al. 1966; Browning and Harrold 1969; Browning 1971) that identify various airstreams associated with cyclones: the warm conveyor belt, the cold conveyor belt, and the dry airstream. A schematic diagram that illustrates the primary components of Carlson's (1980) model in relation to a hypothetical heavy snow situation in the Northeast is shown in Fig. 4-23. Evans et al. (1994) suggest that this conveyor belt model is most appropriate for the mature stage of the third class of cyclone development described by Browning (1990): the frontal wave cyclone.

The warm conveyor belt (Harrold 1973; Browning 1986) is a stream of air that originates in the warm sector of a cyclone that forms along a major confluent zone and ascends as it moves northward, achieving saturation near the warm front. It then turns anticyclonically and joins the upper-level westerly flow north of the sea level cyclone (see Fig. 4-23). In its developing stages, this belt has been described as a baroclinic leaf cloud, which is often a precursor of cyclonic development when there may be no identifiable surface cyclone (Browning 1990). While isobaric charts are incapable of capturing the substantial vertical displacements that characterize the warm conveyor belt and the other airstreams, the warm conveyor belt may be represented on the surface

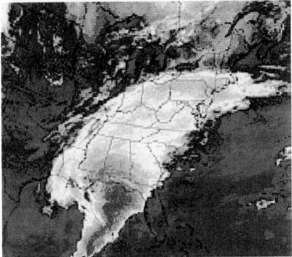

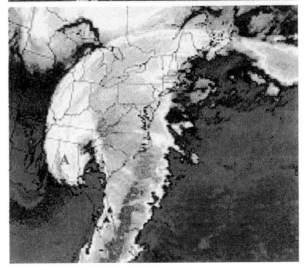

FIG. 4-21. Emergence of "comma head" (Evans et al. 1994) during the Mar 1993 Superstorm. Infrared images are at approximate 12-h intervals on 12–13 Mar 1993.

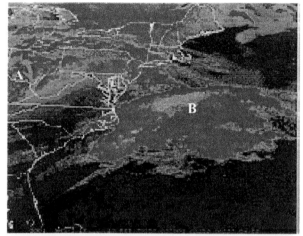

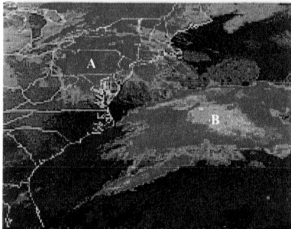

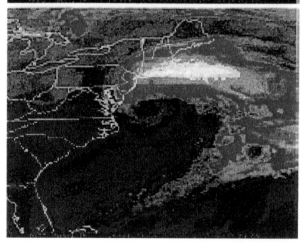

FIG. 4-22. Example of "left exit" (Evans et al. 1994) development of a comma cloud in which two separate cloud features merge, during the Presidents' Day Storm of Feb 1979. Infrared images are from 18 to 19 Feb 1979.

charts by the southerly flow ahead of the surface cold front, and on the upper-level charts (i.e., the 500-, 300-, and 200-hPa analyses) by the southwesterly flow down-wind of the trough axis. The warm conveyor belt typ-

ically forms the comma "tail" and may be bounded by a distinct anticyclonically curved shield of cold upper-level cirrus clouds and a relative-wind jet stream (Browning 1990) as the airstream ascends and veers with height [see Carlson (1980), Browning (1990), and the model diagnostic study by Whitaker et al. (1988)].

The cold conveyor belt (Carlson 1980) may be an especially significant component of snowstorms that af-fect the Northeast coast. It originates within the anti-cyclonic flow around the surface high pressure cell poised to the north of the developing storm system. The air within this stream moves westward within the ocean-influenced boundary layer, toward and to the north of the surface low. This airstream then ascends the steeply sloped isentropic surfaces associated with the upper-level trough–jet system approaching the East Coast (Fig. 4-23), with a portion of the airstream circulating around the surface low. If air parcels ascend to a high enough level, the cold conveyor belt may also turn sharply to the north or northeast as it rises (Carlson 1980; Brown-ing 1990). The model-based trajectories shown by Whi-taker et al. (1988, their Fig. 18, parcel 5) provide evi-dence for this characteristic. The cold conveyor belt is viewed on the operational charts as the surface flow around the southern periphery of the shallow, cold an-ticyclone north of the warm front, and the easterly flow north of the 850-hPa low center, possibly taking the form of a low-level jet. Saturation in the cold conveyor belt results from precipitation falling into it from above as it passes beneath the warm conveyor belt, moisture fluxes from the ocean surface, and the rapid ascent of the airstream as it passes to the north of the developing cyclone center. The cold conveyor belt produces an up-per-level cirriform cloud deck that forms the distinctive comma "head" north of the sea level cyclone, which generally expands westward or northwestward with time.

The cold conveyor belt is quite important since heavy snowfall is located beneath the expanding comma-head cloud mass. Heavy snowfall rates were observed near the transition between the high cloud tops that mark the comma head and the more shallow clouds immediately to its south and southwest (see examples in Fig. 4-24). Similar relationships between the location of heavy snowfall and the distribution and height of the clouds, as inferred from satellite imagery, have been reported by Carlson (1980), Uccellini et al. (1984), Bosart and Sanders (1986), Beckman (1987), and others. No pre-cipitation was reaching the earth's surface under a sig-nificant portion of the cirrus cloud deck that made up the comma head, especially poleward of the highest clouds depicted by the enhancement in the infrared im-agery. These nonprecipitating cloud elements suggest that a relatively thin layer of high clouds was produced by the ascent of the airstream and was then advected northeastward with the upper-level flow.

The dry airstream contains dry air that originates in the lower stratosphere west of the trough axis and de-

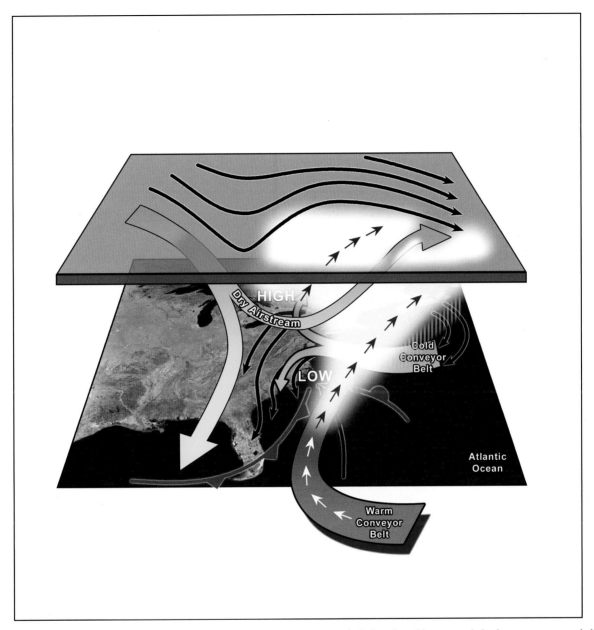

FIG. 4-23. Schematic representation of airflow through a Northeast snowstorm, including the cold conveyor belt, the warm conveyor belt, the dry airstream (see Carlson 1980); surface fronts; sea level cyclone and anticyclone centers; and low- and upper-level streamlines.

scends along an upper-level jet streak toward the developing surface cyclone. Many investigators (Eliassen and Kleinschmidt 1957; Danielsen 1966; Uccellini et al. 1985; Hoskins et al. 1985; Whitaker et al. 1988) have linked the descent of stratospheric air to the evolution of a potential vorticity maximum upstream of the incipient cyclone, a situation that is conducive to cyclogenesis. The descending airstream splits into two branches. One branch descends anticyclonically toward the lower troposphere behind the surface cold front. The other descends initially upstream of the trough axis, then crosses the trough axis, and rises roughly parallel to the

western edge of the warm conveyor belt (see Fig. 4-23). This intrusion of dry air is relatively cloud free and helps to shape the comma cloud (e.g., see the dark blue area along the coast from the Carolinas northward to Delaware and immediately offshore between the comma head and tail in the infrared image for the March 1993 Superstorm in Fig. 4-20). The interaction of the dry airstream and the moisture-laden warm and cold conveyor belts is largely responsible for the asymmetric cloud distribution that marks these storms. The surface low is generally located along the boundary between the dry airstream approaching from the west and the

LOCATIONS OF HEAVY SNOW

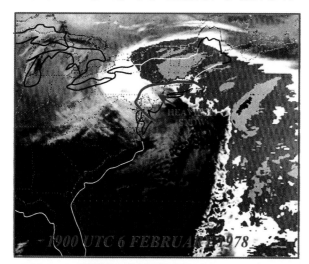

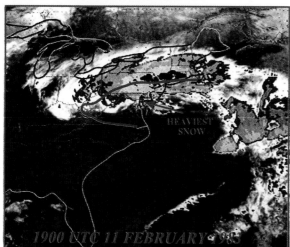

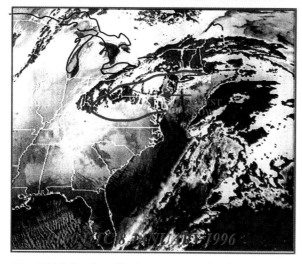

FIG. 4-24. Infrared satellite images representing the occurrence of moderate to heavy snows along the transition between lower and higher cloud features: (top) 1900 UTC 6 Feb 1978, (middle) 1900 UTC 11 Feb 1983, and (bottom) 0100 UTC 8 Jan 1996.

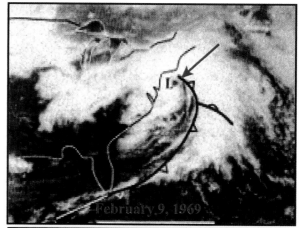

FIG. 4-25. Applications Technology Satellite (ATS-III) and GOES visible imagery of eye-like cloud features associated with three Northeast snowstorms.

moist warm and cold conveyor belts approaching from the south and east, a point emphasized in the review on rapid cyclogenesis by Uccellini (1990).

Modifications to the conveyor belt model have been provided by Martin (1999) and other studies (e.g., Mass and Schultz 1993; Schultz and Mass 1993; Reed et al. 1994; Schultz 2001) that suggest that the model can be more complicated. For example, the comma head, often the site where the heaviest snows fall, can occur from warm sector air (Shultz 2001) originating within the warm conveyor belt region and not necessarily from ascent of the cold conveyor belt as described by Carlson (1980) and others.

In the later stages of cyclone development, when the systems have moved out over the Atlantic Ocean, satellite images often reveal the formation of an "eye"-like cloud-free region near the storm centers, somewhat similar to that observed in tropical cyclones [also described by Sanders and Gyakum (1980), Bosart (1981), and Uccellini et al. (1985)]. Visible satellite imagery from the February 1969 "Lindsay Storm" and the snowstorms of February 1978 and February 1979 (Fig. 4-25) illustrate the appearance of this feature during or toward the end of each cyclone's rapid development phase. An examination of the satellite imagery for cases exhibiting the eye-like feature indicates that the height distribution of the clouds that surround the centers is asymmetric. Deeper clouds were observed to the north, with relatively shallow clouds south of and immediately surrounding the storm center reflecting the downward intrusion of stratospheric air southwest and over the cyclone center. Although convective elements may have been present, the asymmetric cloud height distribution does not resemble the deep convective cloud patterns that typically surround the centers of mature, intense tropical cyclones.

New York City in 1893 (photo by Alfred Stieglitz,
Library of Congress).

The Flatiron Building, New York City, Jan 1905 (photo courtesy of Library of Congress).

Times Square, New York City, 19 Dec 1948 (courtesy of the
Bettmann archive).

Abandoned cars in Woburn, Massachusetts, 10 Feb 1969 (photo from *Weatherwise*, American Meteorological Society, 1969, p. 230).

Clockwise (from top left): Queensberry Street, Boston, Massachusetts, Blizzard of 1978 (photo courtesy of Eric Pence, penceland.com); a Back Bay sidewalk after shoveling, Boston, Massachusetts, Blizzard of 1978 (photo courtesy of Eric Pence, penceland.com); looking east on Boylston Street, Boston, Massachusetts, Blizzard of 1978 (photo courtesy of Eric Pence, penceland.com).

The Long Island Expressway, 11 Feb 1983 (photo reprinted courtesy of *Newsday*).

Dupont Circle, Washington, DC, following the Blizzard of Jan 1996 (photo by Hiram Ruiz).

Street scene in Baltimore, Maryland, following the snowstorm of 25 Jan 2000 (photo courtesy of Bill Swartwout and SouthBaltimore.com).

Waves lash the seawall in Hull, Massachusetts, 7 Mar 2001 (copyright Reuters/Corbis).

True color image of snowcover following record Feb 2003 snowstorm taken by the Moderate Resolution Imaging Spectroradiometer (MODIS) aboard NASA's *Terra* satellite (image courtesy of Jacques Descloitres, MODIS Land Rapid Response Team at NASA Goddard Space Flight Center, Greenbelt, MD).

Street scene in Baltimore, Maryland, following the record-breaking snowstorm of Feb 2003 (photo courtesy of Bill Swartwout and SouthBaltimore.com).

Skiers at Radio City Music Hall, Feb 2003 (courtesy of Rob Gardiner, www.nyclondon.com).

Chapter 5

"NEAR MISS" EVENTS IN THE URBAN CORRIDOR

The review and analyses of 30 snowstorms in the previous two chapters indicate that these storms are relatively rare events. Although the East Coast is the site of intense cyclogenesis during the winter months, cold air must be present, entrenched, or reinforced during the storm in order for precipitation to fall as snow rather than rain in the Northeast urban corridor. Furthermore, precipitation must persist for periods of 12 h or greater to allow snowfall amounts to reach 10 in. (25 cm) or greater. In many coastal storms, precipitation is more than likely to fall as rain or mixed precipitation rather than snow, especially south of New York City toward Washington, D.C. Even during those events in which cold surface air is sufficient for frozen precipitation, it is rare that precipitation will fall completely as snow, and will often fall as a combination of snow, ice pellets, freezing rain, and rain in the urban areas.

In this chapter, the subtle differences that distinguish the relatively rare heavy snowstorm from the "near miss" event are identified. By near miss, we are referring to three situations in which heavy snow is, from a forecasting perspective, a possible outcome in the Northeast urban corridor. However, the heavy snow event does not materialize because either 1) a major cyclone develops but the heavy snowband forms too far inland, and snow changes to rain or other frozen precipitation in the urban corridor, significantly reducing snowfall amounts; 2) moderate snowfall amounts, generally 10 in. (25 cm) or less occur, instead of heavy snow; or 3) heavy precipitation falls as freezing rain or ice pellets (sleet), rather than heavy snow.

It is our experience that there are no simple rules that govern the differences between the "hits" and the near misses, and that each forecasting situation must be assessed through a careful diagnosis of the observations, model output, and experience. However, there are often several atmospheric signatures that distinguish between conditions in which a major snowfall is more likely to occur, and conditions that favor either a changeover to rain, freezing rain, or ice pellets or lesser snowfall amounts. Even though numerical models have improved in delineating precipitation type, these situations often pose the greatest challenges to operational meteorologists who predict winter weather for the eastern United States.

1. Precipitation changeovers within major Northeast snowstorms

A significant number of the 30 heavy snow cases described in the previous two chapters were accompanied by a changeover from snow to rain, freezing rain, or ice pellets (see Fig. 3-6), which acted to reduce the total snowfall in some portion of the Northeast urban corridor (some of the more notable cases include the storms of January 1966, December 1969, February 1972, January 1978, 22 January 1987, the March 1993 "Superstorm," and February 1995). In order to identify any features within major Northeast snowstorms that favored a changeover from snow to rain, freezing rain, or sleet, surface and upper-level features for these seven cases were compared to eight other cases (the storms of December 1960, January 1961, January 1964, February 1967, February 1978, February 1979, February 1983 and 26 January 1987) in which precipitation remained primarily as snow.

A comparison of the paths of the surface low pressure center in the seven changeover cases versus the eight all-snow cases (Fig. 5-1) indicates that the surface low pressure centers in the changeover cases took a path closer to the East Coast, or actually passed inland, than in the cases in which precipitation remained as all snow. The main difference between the two sets of cases is that the surface low pressure centers associated with the changeover cases took a more northerly route once they reached the Carolina coastline, while the surface low pressure centers in the all-snow ensemble moved more toward the east-northeast and remained offshore. As a result, many of the surface low pressure centers in the changeover cases passed near or over the Delaware–Maryland–Virginia (Delmarva) peninsula. This movement toward the north, rather than to the northeast, favored warmer air penetrating the coastline, enhancing the likelihood of a change from snow to rain or mixed precipitation in the Northeast urban corridor.

Anticyclone positions were also compared for systematic differences (Fig. 5-2). As noted in chapter 3, the location and movement of anticyclone centers during major Northeast snowstorms tended to group into two categories. The first category included anticyclones that passed eastward across Ontario and Quebec. The second category included anticyclones over south-central Can-

CYCLONE TRACKS
HEAVY SNOW CASES

ALL SNOW VS CHANGEOVERS

10-13 DECEMBER 1960
18-21 JANUARY 1961
11-14 JANUARY 1964
5-7 FEBRUARY 1967
5-7 FEBRUARY 1978
18-20 FEBRUARY 1979
10-12 FEBRUARY 1983
25-27 JANUARY 1987

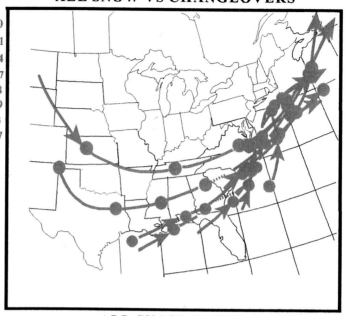

ALL SNOW CASES

29-31 JANUARY 1966
25-28 DECEMBER 1969
18-20 FEBRUARY 1972
19-21 JANUARY 1978
21-23 JANUARY 1987
13-14 MARCH 1993
3-4 FEBRUARY 1995

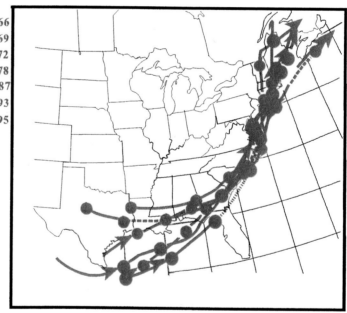

CHANGEOVER CASES

FIG. 5-1. Paths of selected surface low pressure centers derived from the heavy snow sample discussed in chapters 3 and 4 for (top) all-snow cases—eight cases in which much of the precipitation within the Northeast urban corridor fell mainly as snow (10–13 Dec 1960, 18–21 Jan 1961, 11–14 Jan 1964, 5-7 Feb 1967, 5-7 Feb 1978, 18–20 Feb 1979, 10–12 Feb 1983, and 25-27 Jan 1987), and (bottom) changeover cases—seven cases in which the precipitation fell as heavy snow changing to rain within portions of the Northeast urban corridor (29–31 Jan 1966, 25-28 Dec 1969, 18–20 Feb 1972, 19–21 Jan 1978, 21–23 Jan 1987, 13–14 Mar 1993, and 3–4 Feb 1995).

ANTICYCLONE POSTIONS
ALL SNOW VS CHANGEOVER CASES

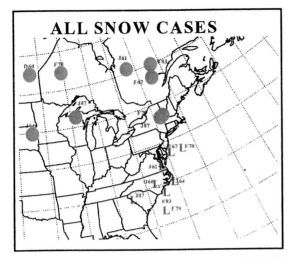

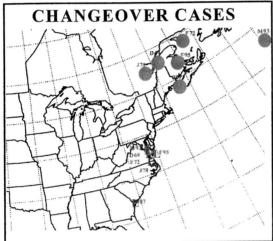

Fig. 5-2. Composite positions of anticyclones for (top) all-snow cases and (bottom) changeover cases, for the same cases shown in Fig. 5-1. Positions (blue circles) were selected at times when the surface cyclone (red L) was moving northeastward along the East Coast.

ada that were associated with a secondary surface ridge or high pressure cell extending eastward into Ontario and Quebec. These two categories are reflected in the locations of anticyclone centers in the all-snow cases shown in Fig. 5-2 (top). The anticyclone centers are located either over south-central Canada, Quebec, or New York. However, for the changeover cases, there is a significant shift of the surface high pressure centers farther east (Fig. 5-2, bottom). The surface high pressure centers are located over southeastern Quebec, near Maine, the Canadian Maritime Provinces, or over the western Atlantic Ocean. The eastward shift of the surface anticyclone allows a longer portion of the cold, low-level airflow that reaches the Northeast urban corridor to pass over and become modified by the mild

waters of the western North Atlantic Ocean, a critical mechanism to change precipitation from snow to rain near the coast.

The upper levels for the all-snow and changeover cases are contrasted in Figs. 5-3 and 5-4. The 500-hPa trough–ridge systems of both all-snow and changeover events involve amplifying troughs characterized by increasing amplitudes, decreasing wavelengths, and the evolution toward a negatively tilted closed low. However, the changeover cases appear to have a larger amplitude–smaller half-wavelength configuration than the all-snow cases. Three examples of each are shown in Fig. 5-3. In the changeover cases, there is evidence of a more amplified upper-trough–ridge system. The significance of a deeper trough and more highly amplified downstream ridge over the western Atlantic Ocean for the changeover cases may indicate that as a trough becomes more highly amplified, the downstream ridge will build due to increased warming east of the trough axis, a process that is marked by warmer air penetrating farther north and west at the 850-hPa level.

Upper-level confluence over southeastern Canada and the northeastern United States is another characteristic of the 30 major urban corridor snowstorms. The associations between upper-level confluence, upper-level jet streaks, and the evolution of the cold surface anticyclones were made in chapter 4 and are explored in more depth in chapter 7. The orientation and location of the surface high pressure systems are important factors for maintaining cold, low-level air in the urban corridor for precipitation to remain as snow and the maintenance of upper-level confluence and its associated jet streaks in the northeast United States are critical to how the surface anticyclone evolves. Upper-level confluence *is* observed in the seven cases of major Northeast snowstorms characterized by snow changing to rain. However, the pattern of confluence is either short lived or not nearly as pronounced as in the all-snow cases. This is demonstrated in Fig. 5-4 where the all-snow cases of the January 1964, February 1967, and February 1983 storms are contrasted with three changeover cases, including December 1969, February 1972, and February 1995.

In general, the maintenance of upper-level confluent flow is lacking in the changeover cases. Furthermore, there is no strong cutoff upper low over eastern Canada in the changeover cases, a feature that is typically associated with a region of confluence that is often associated with major snowstorms. The lack of persistence in the confluence aloft for the changeover cases and the building ridge over the northeast United States during the snow event is consistent with the shift of the surface high pressure to the east as illustrated in Fig. 5-3. All these factors combine to erode the low-level cold air along the coast to the point where snow either mixes with or changes to ice and rain during some portion of the storm.

These differences in synoptic-scale surface and upper-level features among the 30 cases appear to be crucial

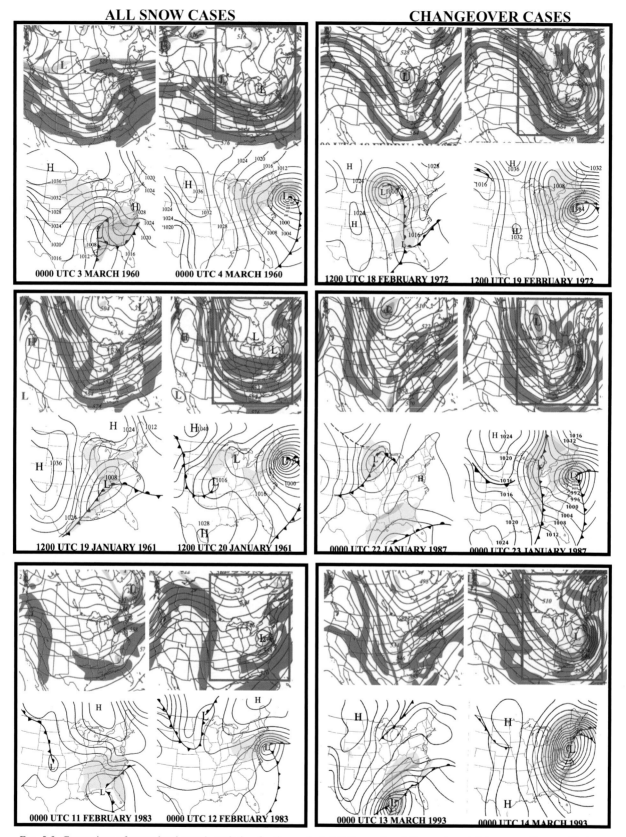

FIG. 5-3. Comparison of upper-level trough evolution for three selected "all-snow" and three selected "changeover" cases. Each case includes successive 12-h upper-level (500-hPa heights and 400-hPa winds; yellow and orange shading represent absolute vorticity (greater than $16 \times 10^{-5} \text{ s}^{-1}$; at $4 \times 10^{-5} \text{ s}^{-1}$ increments), and isotachs of maximum wind speeds observed at 400 hPa (alternating shading represents 5 m s^{-1} increments of wind speed starting with 40 m s^{-1}) and surface frontal analyses (isobars, solid; shading, precipitation). Red box focuses on evolution of the 500-hPa trough.

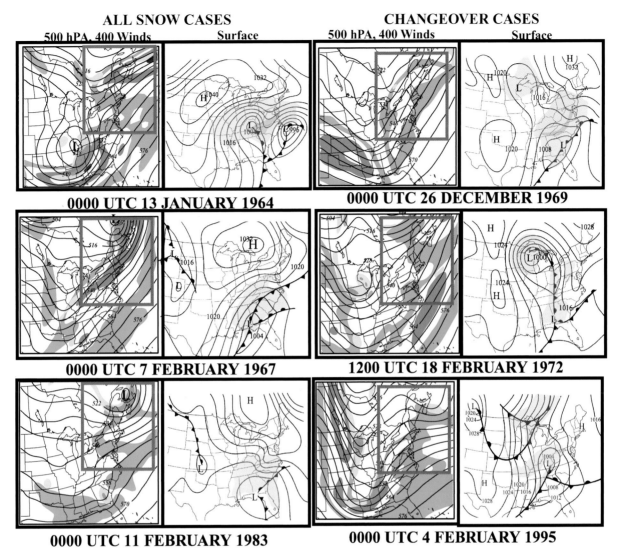

FIG. 5-4. Comparison of 500-hPa confluence and 400-hPa winds between three selected all-snow and three changeover cases. See Fig. 5-3 for details. Red box focuses on 500-hPa confluence.

in differentiating a crippling snowstorm from a potentially less significant event and indicate that there are many elements that must come together to produce a heavy snow event for the urban corridor. When these factors do not come together, the cities of the Northeast are more likely to experience a "changeover" during the snowstorm or a "near miss" rather than a memorable heavy snow event. The next three sections explore some of the conditions that preclude major heavy snow events in the Northeast urban corridor. These conditions are categorized as interior snowstorms, moderate snowstorms, and ice and sleet storms.

2. Interior snowstorms

Fifteen "interior snowstorms" are contrasted with the 30 heavy snow events. These cases are selected because all had the potential to produce heavy snow in the urban

corridor (forecasts appearing in local newspaper reports from Washington to Boston indicated that heavy snow was a potential threat prior to many of these storms), but were marked instead by the heaviest snow band developing farther north and west across interior portions of the middle Atlantic states and New England. In each case, there was a significant synoptic-scale band of snowfall exceeding 10 in. (25 cm), and, often, there were reports of accumulations of 20 in. (50 cm) or greater (Fig. 5-5). The heaviest snows are generally located several hundred kilometers farther inland or farther north than the 30 cases shown in Fig. 3-2, with the major metropolitan areas experiencing more rain than snow. (See Table 5-1 for snowfall amounts.)

Some of the these events are historic storms in their own right, including 1) the 5–7 March 1962 "Ash Wednesday" storm, responsible for some of the most extensive coastal erosion observed during the 20th cen-

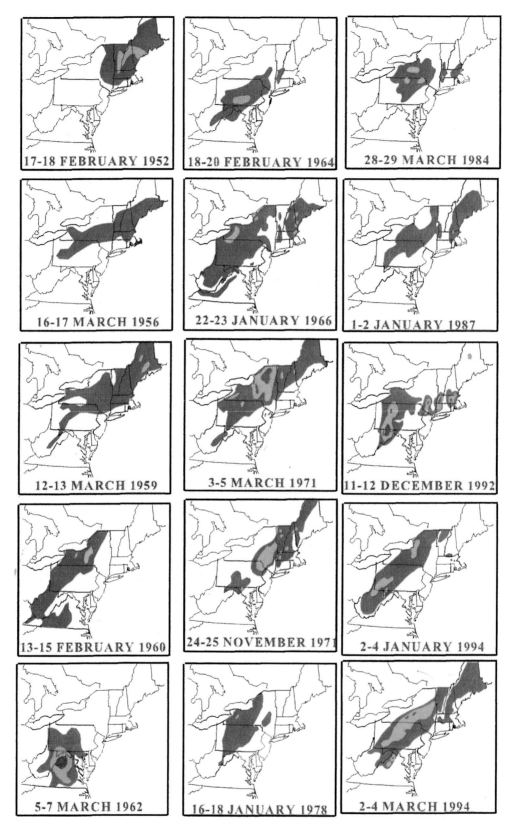

FIG. 5-5. Storm snowfall in excess of 10 in. (25 cm) (blue, >25 cm; green, >50 cm; red, >75 cm) for selected interior snowstorms between 1950 and 2000.

TABLE 5-1. Snowfall amounts (in.) at Boston, MA (Bos); New York City/Central Park, NY (NYC); Philadelphia, PA (Phi); Baltimore, MD (Bal); and Washington, D.C. (DC) for selected "interior snowstorms."

Date	Bos	NYC	Phil	Bal	DC
17–18 Feb 1952	3	1	0	0	0
16–17 Mar 1956	7	5	1	1	0
12–13 Mar 1959	6	5	1	1	0
13–15 Feb 1960	0	1	1	8	6
5–7 Mar 1962	0	1	7	13	4
18–20 Feb 1964	10	7	2	6	0
22–23 Jan 1966	9	2	4	2	0
3–5 Mar 1971	2	2	4	2	0
24–25 Nov 1971	0	0	0	1	0
16–18 Jan 1978	3	2	4	2	1
28–29 Mar 1984	9	3	1	0	0
1–2 Jan 1987	4	1	1	0	0
11–12 Dec 1992	9	1	0	2	1
2–4 Jan 1994	9	1	1	0	1
2–4 Mar 1994	8	5	4	3	3

tury and record-setting snow in Virginia and West Virginia; 2) a powerful nor'easter on 3–4 March 1971 that established low pressure records in New England (Boyle and Bosart 1986); 3) a storm on 28–29 March 1984 that was also responsible for a violent tornado outbreak in North Carolina (Gyakum and Barker 1988; Rozumalski et al. 1998); and 4) a coastal storm on 11–12 December 1992 responsible for severe coastal flooding in New Jersey and New York and record-setting snows in portions of the Appalachian, Catskills, and Berkshire Mountains. These and several other "interior" snowstorms are described in more depth in volume II, chapter 11.

a. Surface features

The individual paths of the surface low pressure centers for 15 interior snowstorms are shown in Fig. 5-6 and summarized in Fig. 5-7. Eleven of 15 cases exhibit a cyclone evolution that is similar to that characterized as the "Gulf of Mexico–Atlantic coast" evolution (see chapter 3). Most of these storms are the typical nor'easter in which the surface cyclones developed along either the Gulf coast, Gulf of Mexico, or southeast United States coasts, and then moved northeastward along the east coast. Only 4 of 15 cases are characterized by a secondary cyclone, or "Atlantic coastal redevelopment" evolution (see chapter 3). These cases are characterized by a primary cyclone that moved northeastward toward the Appalachians, followed by a secondary cyclone developing along and propagating northeastward along the East Coast.

Twelve of the 15 cases (80%) followed a path that took the low center over the Delmarva Peninsula, similar to the changeover cases shown at the bottom of Fig. 5-2, while only 6 of the 30 heavy snow cases (20%; Fig. 3-3) had similar tracks. The large percentage of cases in which the surface low moves near the Carolinas

coastline, or slightly inland, and then crosses the Delmarva Peninsula (Fig. 5-7, top) is highly suggestive that this path is common for severe coastal storms marked by snow changing to rain in portions of the Northeast urban corridor, with heavy snows falling over interior locations.

Surface cyclonic deepening rates for the 15 interior storms are also compared to the 30-case heavy snow sample. Twenty-six of 30 major snowstorm cases (87%) exhibited maximum deepening rates of at least −12 hPa 12 h^{-1} (see chapter 3). Meanwhile, only 8 of the 15 interior cases (53%) exhibited similar rates, indicating that rapid sea level cyclonic development was common, but not as common, for the interior cases as for the major urban snowstorms. In addition, only 3 of the 15 cases (20%) exhibited maximum pressure falls exceeding 10 hPa 6 h^{-1}, the criterion used to select explosive cyclogenesis during the Experiment on Rapidly Intensifying Cyclones over the Atlantic (ERICA; Hadlock and Kreitzberg 1988), while 18 of 30 major snow cases (60%) exhibited similar rates. It appears that the interior snowstorms are characterized by surface cyclones that are less likely to deepen explosively, although some of the storms in the sample obviously did (such as the storms of 15–17 March 1956 and 3–4 March 1971). Perhaps the tendency of these cyclones to deepen along the coastline or slightly inland, rather than offshore, contributes to reduced deepening rates for this sample. With a portion of the cyclone over land, the contributions of oceanic latent and sensible heat flux near the center of the cyclone are apparently diminished.

Characteristics of anticyclones associated with the 15 interior snowstorms were also examined. The positions of the surface anticyclones with respect to the surface low pressure centers along the middle Atlantic coast (Fig. 5-7, bottom) point to a number of distinct characteristics for the "interior" snowstorm sample. The first is the clustering of a portion of the surface high pressure centers over the Maritime Provinces of Canada. This sample is similar to the snow changeover sample of major snowstorms (Fig. 5-2, bottom). The *eastward* locations of these anticyclones result in a more easterly or southeasterly (rather than north or northeasterly) flow over the urban corridor, forcing warmer air near the coastline.

A second characteristic of interior snowstorm anticyclones is a clustering of anticyclones over north-central Canada, fairly far removed from the northeastern United States and significantly farther north than the locations of the surface anticyclones for the all-snow cases (Fig. 5-2). [This configuration is observed in the cases of March 1962 (a 1064-hPa anticyclone), February 1964, January 1966, March 1971, and March 1984.] This characteristic is also observed, but only in a few of the 30 heavy snow cases (including March 1958, January 1966, and late February 1969, which were also cases in which snow changed to rain in some portion of the Northeast urban corridor). The analyses suggest

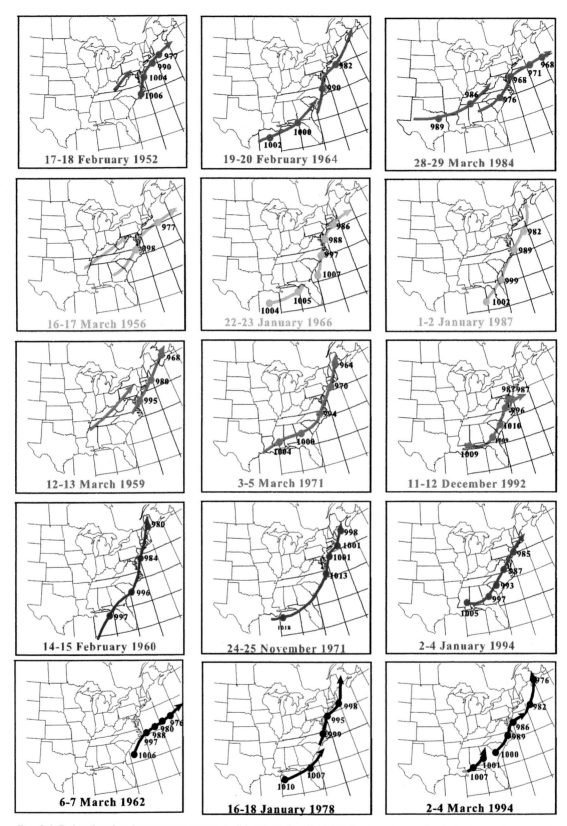

FIG. 5-6. Paths of surface low pressure centers (solid, colors) for all 15 interior snowstorms. Symbols along path represent 12-hourly positions and numerical values are the sea level central pressures (hPa) at 12-hourly intervals.

CYCLONE TRACKS

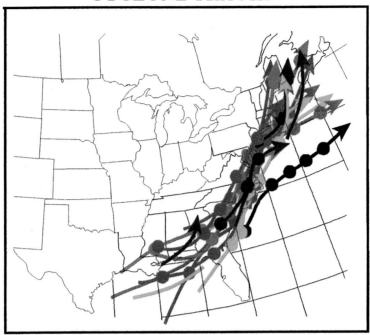

ANTICYCLONE POSITIONS

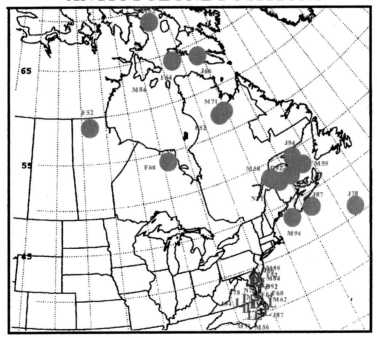

FIG. 5-7. (top) Composite tracks of surface low pressure systems for the interior snowstorms. (bottom) Composite positions of anticyclones associated with interior snowstorms; positions selected at times when the surface cyclone was moving northeastward along the East Coast.

that the surface high pressure systems are so far removed from the Northeast urban corridor that they cannot sustain the transport of cold air into the area needed for snowfall.

b. Upper-level features

The upper-level troughs that generate the interior snow events have distinct characteristics that can be

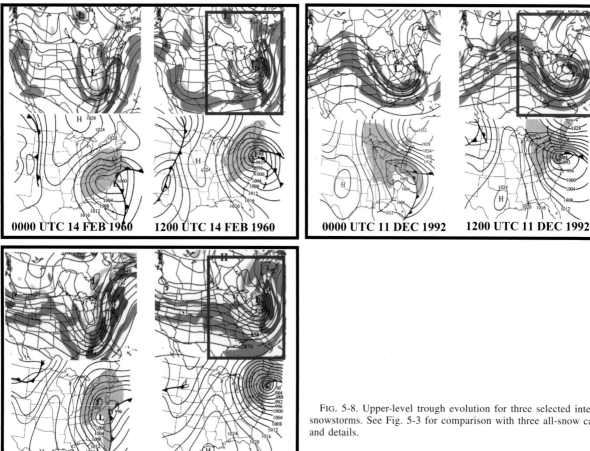

FIG. 5-8. Upper-level trough evolution for three selected interior snowstorms. See Fig. 5-3 for comparison with three all-snow cases and details.

contrasted with the urban corridor snow events. The comparison of the all-snow and interior snowstorms shown in Fig. 5-8 illustrates that for the interior snowstorms 1) the wavelength of the trough–ridge pattern tends to be longer, 2) a ridge axis along the East Coast extends well into southeastern Canada as the surface low develops along the coast, 3) the trough is deeper, and 4) the trough supporting the surface low development immediately along the coast appears to become negatively tilted farther south and west than in the all-snow cases.

A comparison of the tracks of the 500-hPa vorticity maxima for 30 major snowstorms (Fig. 4-8) with the 15 interior snowstorms (Fig. 5-9) indicates the tendency for the vorticity maxima associated with the upper troughs to have a trajectory that is farther south and west than the major snowstorms (cf. the confluence of trajectories in red in Fig 4-8 with those in Fig. 5-9). The vorticity maxima reach the southern United States and then curve northeastward along the east coast. This is consistent with the trough–ridge characteristics noted above and the observations that the surface lows as-

sociated with interior storms track closer to the coastline than those with the heavy snow events.

Another characteristic of the interior snowstorms is either the dissipation or rapid northeastward progression of upper-level confluence over the Northeast during the event. Four examples of interior snowstorms are shown in Fig. 5-10 to illustrate the lack of a strong confluent upper-flow pattern during the evolution of many of these storms. (These examples are a typical subsample of the interior snowstorms that are all summarized in volume II, chapter 11.) Nearly all cases exhibit either an eastward or northeastward progression of the confluence, a weakening of the confluence, or a lack of confluence entirely. Therefore, a significant difference between the interior snowstorms and the coastal snowstorms is that the pattern of upper-level confluence in southeastern Canada or the northeastern United States is weaker or absent in the majority of the interior cases, while prevalent in the heavy snow cases. Without the confluent upper-level pattern, surface anticyclones progress eastward across eastern Canada to a location farther east over the nearby Atlantic Ocean. The eastward move-

500 HPA VORTICITY PATHS
INTERIOR SNOWSTORMS

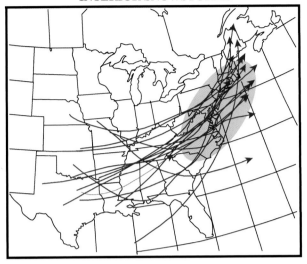

FIG. 5-9. Composite of 500-hPa vorticity maxima for all interior cases. Shaded area represents the location and axis of where the maxima tend to converge.

ment and dissipation of the upper confluence is consistent with the eastward progression of surface anticyclones out over the North Atlantic, resulting in low-level warming and the greater likelihood for a snow changing to rain event within the urban corridor.

Further support for the low-level warming along the coastal zone associated with interior snowstorms is provided by comparing the tracks of the 850-hPa low centers for the interior snowstorms (Fig. 5-11) with the heavy snow cases. The majority of 850-hPa lows track northeastward from Virginia to eastern New England, keeping the urban corridor close to or within the warmer lower-tropospheric air as the heaviest snow falls 100–300 km to the left of the 850-hPa path. Therefore, for the interior snow cases, as the 500-hPa troughs become negatively tilted and the ridge builds into the northeastern United States, the 850-hPa low centers track slightly west and north of their heavy snow counterparts and the surface low moves closer to the coastline, where snow is more likely to change to rain.

c. Summary of differences between major snowstorms and interior snowstorms

A schematic (Fig. 5-12) summarizes the differences between a major Northeast snowstorm and an interior snowstorm, relating the evolution of the surface low, surface high, upper trough, and downstream ridge in association with upper confluent flow over the northeastern United States. With the larger amplification and negative tilt of the upper troughs to the south and west of their heavy snow counterparts, interior snowstorms exhibit a more pronounced downstream ridge along the East Coast and western Atlantic. The building ridge over

the western Atlantic indicates that warmer air is forced toward the coastline, especially if the upper trough attains a negative tilt prior to reaching the Atlantic coast. As the ridge builds, the upper confluence is forced northward and eastward, or diminishes, allowing the cold surface anticyclone to be displaced farther north and/or drift eastward and weaken. The net result is that warmer air at low levels overspreads the coastal zone of the Northeast. The erosion of the confluence over the northeast United States and southeastern Canada, in combination with the more inland track of the east coast cyclone, forces easterly low-level flow throughout the northeast urban corridor, with rain along the immediate coastline and snow falling either farther inland or at higher elevations as the low-level cold air erodes. A comparison of the interior and major snowstorms also shows the following:

1) In major urban snowstorms, the surface cyclone passes immediately offshore east of the middle Atlantic coast and south of southern New England (Figs. 3-3 and 3-4). With the interior snowstorms, a changeover to rain or ice pellets occurs as the surface low often takes a path near the coast of the Carolinas and the Delmarva Peninsula, or slightly inland.
2) In major urban snowstorms the surface highs tend to cluster immediately to the north and northwest of New York and New England (i.e., Figs. 3-13 and 3-14). In the case of interior snowstorms, the anticyclones are either located too far north or east, allowing maritime air to change snow to rain. In some instances, the difference in the location of an anticyclone over Nova Scotia versus Quebec can be a crucial difference between snow or rain in the Northeast urban corridor.
3) Prior to major urban snowstorms, upper confluence is often pronounced and is found downstream of a significant trough, often a closed low, over eastern Canada. With interior snowstorms, the upper confluent flow either does not exist or progresses rapidly eastward or northeastward off the coast of eastern Canada as the ridge builds into the northeast United States dominating the circulation regime.
4) While major snowstorms in the Northeast urban corridor and interior snowstorms can both be associated with negatively tilted upper troughs, the interior snowstorms are marked by troughs that become negatively tilted farther south and west, as the downstream ridge builds into New England and southeastern Canada.

3. Moderate snowstorms

The second category of near-miss events consists of snowstorms that also produce widespread snowfall across the Northeast urban corridor. However, snow amounts of 10–20 in. (25-50 cm) or greater are not nearly as widespread as the 30-case heavy snow sample

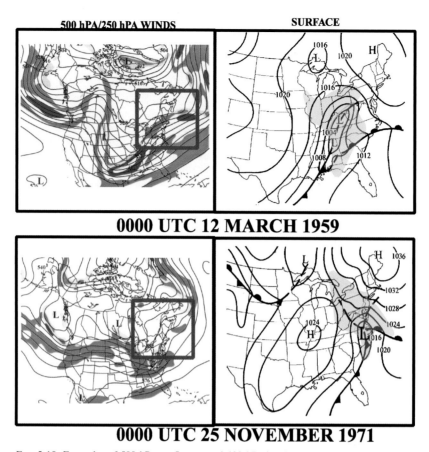

FIG. 5-10. Examples of 500-hPa confluence and 400-hPa jet signatures for selected interior snowstorms. See Fig. 5-3 for details. Red box focuses on regions of 500-hPa confluence.

described in chapters 3 and 4. Fifteen examples of "moderate" snowstorms are selected (Fig. 5-13) and discussed to help contrast the major and moderate snow events (these cases are also shown in more detail in volume II, chapter 11). Snowfall amounts are typically in the 4–10-in. (10–25 cm) range within the Northeast urban corridor, with some local or isolated amounts approaching or exceeding 10–20 in. (25-50 cm; see Table 5-2).

These more moderate snowstorms share several characteristics with the major events. Some of the differences between these events are contrasted to help distinguish conditions that may favor more moderate snowfall over a heavy snowfall. Again, there are no simple rules that can be applied to every case but there are a range of signatures, some distinct and others that are subtle, that separate the major snowstorms from these moderate events.

a. Surface features

All 15 moderate snow events selected for review involve widespread snowfall and are associated with cyclonic development. The paths and central pressures of the surface low pressure centers are shown in Fig. 5-

14. The paths are similar to those of the heavy snow sample (Fig. 3-3) with 10 cases involving secondary cyclone development while only 5 exhibit characteristics more similar to the Gulf of Mexico–Atlantic coast category (see chapter 3).

The maximum deepening rate for the moderate events is generally less than that observed for the heavy snow cases. Deepening rates for only 8 of the 15 moderate cases (53%) exceeded -12 hPa 12 h^{-1} (compared to 26 of the 30 heavy snow cases; 87%). Furthermore, none of the moderate events exhibited deepening rates exceeding -16 hPa 12 h^{-1} (compared to 11 major snow events). In the heavy snow sample, 60% exhibited pressure falls exceeding 10 hPa 6 h^{-1} while none of the moderate cases exhibited similar rates. Therefore, all of the moderate snowstorms are characterized by cyclonic development, but the development does not exhibit a rapid or an explosive development phase, which mark the heavy snow cases.

The presence of a cold Canadian surface anticyclone within the confluent entrance region of an upper- level jet streak is, however, a fairly common element in many of the moderate snowstorms. This characteristic is similar to the 30-case heavy snow sample. Of the 15 cases, 11 were associated with cold surface anticyclones with

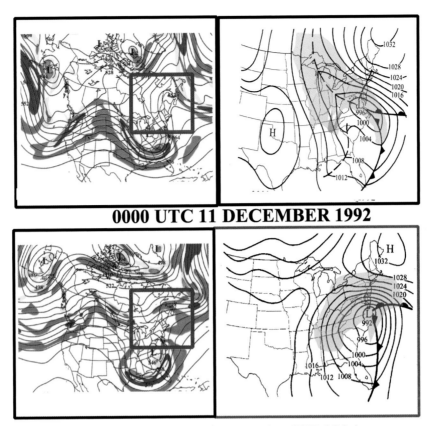

0000 UTC 11 DECEMBER 1992

0000 UTC 3 MARCH 1994

Fɪɢ. 5-10. (*Continued*)

central pressures exceeding 1030 hPa and only three cases (December 1957, January 1971, and 15-16 February 1996), were associated with anticyclones with central pressures less than 1024 hPa.

The positions of the surface anticyclones with respect to the surface low pressure centers along the middle Atlantic coast are shown in Fig. 5-15, which illustrates a clustering of surface high pressure centers that is similar to those associated with the major snowstorms (section 3c). The primary high pressure center is typically located over Ontario and Quebec. High pressure systems may also exist over south-central Canada or the north-central United States, with a secondary high pressure cell or ridge extending over southeastern Canada or the northeastern United States. Therefore, the majority of the moderate snowstorm cases were characterized by relatively strong, cold anticyclones poised to the north of the East Coast that accompany a more modest cyclogenesis.

b. Upper-level features

Modest surface cyclogenesis is more common to the moderate snowstorms than the significant cyclogenesis that often accompanies major snowstorms. Therefore, it is not surprising that the upper-level troughs associated with the low pressure systems are not nearly as pronounced in terms of amplitude, half-wavelength, and negative tilt. In a number of moderate snow events, the surface low and its associated upper trough are weak and nonamplifying. For example, three cases, shown in Fig. 5-16, illustrate the presence of fairly weak short-wave disturbances embedded within gently cyclonic upper-level flow. Note that, at the surface, the surface low pressure systems are either relatively weak or, as in the case of December 1990, reflected by surface troughs rather than a distinct low pressure center.

A significant signature of the moderate cases is the presence of a strongly confluent flow field and associated upper-tropospheric jet streak over the northeast United States and southeast Canada (Fig. 5-16). The moderate snow event tends to be located within the ascending right entrance region of the upper-level jet. In many cases, the snow appears to extend along and to the right of the jet axis as it expands northeastward from the mid-Atlantic region toward New England. The presence of the confluent jet flow over the Northeast is consistent with cold surface anticyclones usually observed for these cases locked into the area just north and east of the track of the surface low.

Another scenario for moderate snowstorms is the presence of two or more separate upper-level troughs

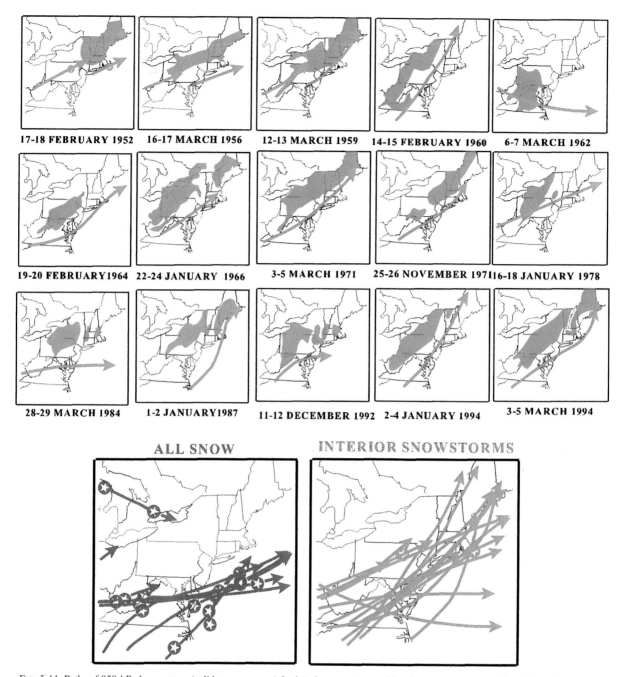

FIG. 5-11. Paths of 850-hPa low centers (solid arrows, green) for interior snowstorms (blue shading represents region of snowfall exceeding 25 cm). Bottom panels offer a comparison of the 850-hPa low center paths for the all-snow cases (blue arrows) described in Fig. 5-1, and all 15 interior snowstorms (green arrows).

that fail to merge [see section 4c(1)]. In some cases, these features may even weaken or damp out. Three examples are presented in Fig. 5-17: 16 January 1965, 14 January 1982, and 20 December 1995. In the January 1965 case, the presence of an amplifying upper trough and intense Canadian anticyclone appear to be precursors to a major snowstorm, but a major snowstorm does not develop since separate upper-trough features failed to merge, resulting in separate cyclonic developments

that did not intensify appreciably. The January 1982 case is also an example of two separate troughs that each produced significant snowfall but failed to merge, precluding a more significant snowfall event. The December 1995 case is an example of multiple trough interactions in which one well-defined trough nearing the East Coast encounters a separate, amplifying trough over southeastern Canada. This interaction results in a deamplification, rather than an intensification, of the

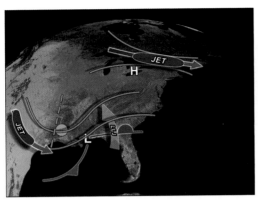
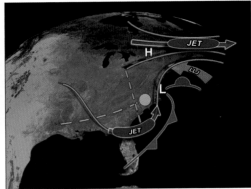

URBAN CORRIDOR SNOWSTORMS

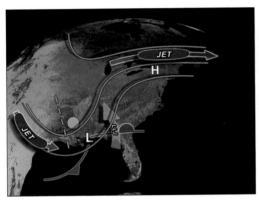
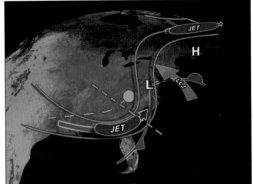

INTERIOR SNOWSTORMS

FIG. 5-12. Schematic comparison of the evolution of the upper-level trough, confluence, and associated jet streaks (upper-level and low-level jet streaks) for (top) major snowstorms in the Northeast urban corridor and (bottom) interior snowstorms. Color gradation on map represents evolution of cold and warm air masses. Violet solid lines represent 500-hPa height contours, a dashed violet line represents the upper trough axis, and the circle and attached dashed lines represent the track of the 500-hPa vorticity maxima.

upper trough moving toward the East Coast. The snowfall produced by this system was significant but did not produce the impressive snowfall rates observed in more significant snowstorms.

The speed of the upper-level disturbance–sea level cyclone can also play a role in precluding more significant snowfall accumulations. The moderate snowstorm of March 1984 was a small and rapidly moving cyclone and heavy snow fell for only a few hours. Especially fast-moving systems will typically produce less snowfall for obvious reasons. In other cases, the evolution of the upper trough–ridge appears to resemble that observed with the heavy snow sample (Fig. 5-18) with a significant amplification of the upper trough and development of a negatively tilted closed 500-hPa low. However, in both the cases of December 1957 and January 1971, rapid cyclogenesis did not occur in conjunction with these upper amplifications. This factor and the association with relatively weak anticyclones may have led to smaller moisture transports for any period of time and, therefore, lower snowfall amounts.

There are also differences in the 850-hPa height and wind fields (described on a case-by-case basis in volume II, section 11b) between the major snowstorms and the moderate snowstorms. The 850-hPa lows associated with moderate snowstorms tend to deepen much less than their heavy snow counterparts. Commensurately, the evolution of the low-level winds is typically weaker than is that of their heavy snow counterparts and does not develop the strong easterly jet directed from the Atlantic Ocean toward the region of heavy precipitation. The moderate cases exhibit 850-hPa winds that are weaker and remain more from a southerly direction. The weaker southerly lower-tropospheric winds are also indicative of smaller moisture transports that result in lower snowfall amounts.

c. Summary of differences between major snowstorms and moderate snowstorms

The moderate snowstorms described in this section tend to produce widespread but lesser snowfall amounts

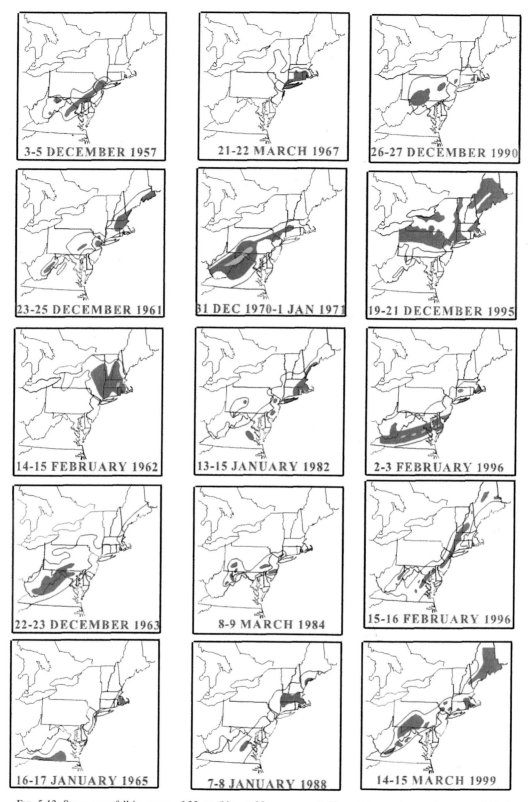

FIG. 5-13. Storm snowfall in excess of 25 cm (blue, >25 cm; green, >50 cm; blue contour encloses snowfall of greater than 15 cm) for moderate snowstorms between 1950 and 2000.

TABLE 5-2. Snowfall amounts (in.) at Boston, MA (Bos); New York City/Central Park, NY (NYC); Philadephia, PA (Phi); Baltimore, MD (Bal); and Washington, D.C. (DC) for selected "moderate snowstorms."

Date	Bos	NYC	Phi	Bal	DC
3–5 Dec 1957	0	8	7	8	11
23–25 Dec 1961	10	3	3	1	1
14–15 Feb 1962	8	3	4	3	1
22–23 Dec 1963	5	7	5	8	6
16–17 Jan 1965	8	5	4	3	3
21–22 Mar 1967	2	10	4	2	0
31 Dec–1 Jan 1971	5	6	5	6	9
20–22 Dec 1975	18	2	1	0	0
13–15 Jan 1982	13	9	9	8	10
8–9 Mar 1984	2	6	7	5	2
7–8 Jan 1988	9	5	4	8	8
26–27 Dec 1990	1	7	6	5	3
19–21 Dec 1995	12	10	2	1	0
3 Feb 1996	5	8	7	8	10
15–16 Feb 1996	7	11	8	11	8

anticyclones are common to these moderate events, as they are to the major snow events. Therefore, the presence of cold anticyclones and their associated upper-level conditions are common to many snowstorms in the Northeast urban corridor whether they be of moderate or heavy intensity.

3) Factors that diminish the chances for rapid cyclogenesis appear to significantly reduce the likelihood for heavy snow, which should come as no surprise. However, the complex wave interactions that influence trough development are often poorly understood and sometimes poorly forecast. The feedbacks that allow the "self-development" of cyclones (discussed in chapter 7) are more often retarded than are those interactions that allow a cyclone to develop rapidly.

4) Speed is also be an important consideration, with some moderate snowstorms noted as fast-moving cyclones that cannot sustain heavy snowfall rates for long periods of time at any given location.

4. Ice and sleet storms

Many significant snowstorms include regions that experience precipitation in the form of freezing rain ("ice storms") or ice pellets (sleet). The presence of a warm layer above the earth's surface with temperatures above 32°F (0°C) for a significant depth to allow snow to partially or fully melt is a well-understood phenomena (Forbes et al. 1987; Stewart and King 1987; Huffman and Norman 1988; Ramamurthy et al. 1991; Rauber et al. 1994). Even very small accumulations of freezing rain are often sufficient to cause catastrophic effects. The presence of freezing rain or sleet can have a major impact to roadways while heavier accumulations of freezing rain present a significant threat to trees and electrical lines, resulting in potentially massive power outages.

Ice pellets may present less of a threat than freezing rain or snow since the pellets can add some traction to roadways in contrast to the slick road surfaces associated with freezing rain. In addition, large amounts of sleet do not accumulate to depths observed with snowfall composed of similar liquid amounts. Nevertheless, significant accumulations of ice pellets require removal for transportation and are still a significant winter weather threat.

In many instances, the region of freezing rain and sleet is found between those areas of significant snow and rain, and there is often an overlap over a considerable area that receives some combination of these elements. In some winter storms, freezing rain or ice pellets can be the predominating winter weather problem. Of the 30 major snowstorms examined in the previous two chapters, freezing rain and sleet were observed to some degree with all storms.

For example, the March 1993 Superstorm produced

than the major snowstorms discussed in the previous two chapters. In these more moderate events, snowfall amounts of 4–10 in. (10–25 cm) were common in the Northeast urban corridor, with some regions receiving greater than 10 in. (25 cm).

A common scenario that marks these more moderate events is that the upper-level trough associated with the surface cyclogenesis is weak or less pronounced in terms of amplitude and wavelength than in the major snowstorms (see schematic in Fig. 5-19). However, the confluent upper-level patterns in the northeast United States and southeastern Canada associated with the cold surface anticyclone are present. In these cases, the cold surface anticyclone and its associated jet entrance region within a region of upper-level confluence are dominant features, but the short-wave trough associated with cyclogenesis is much weaker than those associated with the major snowstorms. In nearly all the examples shown in Figs. 5-16–5-18, upper-level confluence and an associated jet streak are present and appear to act to maintain cold temperatures for the snow event. However, the lack of an amplifying trough in the southern stream appears to be an important factor in limiting the potential of these storms to grow into major snow-producing systems.

1) In general, the major differences between major and moderate snowstorms involve differences in the cyclogenesis supporting the snow event. The surface lows tend to be weaker, develop less intensity and do not exhibit the rapid deepening observed for many major snowstorms. Commensurately, the upper troughs also tend to be weaker (smaller amplitude, do not evolve into deepening closed circulations) and, generally, do not amplify very much. Commensurately, the 850-hPa lows are also weaker and accompanied by a weaker southerly (rather than easterly) low-level jet streak.

2) The presence of upper confluence and strong surface

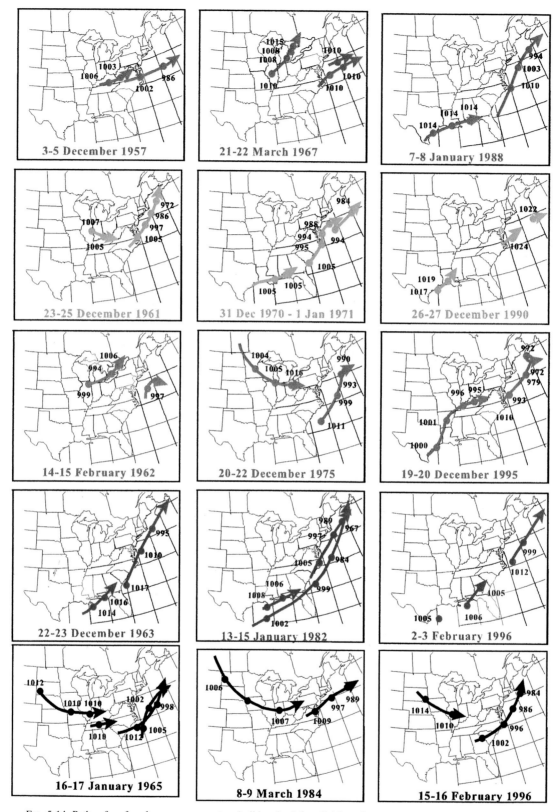

FIG. 5-14. Paths of surface low pressure centers (solid, colors) for all 15 moderate snowstorms. Symbols along path represent 12-hourly positions and numerical values are the sea level central pressures (hPa) at 12-hourly intervals.

MODERATE SNOWSTORMS
CYCLONE TRACKS

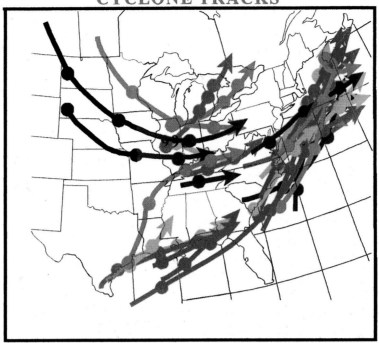

ANTICYCLONE POSITIONS

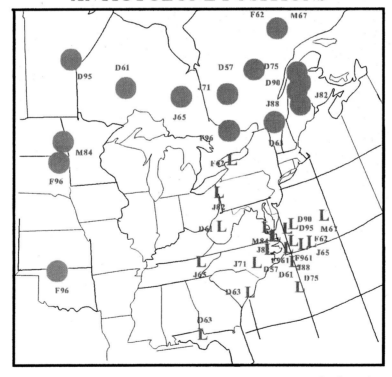

FIG. 5-15. (top) Composite tracks of surface low pressure systems for the moderate snowstorms. (bottom) Composite positions of anticyclones associated with moderate snowstorms; positions (blue circles) selected at times when the surface cyclone (red L) was moving northeastward along the East Coast.

500 hPa/400 hPA WINDS   **SURFACE**

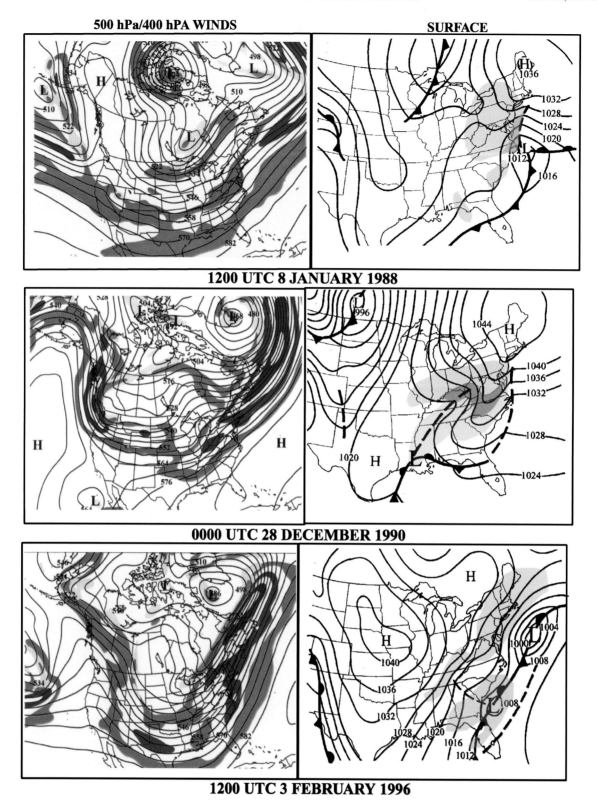

1200 UTC 8 JANUARY 1988

0000 UTC 28 DECEMBER 1990

1200 UTC 3 FEBRUARY 1996

FIG. 5-16. Examples of moderate snowstorm evolution; three cases exhibiting strong surface anticyclones, 500-hPa confluence, and relatively weak shortwave troughs. Panels in the left column show 500-hPa heights and 400-hPa winds; panels in the right column show surface frontal analyses. See Fig. 5-3 for specific details.

500 hPa/400 hPa WINDS

SURFACE

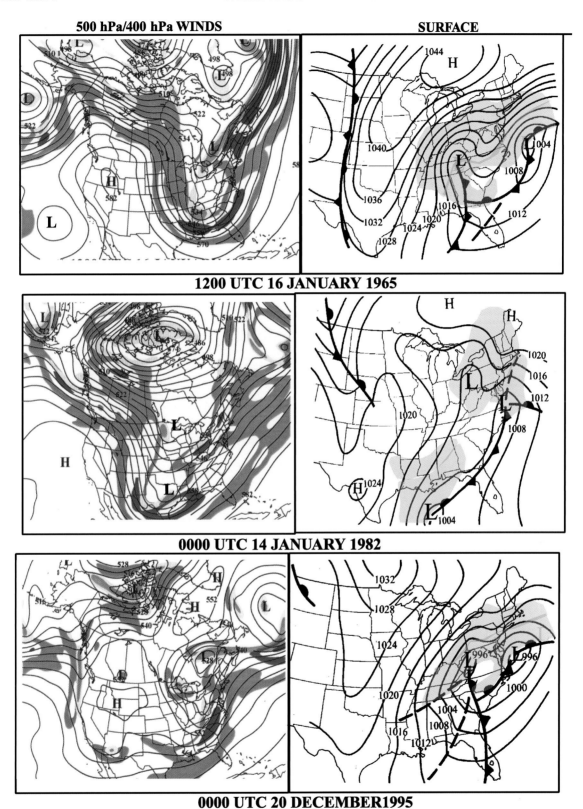

1200 UTC 16 JANUARY 1965

0000 UTC 14 JANUARY 1982

0000 UTC 20 DECEMBER 1995

FIG. 5-17. Examples of moderate snowstorm evolution; three cases exhibiting upper-level trough "fractures" or "nonconsolidation" (see text). Left column shows 500-hPa heights and 400-hPa winds; right column shows surface frontal analyses. See Fig. 5-3 for specific details.

500 hPA/400 hPa WINDS   **SURFACE**

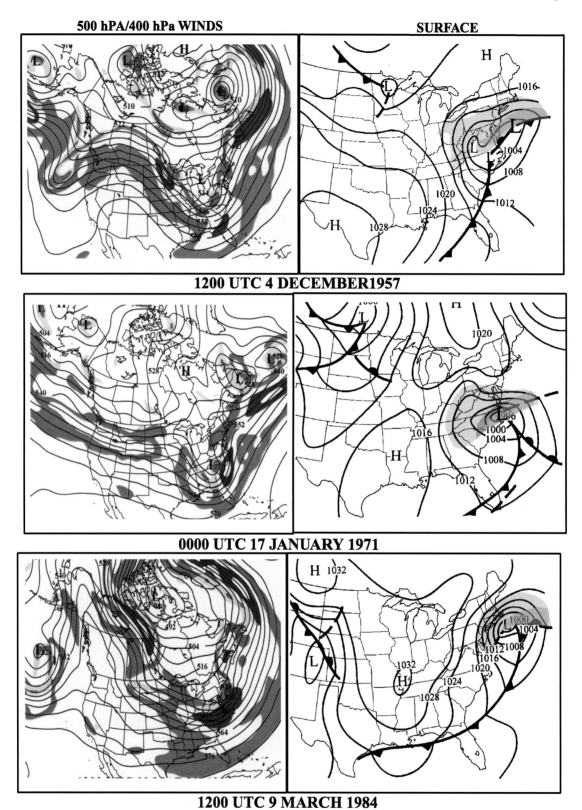

1200 UTC 4 DECEMBER 1957

0000 UTC 17 JANUARY 1971

1200 UTC 9 MARCH 1984

FIG. 5-18. Examples of moderate snowstorm evolution; three cases exhibiting surface and upper-level structure similar to major snowstorms. Left column shows 500-hPa heights and 400-hPa winds; right column shows surface frontal analyses. See Fig. 5-3 for specific details.

URBAN CORRIDOR VS MODERATE SNOWSTORMS

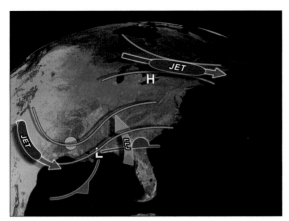 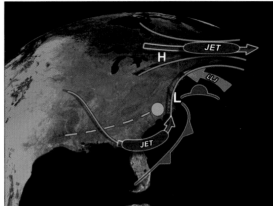

URBAN CORRIDOR SNOWSTORMS

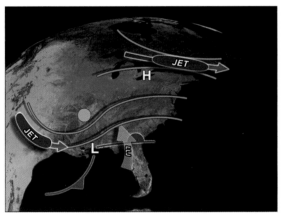 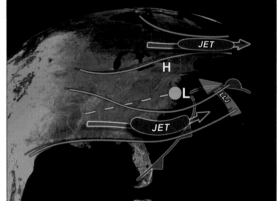

MODERATE SNOWSTORMS

FIG. 5-19. Schematic comparison of the evolution of the upper-level trough, confluence, and associated jet streaks (upper-level and low-level jet streaks) for (top) major snowstorms in the Northeast urban corridor and (bottom) moderate snowstorms. Color gradation on map represents evolution of cold and warm air masses. Purple lines represent 500-hPa height contours, and the circle and attached dashed lines represent the track of the 500-hPa vorticity maxima.

mostly snow over the Appalachian Mountains, but precipitation was generally a combination of heavy snow, heavy ice pellets, and some freezing rain and rain in the highly populated Northeast urban corridor. Most snowfall amounts in the urban corridor averaged about a foot (30 cm) of snow, but the total precipitation was made up of a significant amount of sleet. Had precipitation remained as all snow, 24 in. (60 cm) accumulations would have been likely since many locations reported 2–3 in. (5-7.5 cm) of liquid water equivalent. While the ice pellets cut down on final snowfall accumulations (the addition of ice pellets can actually cause snow depths to diminish because of compaction), it still added to the weight of the snow/sleet mass, requiring a massive removal effort.

The post-Christmas storm of December 1969 (see volume II, chapter 10.14) is another winter storm that produced severe icing throughout the Connecticut Valley in New England. The storm of 10–11 February 1994 (see volume II, chapter 10.25) produced significant snowfall accumulations from New York City through New England but was also one of the worst ice storms on record across portions of the eastern United States. In Washington and Baltimore, most of the precipitation from this storm fell as ice pellets. Freezing rain accumulations of 0.5-1.5 in. (1.2–3.6 cm) devastated portions of the southeast United States, Virginia, and southeastern Maryland, causing more than a billion dollars in damage.

According to Ludlum (1982) the most damaging ice

storm in American history occurred over a large area from Texas to West Virginia from 28 January through 1 February 1951 (Harlin 1952). Ice formed on exposed objects to thicknesses of 0.5–4 in. (1.2–10 cm), impeding traffic for up to 10 days afterward. Some of these icing events can be as damaging, if not more damaging, than a major snowstorm. A severe ice storm in Massachusetts through Maine in late November 1921 caused enormous destruction and economic disruption (Burnham 1922) with ice coating electrical wires up to 3 in. (7.5 cm) in diameter. As the 20th century drew to a close, one of the worst ice storms of the century affected portions of northern New York and New England and paralyzed much of southeastern Canada on 5–9 January 1998. The personal and economic impact of this storm is well documented (DeGaetano 2000), with damage estimated at $4 billion (Gyakum and Roebber 2001).

In the following sections, seven significant freezing rain and ice pellet events during the last half of the 20th century are examined to contrast the surface and upper-level features with major snowstorms. Examples of vertical temperature profiles associated with freezing rain and ice pellets are presented and possible scenarios are also presented in which above-surface melting would be promoted. The heavy snowfall and icing patterns of these seven significant icing events are shown in Fig. 5-20, illustrating the locations of icing within the Northeast urban corridor. Several of these events are also associated with significant snowfall. However, icing rather than snow dominated a few of the cases.

a. Surface features

The surface features associated with some of the ice and sleet storms discussed in this section exhibit many similarities to their heavy snow counterparts. One case in particular, the ice storm of 7 January 1994 (see volume II, chapter 11.3), was also a significant snowstorm over much of New England and its surface features, including low pressure moving eastward from the Ohio Valley and redeveloping near the mid-Atlantic coast, and high pressure to the northwest of New England, are very similar to many of the heavy snow cases. However, there are a number of significant differences in the other cases that may be indicative of conditions that are more conducive to freezing rain rather than snow. Although only seven cases are shown, they do point to some of the factors that support freezing rain or sleet rather than snow.

A summary of the tracks of the surface low pressure systems for this small sample of cases (Fig. 5-21, top) shows a wide variety of paths. A number of storms parallel the East Coast, somewhat similar to those cases of interior snowstorms rather than the major snowstorms as the surface low pressure centers remain relatively close to the coastline. There are also several examples of surface low pressure systems that track northeastward across the Ohio Valley or Great Lakes, similar to the primary low pressure systems associated with several of the major Northeast snowstorms (see chapter 3). But the paths of the ice storms track much farther to the north than those shown in Fig. 3-5. These cases, all occurring in January 1994 and one in January 1999, are associated with secondary low pressure systems along the East Coast with the exception of the storm of 27–28 January 1994 (in which no defined secondary low developed along the East Coast). The storm of January 1956 (McQueen and Keith 1956) is an anomalous case of an oceanic low pressure system that first moves southeastward off the Carolina coastline and then north and northwestward over the western Atlantic Ocean. Hart and Grumm (2001) rated this storm system as the highest-ranked anomalous weather event over a 53-yr period, based on a normalized departure from climatology of tropospheric values of height, temperature, wind, and moisture computed from the National Centers for Environmental Prediction–National Center for Atmospheric Research (NCEP–NCAR) reanalysis dataset (Kalnay et al. 1996).

The surface anticyclones (Fig. 5-21, bottom) also exhibit some similarities and differences to the heavy snow sample. Some of the anticyclones are located over central or northern Canada, mostly over Ontario, delivering cold air at low levels to the northeast United States (one of the anticyclones, associated with the January 1956 ice storm, is located over extreme northern Quebec, while another associated with an ice storm in January 1999 is located over northern Maine). These locations are similar to those observed with some of the heavy and moderate snowstorms, as well as the interior snowstorms (cf. Figs. 3-5, 5-7, and 5-15). However, anticyclones associated with the ice storms of mid- and late January 1994 are located over the western Atlantic Ocean, clearly much farther east than most of the other systems associated with the snow events.

The freezing rainstorms of January 1956, December 1973, and January 1978 exhibit some similarities to each other (Fig. 5-22). These three events are all associated with significant East Coast cyclones associated with large cold anticyclones located over central or northern Canada (Fig. 5-21, bottom). One feature that characterizes all three cases is a frontal zone extending northward from the surface low located over the western Atlantic Ocean. This north–south surface frontal zone reflects the effect of a very strong upper ridge located to the east or northeast of the surface low. The frontal boundary may also be indicative of the tendency of warmer air to the north or northeast of the surface low to move above the cold surface air over the Northeast states. Another key feature is the presence of a cold anticyclone north of the storm system, which is strong enough to sustain at least a shallow layer of cold air over a large portion of the Northeast as the precipitation falls on the urban region.

The ice storms of 17 and 27–28 January 1994 and January 1999 (Fig. 5-23) are three examples that clearly

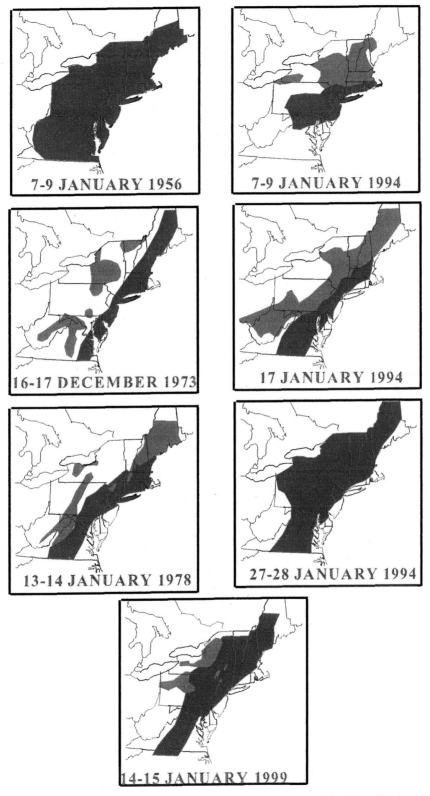

FIG. 5-20. Storm snowfall in excess of 10 in. (25 cm; blue shading) and locations of icing and sleet occurrences (violet shading) of seven selected ice storm cases.

CYCLONE TRACKS

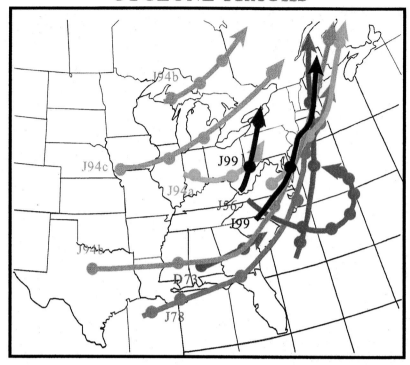

ANTICYCLONE POSITIONS

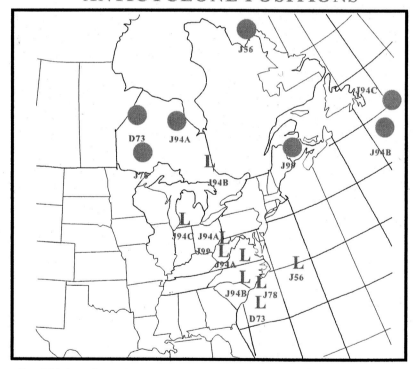

FIG. 5-21. (top) Composite tracks of surface low pressure systems for the seven selected ice storm cases. Symbols along path represent 12-hourly positions. (bottom) Composite positions of anticyclones associated with ice storms; positions (blue circles) selected at times when the surface cyclone (red L) was moving northeastward along the East Coast.

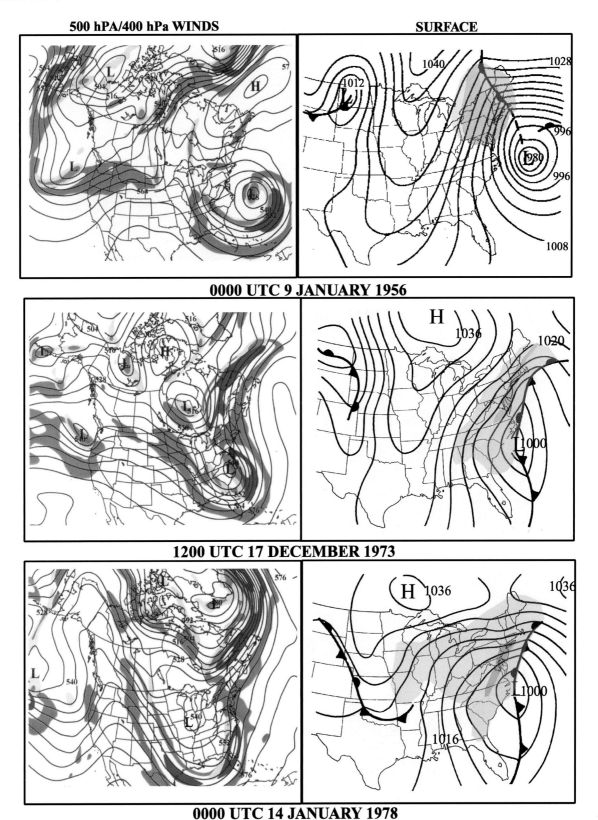

500 hPA/400 hPa WINDS   **SURFACE**

0000 UTC 9 JANUARY 1956

1200 UTC 17 DECEMBER 1973

0000 UTC 14 JANUARY 1978

FIG. 5-22. Examples of ice storms exhibiting a strong north–south frontal boundary located over the western Atlantic Ocean. Left column shows 500-hPa heights and 400-hPa winds; right column shows surface frontal analyses. See Fig. 5-3 for specific details.

500 hPA/400 hPa WINDS   **SURFACE**

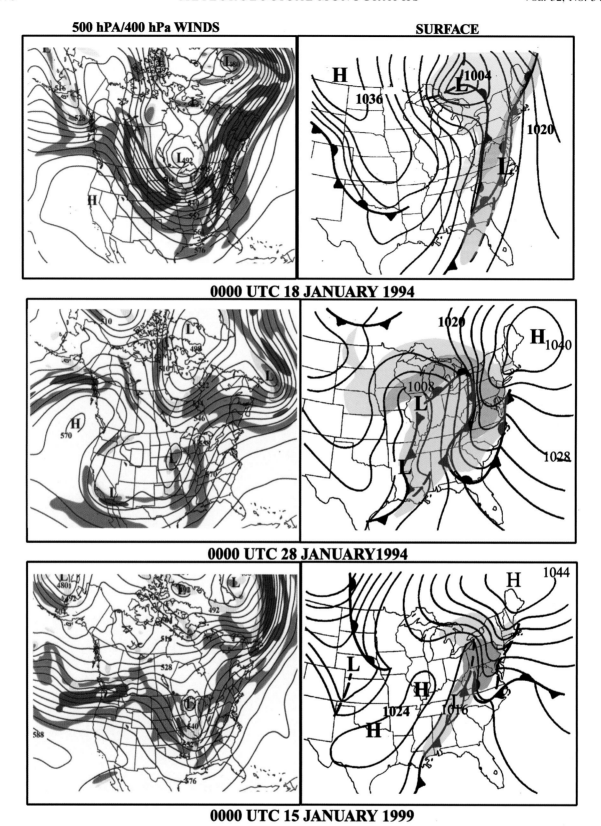

0000 UTC 18 JANUARY 1994

0000 UTC 28 JANUARY1994

0000 UTC 15 JANUARY 1999

FIG. 5-23. Examples of ice storms exhibiting a strong wedge of cold-air damming prior to the onset of the ice storm. Left column shows 500-hPa heights and 400-hPa winds; right column shows surface frontal analyses. See Fig. 5-3 for specific details.

do *not* resemble the significant snowstorm cases since there is no significant East Coast cyclone. Instead, these cases are dominated by the effects of retreating cold surface anticyclones and low pressure systems that move to the west-northwest of the Northeast urban corridor. Typically very pronounced wedges of cold air remain dammed east of the Appalachians. For example, the 28 January 1994 and January 1999 cases are each associated with very strong anticyclones associated with very cold air. In the case of January 1999, the anticyclone is poised north of Maine and very cold temperatures extend southwestward into central North Carolina. This event was marked by a major ice storm in Virginia and Maryland. While the surface high associated with the 28 January 1994 ice storm had already moved east of the Maine coast, cold surface air remained east of the Appalachian Mountains southward to southern Georgia, again providing a shallow pool of air cold enough to produce freezing rain throughout the mid-Atlantic region. Meanwhile, in both cases, the surface low was located well to the west-northwest of the coastline.

The ice and ice pellet storm of 17 January 1994 is less indicative of the influence of a strong cold anticyclone but still exhibits the pronounced wedge of cold air just west of the Atlantic coastline. Although the surface high had already retreated eastward over the Atlantic Ocean, this anticyclone had been associated with some of the coldest temperatures of the 1994 winter season and the cold temperatures were slow to modify east of the Appalachians. The only areas to escape the significant icing were the immediate mid-Atlantic and New England coasts where warmer oceanic air was able to penetrate immediately inland.

b. Upper-level features

One should expect that a major difference between the heavy snow cases and the ice storm cases is the presence of a layer of relatively mild air above the surface with temperatures above 0°C in the ice storm cases versus temperatures that remain below freezing for the snowstorm cases. In this section, the weather patterns that promote warming in the lower troposphere while temperatures remain below freezing near the earth's surface are contrasted with those factors that produce major snowstorms. A sample of soundings is shown in Fig. 5-24 to contrast the temperature profiles in the icing events from the heavy snow cases. In general, it appears that the warming is typically found in the layer between 900 and 700 hPa.

An examination of the seven cases indicates that the presence of a warm layer above the earth's surface can arise in a number of ways. For example, the icing event of January 1956 occurred during an anomalous blocking pattern with a distinct cutoff low over the western Atlantic and an impressive cutoff ridge extending from south of Greenland toward eastern Canada. The pres-

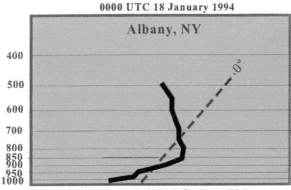

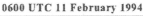

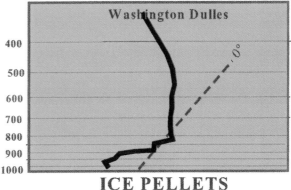

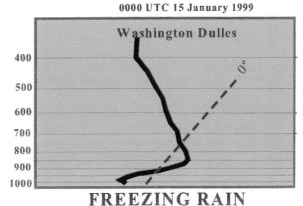

FIG. 5-24. Schematic skew *T*–log*p* vertical temperature profiles for two freezing rain soundings and one ice pellet sounding. Dashed line is the 0°C isotherm.

ence of the huge ridge north of the United States allowed very mild oceanic air to penetrate westward across eastern Canada into the northeastern United States before cold air was dislodged at the surface. The presence of a very intense ridge over eastern Canada is a significant deviation from many Northeast snowstorms often characterized by a deep trough over eastern Canada and a strong ridge farther north and east across Greenland and Iceland.

The December 1973 and January 1978 ice storm cases (Fig. 5-22) are two examples of slowly moving negatively tilted troughs, their significant downstream ridges, and strong surface anticyclones located to the northwest of the Northeast urban corridor that supply cold low-level air. These surface anticyclones remain to the northwest of the urban corridor. But, at upper levels, the development of a strong negatively tilted trough south and west of the Northeast urban corridor is also associated with the development of a strengthening ridge over the western Atlantic. This scenario produces a strong meridional upper-level flow along the East Coast that establishes a strong north–south frontal boundary near the coast, with the boundary separating very cold air to the west from much milder oceanic air. Cold air remains at the surface near the coast but much warmer air rises above the shallow cold air immediately to the west of the north–south frontal boundary. In the case of December 1973, there is no upper trough or cutoff low over eastern Canada, conditions typically found with heavy snow systems. Instead, the upper low is located farther west over Hudson Bay. With no confluent trough over eastern Canada, warmer air can advect rapidly northward in conjunction with the building ridge. In the case of 13–14 January 1978, a strong region of upper-level confluence over eastern Canada lifts northward as a trough nearing the East Coast deepens and becomes negatively tilted. With cold high pressure remaining to the northwest of the urban corridor, the upper ridge builds dramatically over the western Atlantic and a north–south frontal boundary forms over the western Atlantic, similar to that associated with the December 1973 storm.

In these two cases, rather than having the cold air located to the north of New England, as is common for the major snowstorms events, the cold air is located farther to the west and a building ridge to the east. The building ridge appears to be a significant factor that would promote enough warming aloft to change snow to rain, similar to that observed with the interior snowstorms discussed earlier. But, the presence of an anticyclone over south-central Canada still sustains cold low-level air in portions of the Northeast urban corridor. Therefore, this configuration tends to produce a north–south-oriented frontal zone with cold low-level air to the west of the boundary and warmer air to the east that then rises above the shallow colder air as it is advected to the west.

An additional subset of cases, shown in Fig. 5-23, is associated with surface anticyclones responding to *rapidly moving* progressive upper-level troughs over eastern Canada. These events are contrasted with the heavy snow cases, which are characterized by the slow-moving cutoff lows that tend to "lock in" the surface high pressure system and cold air. In these ice storm cases, very cold Canadian anticyclones supply cold air, but the high pressure systems continue to move toward the east. Even though the surface high has drifted over the Atlantic

Ocean, a narrow wedge of shallow cold air remains trapped east of the Appalachians. As the upper troughs move east of eastern Canada, warming and rising heights follow in their wake. Meanwhile, a separate upper trough nearing the East Coast propagates toward the northeast, forcing warmer air above the entrenched cold, surface air, resulting in widespread sleet and freezing rain.

In the case of 17 January 1994 (Fig. 5-23) a surface high pressure system accompanied by extremely cold air retreated well off the East Coast, leaving a narrow wedge of cold air between the coast and the Appalachian Mountains. With a broad upper trough to the west, a surface low moved well to the northwest of the Northeast urban corridor, associated with strong south to southwesterly winds in the lower troposphere, producing a narrow wedge of warm air above the cold surface air. Significant ice and sleet occurred across much of the interior mid-Atlantic states and Northeast. A similar scenario is found just 10 days later (Fig. 5-23), when another strong surface anticyclone retreated east off the Northeast coast. An upper trough is found over the upper Midwest, moving toward the northeast, accompanied by strong ridging over the eastern United States, representing warming occurring throughout the atmosphere, except for the remaining wedge of very cold air near the surface along the East Coast that resulted in significant icing, especially in the interior portions of the mid-Atlantic states and New England. The final example of 15 January 1999 (Fig. 5-24) is also associated with a very strong, yet retreating, surface anticyclone, rapidly rising 500-hPa heights across the eastern United States, and a significant upper-level trough moving northeastward toward New England. Very cold air remained entrenched as far south as Virginia as the surface high drifted eastward across southeastern Canada but the northeastward-moving trough over the Midwest was associated with lower-tropospheric warming and strong warm-air advection above the cold surface air, resulting in one of the worst ice storms on record in the Washington metropolitan area.

These three examples are all associated with retreating anticyclones and upper-level troughs that moved to the northwest of the Northeast urban corridor as strong upper-level ridging and related high geopotential heights were evident off the Southeast coast. In each case, surface warming occurred along the western slopes of the Appalachians and over the immediate coastline as surface winds shifted to a southeasterly direction off the relatively mild Atlantic Ocean. As a result, icing conditions eased along the immediate coastline as temperatures warmed well above freezing. In these cases, upper-level confluence retreated from the northeast United States and was replaced by rising heights and an upper-level trough that lifted northeastward partly in response to the rising heights. Thus, these cases typically resulted in a changeover to rain along the immediate coastline, but temperatures inland remained below freezing. As

850 hPA Analysis
850 0° vs. Surface 0°

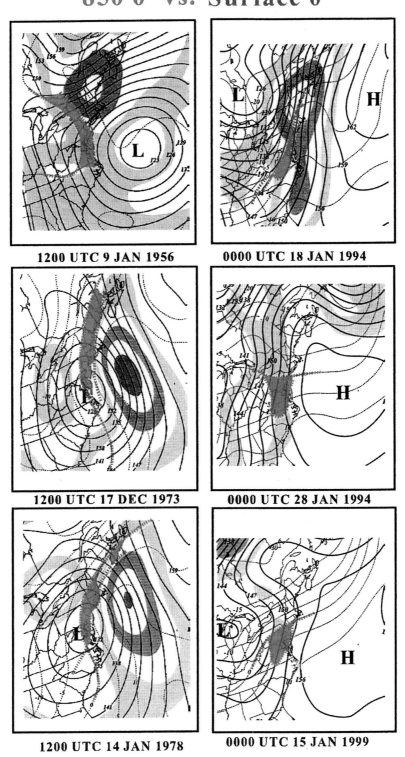

FIG. 5-25. The 850-hPa analyses for six selected ice storms including heights (solid, dm), winds (alternated shading at 5 m s⁻¹ increments above 15 m s⁻¹), and 0°C isotherms (850 hPa, dotted red; surface, dotted blue). Blue region represents area where 850-hPa temperatures are above 0°C, while surface temperatures are below 0°C.

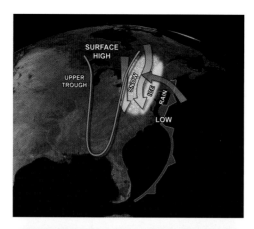

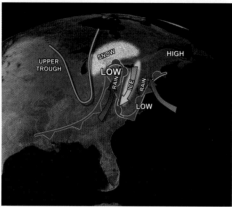

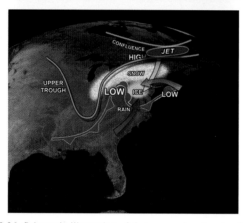

FIG. 5-26. Schematic illustration of three scenarios showing the surface and upper-level features associated with Northeast ice storms.

such, the precipitation changed to rain in cities very close to the Atlantic coast (New York City, Boston, and Philadelphia) while the more inland cities (Baltimore and Washington are located farther west of the Atlantic Ocean) experienced a longer period of frozen precipitation.

Selected 850-hPa analyses for six events (Fig. 5-25) show the presence of warmer air aloft, superimposed with locations of the corresponding surface freezing contour. The color-shaded regions represent the regions with surface temperatures below freezing and 850-hPa

temperatures above freezing, denoting the areas where freezing rain and ice pellets were occurring. These rather narrow regions, often located between areas of rain and snow, are obviously a difficult forecast problem due to their localized nature and the warming aloft, which is often located between rawinsonde stations. Analysis of vertical profiles from numerical model data is presently the best method to assess the potential for freezing rain and sleet versus rain and snow.

c. Summary of differences between major snowstorms and ice storms

The major difference between ice storms and snowstorms is the appearance of a warm layer in the lower troposphere, especially between 900 and 700 hPa, where the temperatures rise above freezing, overlaying a shallow pool of cold air near the surface. The depths of both the warm layer and the surface-based cold layer determine whether precipitation reaches the ground as either ice pellets or snow. Ice pellets occur when partial melting of snow occurs in the warm layer and colder layers and/or drier layers (which promote evaporative cooling) beneath allow precipitation to reach the ground as ice pellets. Freezing rain occurs as the depth of the warm layer is great enough to completely melt the snow. The rain reaches the ground and freezes upon contact. It is the authors' experience that freezing rain has been observed in the northeast United States with temperatures as low as the single digits (°F; −15° to −12°C). More typically, within the Northeast urban corridor, temperatures are observed to range between the 20°s and low 30°s (°F; −5° to 0°C) during most freezing rain events.

Within the seven ice storm cases selected to contrast the major snow events, there are three scenarios that tend to favor icing events as opposed to major snowstorms (Fig. 5-26). In all three scenarios, the presence of warm air over the shallow pool of cold surface air is emphasized and the role of the Appalachian Mountains in maintaining cold, surface temperatures is noted.

Scenario 1 (Fig. 5-26a): This scenario has a deep upper trough along the East Coast and an offshore surface low along a north–south surface frontal zone. Cold air is supplied by an anticyclone located to the north-northwest of the urban corridor. Strong upper-level ridging over the western Atlantic Ocean contributes to a north–south-oriented baroclinic zone characterized by a strong surface front that separates cold air from the west and warmer air over the Atlantic. As warmer air rises above the shallow cold surface air, freezing rain or sleet is the result. The storms of December 1973 and January 1978 are two examples of this scenario that resulted in widespread, damaging ice storms.

Scenario 2 (Fig. 5-26b): This scenario is perhaps a more common one in which a very strong anticyclone accompanied by very cold air moves into the Northeast urban corridor but continues to retreat eastward across eastern Canada or the western Atlantic Ocean. As the high pressure center retreats eastward, warming occurs, except near the earth's surface where the shallow remnants of the cold air mass remain wedged to the east of the Appalachian Mountains. This scenario is often accompanied by surface ridging along the East Coast and the movement of an upper-level trough to the northwest of the urban corridor, resulting in significant warm-air advection just above the cold surface air trapped to the east of the Appalachian Mountains. Since the cold air is wedged in between the mountains and coastline, the cold air is eroded by warming above and over the coastline and eastward, where southeasterly flow behind the retreating anticyclone quickly modifies over the ocean and begins to move inland. In these cases (including 17 and 27–28 January 1994 and an ice storm on 14–15 January 1999), regions along the immediate coastline, including New York City and Boston, experience a changeover from freezing rain to rain. Meanwhile, more interior locations remain as freezing rain for a longer period of time, or do not change over at all, sometimes allowing the metropolitan areas of Baltimore and Washington to experience more significant icing that its north-

ern but more coastal metropolitan centers. In general, the more northern and western regions of the Northeast urban corridor are more susceptible to icing than the eastern and southern portion.

Scenario 3 (Fig. 5-26c): Scenario 3 is one that often occurs as the transition region between snow and rain. This scenario is accompanied by a very strong anticyclone to the north-northwest of the urban corridor, similar to that found in major snowstorms, which provides the cold air at the surface. However, the cold outbreak does not extend as far south as those in the heavy snow sample and portions of the Northeast urban corridor lie along the southern fringes of the cold outbreak where shallow, subfreezing air undercuts milder air aloft. These weather systems often have a gradation from snow to sleet and freezing rain to rain on a north–south axis. The case of 7 January 1994 is an example where heavy snow mostly fell in Boston; snow, sleet and freezing rain fell in New York City; freezing rain occurred in Philadelphia; and a combination of rain and freezing rain fell in the Washington and Baltimore metropolitan areas, with more rain than freezing rain in the southeastern suburbs.

Chapter 6

MESOSCALE ASPECTS OF NORTHEAST SNOWFALL DISTRIBUTION

The descriptions of 30 major Northeast snowstorms in chapters 3 and 4, and 37 "near miss" cases in chapter 5, provide a basis for understanding the evolution of the cyclones responsible for the heavy snow-producing systems in the Northeast, while describing atmospheric features and processes that operate on a continuum of scales from the very large, or synoptic scale, to very small, or mesoscale. The snowfall distributions exhibited by the many cases shown in Figs. 3-2, 5-5, and 5-13 cover large, synoptic-scale areas, but all contain mesoscale detail. Mesoscale regions of enhanced snowfall rates are often associated with snowbands that can occur within larger synoptic-scale areas of heavy snow or they can exist as isolated phenomena. These bands are often less than 100 km wide and can extend over lengths of several hundred kilometers, and in some cases, the bands may extend over lengths approaching 1000 km.

An example of distinct mesoscale structure within a synoptic-scale area of snow is shown in a visible satellite image of the mid-Atlantic states following the 31 March–1 April 1997 snowstorm (Fig. 6-1). This storm primarily affected southeastern New York and southern New England with very heavy snows (Fig. 6-1a; also see volume II, chapter 10.28). However, an area of heavier snow also fell across a very small portion of southeastern Pennsylvania, northeastern Maryland, and northern Delaware, as observed in the visible satellite image (Fig. 6-1b). The morning following the snowstorm, the town of Christiana in eastern Lancaster County, Pennsylvania, reported 18 in. (45 cm) of snow, while whatever snow that had fallen in western Lancaster County had already melted. The mesoscale structure noted for the April 1997 snowstorm can be found in virtually every case described throughout this monograph.

In this chapter, mesoscale characteristics of snowfall are addressed and related to features such as conditional symmetric instability and frontogenesis, inverted surface pressure troughs and other quasi-frontal boundaries, gravity waves, elevation, thundersnow, and "bay effect" or "ocean effect" snowfall. Storm Data and Climatological Data dating from 1960 to 1999 were examined to find cases, using a methodology similar to that described by Branick (1997), who examined winter storm events across the entire United States for a 12-yr period from 1982 to 1994.

Orlanski (1975) provides a basis for defining the scales of weather systems, which is shown in Fig. 6-2 to include Northeast snowstorms. Synoptic, or "macro β," scales of motion cover distances between 2000 and 10 000 km and time periods of a day to several days. Mesoscale features are defined as spanning distances of 2–2000 km and time periods of 1 h to 1 day. Orlanski subdivides mesoscale into three subscales: meso-α (200–2000 km; around 1 day), meso-β (20–200 km; several hours to 1 day), and meso-γ (2–20 km; 30 min to a few hours). This chapter will examine the variety of mesoscale snowfall detail found within Northeast snowstorms, spanning the meso-α to meso-γ scales.

The development of narrow, singular, or multiple snowbands with meso-α to meso-γ scales remains a perilous prospect for forecasters and is rarely resolved by current operational models. When these historically unpredicted events affect a populated urban environment, such as the Northeast urban corridor, its societal and economic impacts are large, and public perceptions of forecast capability are severely affected. A statement released by the New York City forecast office following a mesoscale snow event that left up to a foot (30 cm) of snow over heavily populated regions of Long Island and western Connecticut on 13 December 1988 that was not predicted points to frustration associated with the forecast challenges that remain.

> Events like this cause weather forecasters to lose their hair at an early age and in some cases to develop stomach ulcers. We have to admit the existence of many weather events that are...at this stage of our weather science...unforecastable. There are situations that slip between the cracks and I am afraid this has been one of them. We missed it and so did the computers. We could only sit here at the Weather Service Office and watch the situation go from bad to worse. We became observers as opposed to forecasters.

1. Examples of snowbands with possible linkage to midlevel frontogenesis and symmetric stability

The roles of moist symmetric stability and frontogenesis in the development of precipitation bands have been described in numerous articles over the past two decades (Bennetts and Hoskins, 1979; Emanuel 1979, 1983, 1985; Sanders and Bosart 1985a; Sanders 1986b;

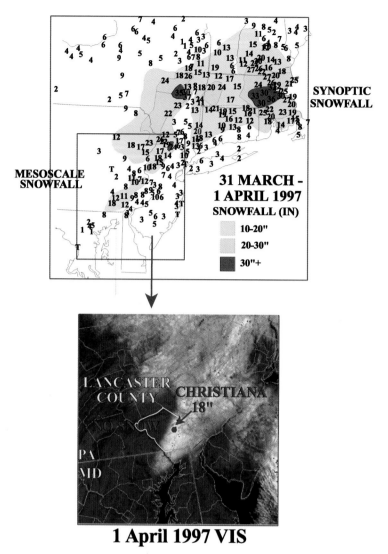

FIG. 6-1. Example of synoptic vs mesoscale detail within the snowfall of 31 Mar–1 Apr 1997: (top) total snowfall (in.) and (bottom) visible satellite image taken the morning following the snowstorm.

Wolfsberg et al. 1986; Nicosia and Grumm 1999; Graves et al. 2003). From these studies, a relationship between frontogenesis and symmetric stability can be applied to describe the narrow regions of ascent that lead to the banded structure often observed with snowfall. The ascent regions are enhanced and contracted to meso-α and meso-γ scales when the symmetric stability is low and where the frontogenesis is maximized. Nicosia and Grumm (1999) examine heavy snowbands located along the northwestern edge of the synoptic-scale snowband for several recent snow events in the eastern United States, including the cases of 3–4 February 1995 (see volume II, chapter 10.26), 19–20 November 1995, 3–4 March 1994 (also see volume II, chapter 11.1), and 12 January 1996 (Fig. 6-3). The snowbands are often west of the heaviest precipitation predicted by current numerical models but occur near model-forecast front-

ogenesis at 700 hPa. The frontogenesis process is enhanced as a midtropospheric low develops and intensifies (as the cyclonic circulation extends upward during cyclogenesis) and as the exit region of an upper-level jet streak/dry intrusion encounters the region.

The Next-Generation Doppler Radar (NEXRAD) imagery for cases provided by Nicosia and Grumm (1999) and Graves et al. (2003) shows that the development of one or more snowbands are often embedded within a larger synoptic-scale region of snow. These bands are typically oriented from southwest to northeast and appear to move slowly northeastward, or to remain stationary and "pivot" slowly in a counterclockwise direction as individual radar elements merge into the band from the east, southeast, or northeast.

A series of images from the 25 January 2000 snowstorm in Washington, D.C., illustrates the banded struc-

NORTHEAST SNOWSTORMS
SYNOPTIC TO MESOSCALE PHENOMENA

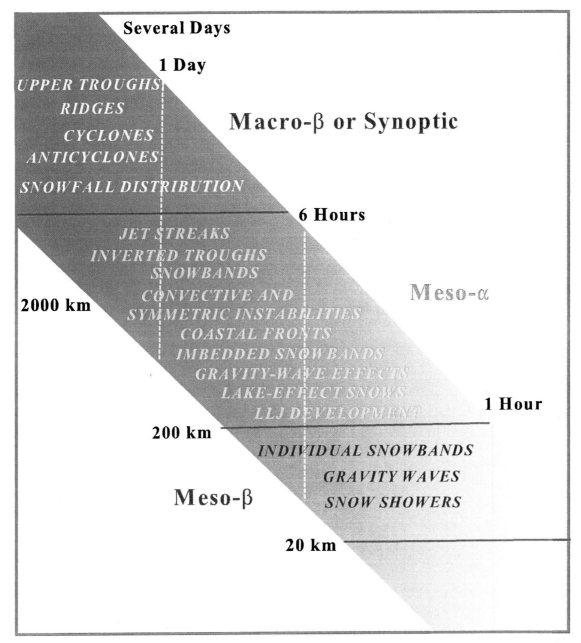

FIG. 6-2. Classification of scales of motion involved in Northeast snowstorms, as described within this monograph. Based on a figure in Orlanski (1975).

ture of the snowfall radar echoes. The heaviest bands, indicated by the green shading (Fig. 6-4), are located just west of Chesapeake Bay and are relatively stationary, while individual elements within the band are rotating cyclonically toward the south. Note that surrounding bands appear to be directed toward the main band over eastern Maryland, with convergence implied by the red arrows. This converging area of snowbands pivoted slowly toward the east-northeast, where some of the greatest snow totals, exceeding 15 in. (38 cm), were reported (Fig. 6-4), right along Chesapeake Bay.

An examination of Climatological Data and Storm Data provides many examples of narrow, mesoscale snowbands, several of which are presented here.

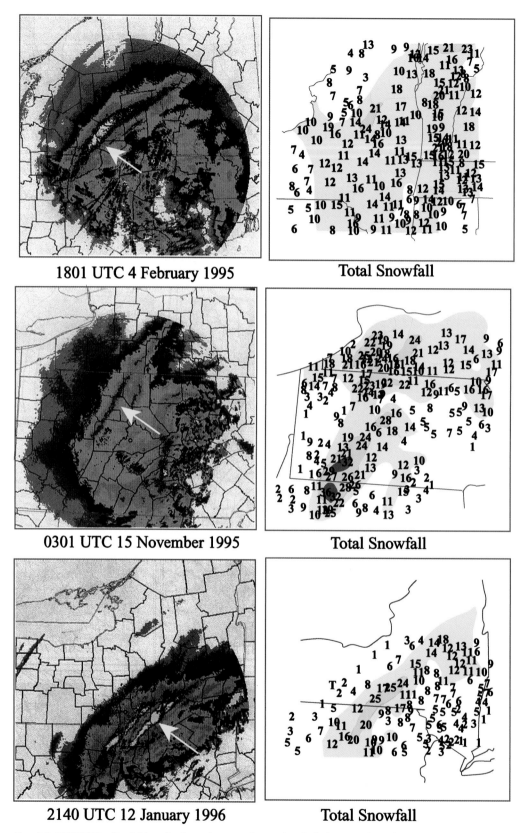

FIG. 6-3. NEXRAD reflectivities of enhanced mesoscale snowbands during a variety of Northeast snow events on 4 Feb 1995, 15 Nov 1995, and 12 Jan 1996. [From Nicosia and Grumm (1999).]

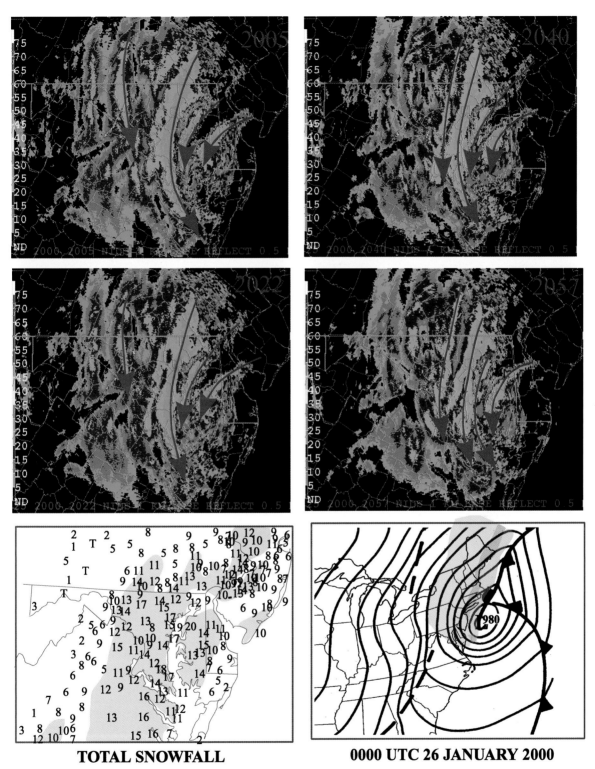

TOTAL SNOWFALL   **0000 UTC 26 JANUARY 2000**

FIG. 6-4. NEXRAD depiction of time evolution of snowbands embedded within the 25 Jan 2000 snowstorm in the Baltimore–Washington metropolitan areas for 2005–2057 UTC. Colors represent gradations in base reflectivity at 0.5° elevation; note scale on left of each figure. Red arrows represent instantaneous motion of snowbands. (bottom left) Total snowfall (in.; contours at 10, 15, and 20 in.) in the middle Atlantic states. (bottom right) Surface analysis at 0000 UTC 26 Jan 2000.

a. 7–8 January 1958

A narrow heavy snowband was associated with a small and explosively deepening cyclone on 7–8 January 1958. A narrow belt of 14–20-in. (35–50 cm) snowfall accumulations was described as 10 mi (16 km) wide[1] from New Haven, Connecticut, to northwestern Rhode Island. An analysis of available snow data indicates that the length of the band was much greater than its width, which extended from the lower Chesapeake Bay into western Maine (Fig. 6-5a). The mesoscale band of snow was located within a larger region of snowfall associated with a low pressure system that evolved from the merger of two separate systems (see chapter 4). Eventually, the storm underwent a period of rapid deepening over the western Atlantic Ocean, characterized by pressure falls exceeding -20 hPa per 3 h in extreme southeastern Massachusetts.

b. 2–3 March 1969

Another interesting example of mesoscale areas of heavy snow occurred on 2–3 March 1969, when a slow-moving coastal cyclone produced a large area of snowfall from Virginia to Maine. Embedded within this synoptic-scale event were a number of smaller regions with significant amounts of snow, including southern Virginia, southern New Jersey, and southeastern Massachusetts. However, the most anomalous area of snow was located along a narrow 30-mi- (50 km) wide southwest-to-northeast band extending from eastern Pennsylvania through southeastern New York into southern Vermont. Within this narrow band, accumulations of 20 to greater than 30 in. (50–75 cm) of snow fell across extreme northeastern Pennsylvania, northwestern New Jersey, and into portions of southeastern New York (Fig. 6-5b). Port Jervis, New York, recorded more than 33 in. (83 cm) of snow, while Liberty, New York [only 30 mi (50 km) to its northwest], received only an inch (2.5 cm).

c. 11–12 November 1987

Sometimes, the heavy snowbands occur over isolated, relatively small regions that are separate from a larger-scale snowfall pattern. The snowfall during the 10–12 November 1987 Veterans' Day snowstorm (Fig. 6-5c) exhibits a well-defined synoptic band of snow (Fig. 6-6, top; labeled A) associated with a weak area of low pressure on 10 November. A synoptic-scale cyclone developed early on 11 November and produced a significant band of snow across southeastern New England (Fig. 6-6, top; labeled B). In addition to these synoptic-scale areas of snow, a 40 km × 200 km mesoscale band of heavy snows left up to 17 in. (42 cm) of snow in the Washington metropolitan area [Fig. 6-6, labeled C; also

---
[1] New England Climatological Data, January 1958.

see Maglaras et al. (1995)]. This mesoscale band of snow formed locally over the Washington metropolitan area early on 11 November during the early stages of cyclogenesis and dissipated by late afternoon. The eastern portion of Washington and nearby suburbs received the heaviest snows (Fig. 6-6, bottom) while the western suburbs reported as little as 2 in. (5 cm). At the height of the storm, snowfall rates of 3 in. (8 cm) h^{-1} were reported and often accompanied by thunder and lightning. The 11.5 in. (29 cm) of snow at National Airport in Washington was the greatest November snowfall on record. The heavy band of snow did not translate northeastward toward the Philadelphia and New York metropolitan areas, where only about an inch (2.5 cm) of snow fell. Rather, the area of heaviest snow seemed to develop and then dissipate locally in the mid-Atlantic region. Heavy snowfall then redeveloped over southeastern Massachusetts (labeled B in Fig. 6-6) while the synoptic-scale cyclone intensified offshore beneath an advancing upper-level trough–jet system translating off the Virginia coast (Fig. 6-5c).

This storm proved to be exceptionally difficult to forecast because of a number of factors. Foremost was the small scale of the event that operational models could not resolve. In addition, the storm occurred unusually early in the season with few historical precedents. Finally, the operational models failed to properly simulate the amount and the mesoscale banded structure of precipitation, and the local enhancement of the snowfall that occurred in the Washington, D.C., area.

d. Three snowstorms during January and February 1996

Only a few days following the January 1996 "Blizzard of '96" (see volume II, chapter 10.27), a new storm system threatened the Northeast with more heavy snow. On 12 January 1996, a modest storm system managed to produce six additional inches (15 cm) of snow in the Washington–Baltimore area, while snow changed to rain from Philadelphia to New York City and Boston after only a few inches of accumulation. While this storm system was much less widespread and intense than the blizzard earlier in the week, a narrow 20–30-km-wide snowband developed along the northwestern side of the storm over northeastern Pennsylvania into southeastern New York (Nicosia and Grumm 1999). The snowband was located within the larger area of 6-12 in. (15–30 cm) of snow, and produced snowfall amounts in northeastern Pennsylvania of 20–25 in. (50–63 cm). There is an isolated report of 36 in. (90 cm) in northeastern Pennsylvania, where snowfall rates approached 6 in. (15 cm) h^{-1}. The mesoscale structure of the snowband was also captured by the NEXRAD base reflectivities (Fig. 6-7a).

Another example of a mesoscale snowband occurred on 2 February 1996, when a narrow band of heavy snow

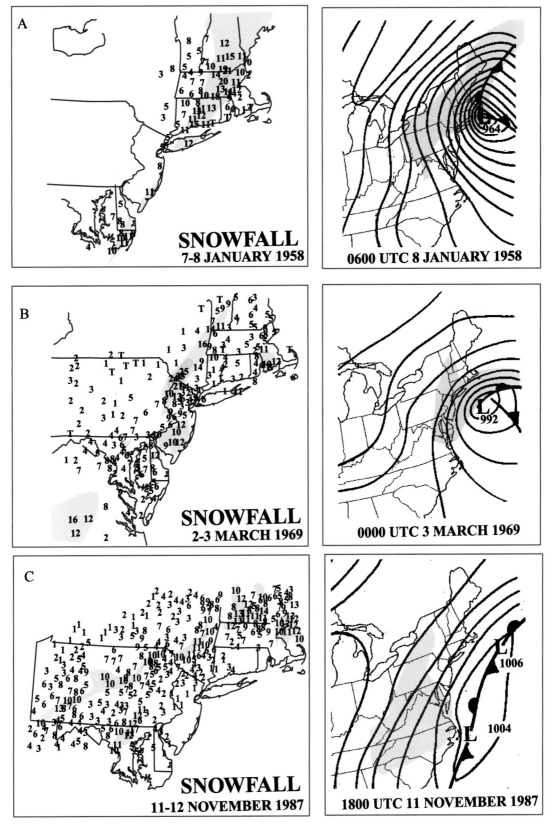

FIG. 6-5. Snowfall distributions and a representative surface analysis for the mesoscale snow events of (a) 7–8 Jan 1958, (b) 2–3 Mar 1969, and (c) 11 Nov 1987.

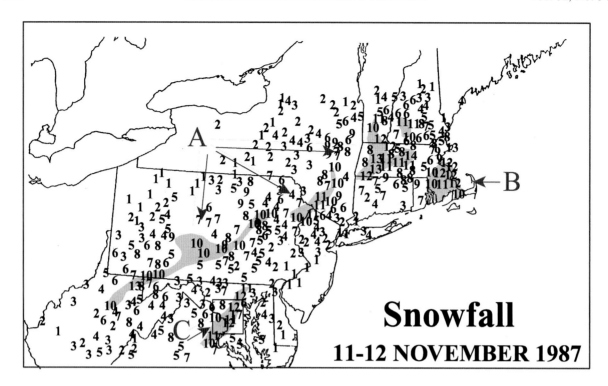

SNOWFALL (INCHES)

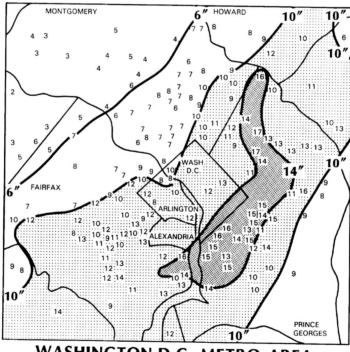

WASHINGTON, D.C. METRO AREA
NOVEMBER 11, 1987

FIG. 6-6. Snowfall distribution for the Veterans' Day snowstorm of 10–11 Nov 1987. (top) A, B, and C refer to separate bands of snowfall attributed to several distinct periods described in the text; (bottom) detailed snowfall analysis for the Washington metropolitan area on 11 Nov 1987 (from Maglaras et al. 1995).

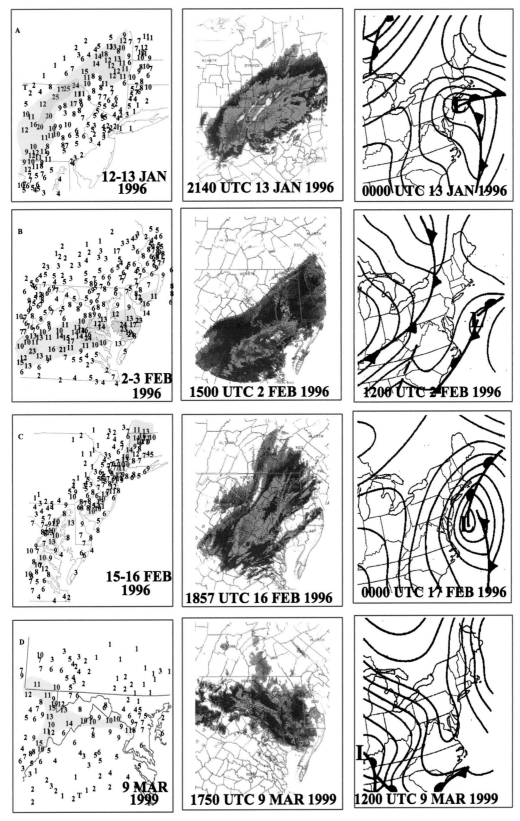

Fig. 6-7. Snowfall distributions, NEXRAD imagery, and a representative surface analysis for the mesoscale snow events of (a) 12–13 Jan 1996, (b) 2–3 Feb 1996, (c) 15–16 Feb 1996, and (d) 9 Mar 1999.

oriented along a southwest–northeast axis produced up to 14 in. (36 cm) of snow along a swath from northern Virginia into southeastern Maryland and southern Delaware (Fig. 6-7b). The event produced snowfall rates of 3 in. (7.5 cm) h^{-1} and was associated with a weak center of low pressure. A second, more significant cyclone the next day produced a larger region of snow, but snowfall rates were not nearly as heavy as the prior day, although they did add to the significant snowfall accumulations (see volume II, chapter 11.2). Note that the smaller, mesoscale event from Virginia through Delaware is very evident in the NEXRAD imagery.

A third example of a narrow snowband during the winter of 1995/96 was associated with a developing cyclone on 16 February 1996. This synoptic-scale snow event (see volume II, chapter 11) was associated with a narrow band of heavier snow with up to 15 in. (38 cm) along the heavily populated Interstate 95 route from northern Virginia to southeastern New York and central New England (Fig. 6-7c). This narrow band (see NEXRAD imagery) was located within a larger-scale snowband that produced 4–8 in. of snow (10–20 cm), on average, from Virginia to New York.

e. 9 March 1999

As a significant and widespread winter storm over the Midwest moved eastward on 8 March 1999, it began to weaken and its large snow-producing capabilities appeared to wane. However, as this system approached the middle Atlantic states on 9 March, a narrow band of heavy snow developed from southwestern Pennsylvania into northern West Virginia, northern Virginia, and into portions of the Washington metropolitan area, leaving up to 11 in. (28 cm) of snow over a very localized area (see Fig. 6-7d). The band was less than 100 km wide within which heavy snowfall rates [up to 2 in. (5 cm) h^{-1}] left more than 6 in. (15 cm) of snow.

The cases briefly described above reveal some of the complexities that occur in Northeast snowstorms, including spatial and temporal variations of snowfall that challenge simple explanations based on synoptic-scale principles, and routinely have confounded forecasters. The ability to forecast these events will be a crucial issue to resolve (as described in chapter 8), in order to predict the snowfall with acceptable reliability.

2. Examples of snowbands linked to surface fronts and inverted pressure troughs

Surface fronts are occasionally associated with locally enhanced snowfall amounts in major snowstorms. The fronts represent sites of localized baroclinic zones that can act to focus strong ascent in relatively narrow bands. The March "Blizzard" of 1888 (Fig. 6-8a; also see volume II, chapter 9) is a well-known example of a major cyclone associated with a pronounced surface front extending to the north of the low pressure center. The

heaviest snows, up to 55 in. (140 cm), fell in the Hudson River valley of New York, just west of and parallel to the frontal boundary, with the heaviest snowfall occurring on the cold side of the front. The November 1950 "Appalachian Storm" (Fig. 6-8b; see volume II, chapter 9) is another example where the heaviest snows, exceeding 40–50 in. (100–125 cm) in West Virginia, occurred just west of a strong frontal boundary oriented to the north of a strengthening surface low. The December 1962 case (Cunningham and Sanders 1987), marked by record-setting snowfall of up to 46 in. (117 cm), occurred over central and eastern Maine, west of a north–south frontal boundary lying to the north of an intensifying cyclone (Fig. 6-8c).

These frontal features act to focus the heavy snowfall in areas along and west of the frontal boundary. The large-scale circulation often exhibits a pronounced easterly low-level flow east of the boundary that ascends the frontal boundary and then moves more north and then northeastward as the rising air encounters the more southwesterly flow aloft. Since these three snowstorms produced record-setting snowfall, the interaction of a developing cyclone with a north–south front appears to be a factor in producing some of the most significant snowstorms to ever affect the Northeast.

Coastal fronts (Bosart et al. 1972; Ballentine 1980; Bosart 1981; Bosart and Lin 1984; Forbes et al. 1987; Stauffer and Warner 1987; Bell and Bosart 1988, 1989; Nielsen 1989; LaPenta and Seaman 1990, 1992; Nielsen and Neilley 1990; Doyle and Warner 1993a,b; also see chapters 3 and 7) are often associated with localized heavy snowfall amounts, particularly on the cold side of the boundary (Bosart et al. 1972; Marks and Austin 1979), similar to the larger-scale fronts. Low-level air originating over the warmer ocean rises above the frontal boundary, resulting in precipitation falling on the cold side of the boundary. A number of the heavy snow events across eastern New England were associated with the development of a significant coastal front off the southeast New England coast (Fig. 6-9) including one described by Bosart et al. (1972) that may have enhanced snowfall rates. Weak centers of low pressure, sometimes called zipper lows (Keshishian and Bosart 1987), can develop along the coastal front and are associated with low-level warm-air advection and intense frontogenesis, and can also enhance the development of heavy snow along the cold side of the coastal boundary. Three examples of New England snowstorms are shown in Fig. 6-9 to illustrate several coastal fronts and the mesoscale snowfall distribution associated with each event.

An inverted surface pressure trough is another feature that is occasionally associated with enhanced snowfall amounts (see, e.g., Keshishian et al. 1994). While these features do not exhibit the well-defined temperature gradients observed with coastal fronts and larger-scale frontal boundaries, they can eventually exhibit well-defined frontal boundaries as the storm system intensifies. One

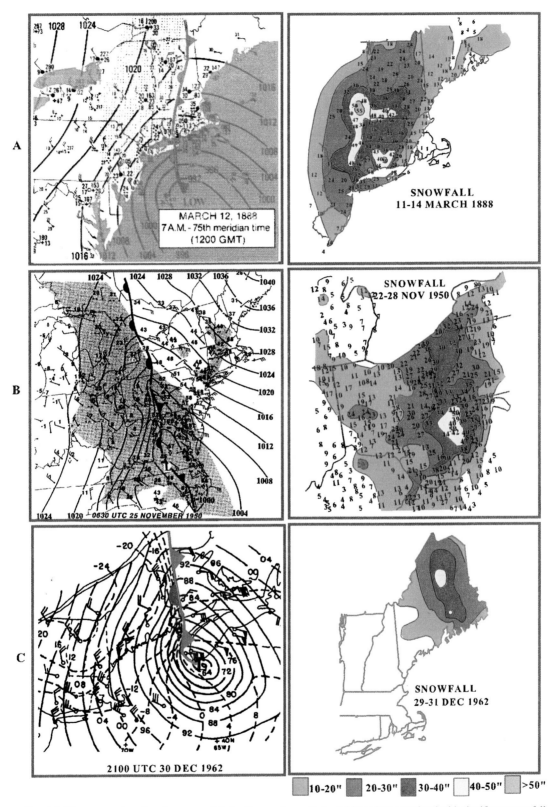

FIG. 6-8. Three examples of cyclones associated with a north–south frontal boundary associated with significant snowfall accumulations: (a) 1200 UTC 12 Mar 1888, (b) 0630 UTC 25 Nov 1950, and (c) 2100 UTC 30 Dec 1962 (derived from Cunningham and Sanders 1987). Total snowfall for each storm is shown at right.

SURFACE   SNOWFALL

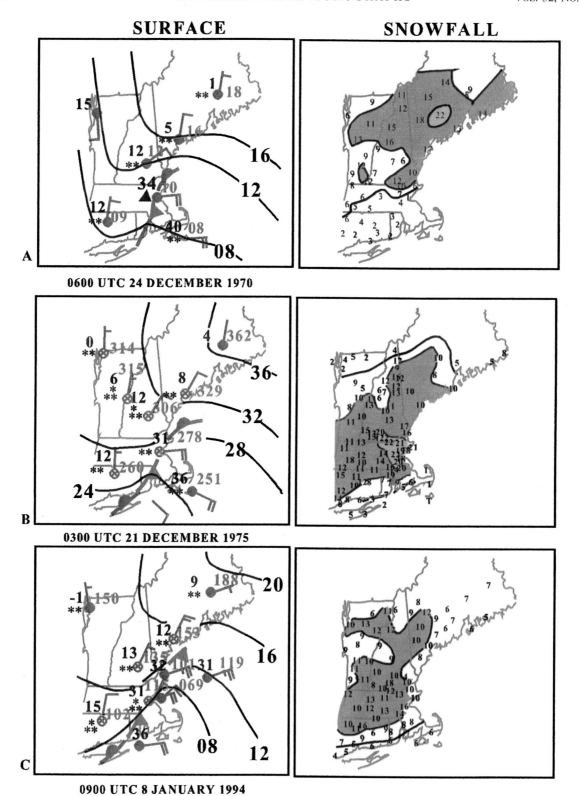

0600 UTC 24 DECEMBER 1970

0300 UTC 21 DECEMBER 1975

0900 UTC 8 JANUARY 1994

FIG. 6-9. Examples of mesoscale snowfall enhancement associated with coastal fronts along the New England coastline. Surface maps at (a) 0600 UTC 24 Dec 1970, (b) 0300 UTC 21 Dec 1975, and (c) 0900 UTC 8 Jan 1994. Total snowfall amounts are included on right.

localized snowfall event associated with a coastal pressure trough left only 1–2 in. (2.5–5 cm) of snow in the Boston metropolitan area but produced more than a foot (30 cm) of snow just north of the city on Cape Ann (Rosenblum and Sanders 1974). Only 6 mi (10 km) separated the heaviest snows from nearly no snow. A synoptic-scale cyclone developed out of a broad trough of low pressure over New England that consolidated into a synoptic-scale cyclone that moved well east-southeast of southeastern New England early on 18 December. As this cyclone developed offshore, an inverted surface pressure trough developed across eastern New England (see Rosenblum and Sanders 1974, their Fig. 4) with northerly and westerly flow over the interior of New England and onshore flow near the coastline. An examination of Rosenblum and Sanders's mesoanalyses indicates the presence of a small mesoscale cyclone within the coastal trough.

Below are examples of several other significant snow events that appear to be linked to the presence of an inverted trough.

a. 28–29 March 1942

One of the most anomalous, late season surprise snowstorms affected a relatively narrow area (around 100–150 km in width), extending from the nation's capital and Baltimore, Maryland, northward into central Pennsylvania on 28–29 March 1942. Eleven inches of snow (28 cm) fell in downtown Washington, with 22 in. (55 cm) in Baltimore, and snowfall amounts exceeded 30 in. (75 cm) from north-central Maryland into central Pennsylvania (Fig. 6-10). Meanwhile, much lighter snowfall totals surrounded the band to its east and west, over western and eastern Pennsylvania and Maryland.

This snowstorm was unusual for several reasons. First, it occurred relatively late in the season, during early spring, and was not characterized by many of the features found in other Northeast snowstorms described in chapters 3 and 4. The unusual synoptic setting for the storm involved a primary low moving toward the Midwest, farther to the west than is typical for other snowstorms; a surface high located over eastern Canada, typically farther east than often observed (see chapter 3); and a secondary low pressure system that moved eastward off the mid-Atlantic coast, but did not intensify appreciably. The heaviest snow fell to the west of a slowly moving occluded frontal boundary separating the weakening primary low pressure system over the Midwest from the developing low off the mid-Atlantic coast. This boundary later took the form of an inverted trough over eastern Maryland and Pennsylvania with the heaviest snows located to its west (Fig. 6-10). This anomalous storm produced some of the heaviest snows ever recorded in central Maryland and central Pennsylvania and was understandably not predicted with any accuracy.

b. 26 December 1947

New York City's greatest snowstorm, in December 1947, also appears to be associated with the development of a pronounced inverted trough extending north and west of the cyclone center. This feature appears to be a critical factor in the production of record-breaking snowfall in the city's metropolitan area. While the heavy snow was associated with a developing cyclone, including the merger of two separate upper-level troughs (see volume II, chapter 9), a mesoscale region of very heavy snow developed along and to the west of a localized region of rapid pressure falls over southern New England that created an inverted trough to the northwest of the main cyclone near Long Island and New York City. The development of this trough alters a primarily northeasterly wind field over much of the Northeast and generated a localized region of northwesterly winds over New Jersey and southeastern New York. As such, the trough axis represented a region of low-level convergence that focused the heaviest snow (Fig. 6-10) immediately west of the trough axis. Thus, the mesoscale features related to the inverted trough helped transform a significant synoptic-scale event into a historical snowfall in New York City and its surrounding suburbs.

c. 13 December 1988

A mesoscale snow event on 13 December 1988 that produced heavy snow on Long Island is marked by an inverted surface pressure trough that evolved into a surface coastal front and led to the issuance of the forecaster lament quoted at the beginning of the chapter. As in other cases marked by a mesoscale distribution of snow, this case is also associated with a major cyclone that developed over the Atlantic Ocean west of Bermuda, one of a number of cyclones to occur during the Experiment on Rapidly Intensifying Cyclones (ERICA; see Hadlock and Kreitzberg 1988). In this case, a narrow band of precipitation developed far to the north and west of the cyclone along and near a coastal trough just east of the mid-Atlantic coast (Fig. 6-11). While New York City received only 1 in. (2.5 cm) of snow or less, a north–south band of snowfall (Fig. 6-11a) exceeding 6 in. (15 cm) and no wider than 70 km, developed across central Long Island and extended north into southern New England. Maximum snowfall on Long Island exceeded 12 in. (30 cm), much of it falling within a 4–6-h period during the evening rush hour, creating a commuting nightmare. As the band extended northward into southwestern Connecticut, up to 10 in. (25 cm) of snow fell in that area.

The inverted trough appeared to develop in response to an advancing upper-level trough. However, the primary surface low developed nearly 1000 km southeast of New York, with the main upper vorticity maximum passing far to the south over eastern North Carolina. The surface inverted trough developed from the cyclone

HEAVY SNOW with INVERTED TROUGH

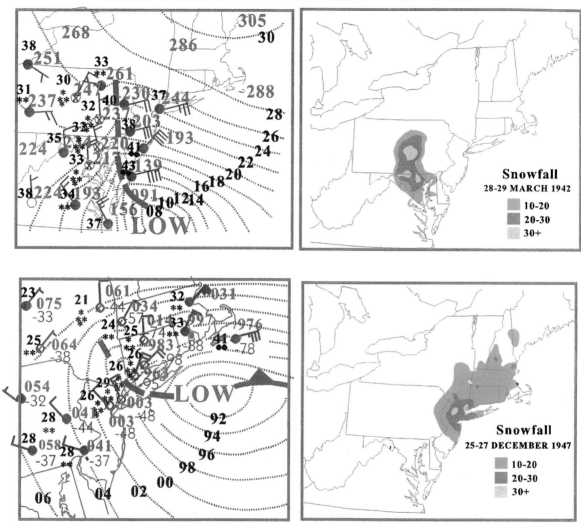

FIG. 6-10. Two examples of heavy snowfall associated with a surface low and an inverted surface trough. Surface analyses at (top) 1230 UTC 29 Mar 1942 and (bottom) 1830 UTC 26 Dec 1947, including surface temperature (°F), winds (conventional notation), pressure (064 = 1006.4 hPa), and pressure tendency (−41 = −4.1 hPa per 3 h). Total snowfalls for each storm are shown at right.

center northward into southern New England, while a coastal front (see Sanders 1990; Maglaras et al. 1995; Roebber et al. 1994) developed within the trough across eastern Long Island, separating oceanic-modified air with temperatures in the 40°s (°F; 5°–10°C) across eastern Long Island from temperatures in the 20°s (°F; −5°C) across western Long Island (Fig. 6-11b). Melted precipitation amounts in the area of heavy snow exceeded 1.5 in. (3.7 cm). Operational numerical models hinted at the development of an inverted trough but greatly underestimated precipitation amounts associated with this feature.

d. 21 March 1992

A very heavy, localized snowstorm affected Portland, Maine, and surroundings on the morning of 21 March

1992 (Bosart and Bracken 1996). Over 25 in. (60 cm) of snow fell in the area within 12 h and was associated with the development of an inverted trough extending north of the developing surface cyclone over the western Atlantic (Fig. 6-12) associated with a mesoscale disturbance aloft. Bosart and Bracken (1996) have described the presence of cyclonic development to the northwest of an oceanic cyclone as a source for "surprise" mesoscale snowfall phenomena. Bosart and Bracken hypothesize that the snowband is a mesoscale response to enhanced baroclinicity and reduced static stability near the coast coupled with the passage of a cyclonic vorticity maximum in very cold air aloft. A confluent airstream, with a moist northeasterly flow warmed and destabilized by the underlying ocean and a cold north-northwesterly flow with an overland trajectory to the west of the boundary focus the baroclin-

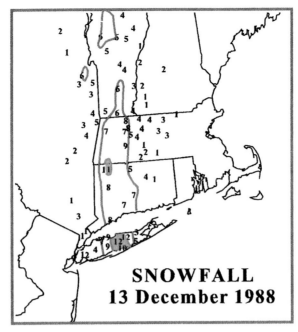

**SNOWFALL
13 December 1988**

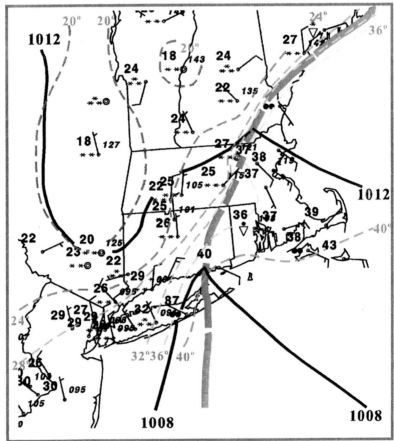

2200 UTC 13 DECEMBER 1988

FIG. 6-11. (top) Total snowfall for 13 Dec 1988. (bottom) Surface analysis at 2200 UTC 13 Dec 1988. Red and blue line represents coastal trough/front, dashed lines are isotherms, and surface observations are conventional notation.

Snowfall

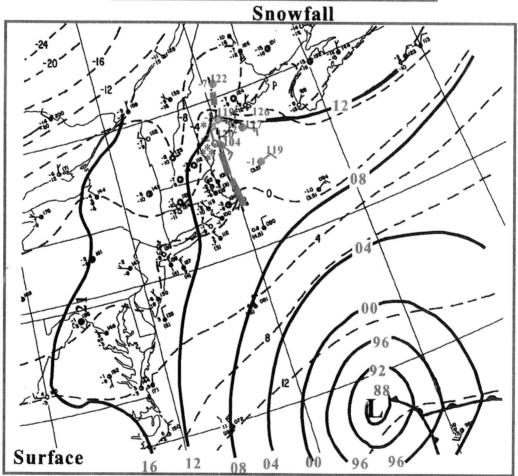

Surface

FIG. 6-12. (top) Total snowfall for 21 Mar 1992. (bottom) Surface analysis at 0900 UTC 21 Mar 1992 (derived from Bosart and Bracken 1996).

icity into a narrow zone, where low-level convergence and associated ascent in a very narrow region resulted in the heavy snowfall around Portland. Operational forecast models failed to predict any precipitation over eastern Maine prior to the event.

These examples of heavy snow events associated with surface inverted pressure troughs extending north or west from an oceanic surface low appear to have several characteristics in common. First, the heaviest snow is located to the north, northwest, and west of the trough axis. Second, the surface trough appears to be associated with a distinct upper-level trough although the main cyclone associated with the upper trough can be located well to the south and east of the surface trough axis. In some cases, there is evidence of two distinct troughs and associated vorticity maxima, with the northern upper-level system related specifically to the surface trough while the southern upper-level system is associated with the more pronounced cyclone. Third, the snow developed fairly rapidly and remained localized near the apex of the inverted trough. Finally, the operational numerical models seem to capture the upper-level features associated with these events and at times provide indications of the inverted surface troughs in the sea level pressure forecasts. However, the models still have problems capturing the mesoscale ascent patterns and associated precipitation rates, tending only to predict the primary surface low far to the south and east of the areas affected by the local snow event.

In this and the previous section, some of the mesoscale snowbands also appear to have been associated with the process of frontogenesis, either aloft or near the earth's surface. Nicosia and Grumm (1999) emphasized the role of frontogenesis in combination with other factors in a study of three mesoscale snowbands associated with Northeast snowstorms. In this section, especially in the case of December 1988, the presence of an inverted surface trough was sometimes indicative of surface frontogenesis, occurring in conjunction with the development of heavy snow. Therefore, any factor that would contribute to frontogenesis, such as shearing or confluent deformation, might be an important contribution to mesoscale snow phenomena in a significant number of cases, but further research is necessary to clarify this point.

3. Gravity waves

Mesoscale gravity waves are mesoscale perturbations in the surface pressure and wind fields induced by unbalanced flow throughout the troposphere. These mesoscale features can result in enhancements of convergence–divergence patterns, and associated vertical motion fields, that act to enhance snowfall rates in one area while inhibiting precipitation in others, while also generating strong surface winds (see Brunk 1949; Tepper 1954; Bosart and Cussen 1973; Uccellini 1975; Lindzen and Tung 1976; Hooke 1986; Bosart and Sanders 1986;

Uccellini and Koch 1987; Ferretti et al. 1988; Bosart and Seimon 1988; Schneider 1990; Koch and O'Handley 1997; Bosart et al. 1998). These mesoscale pressure perturbations can last for many hours and extend over hundreds to thousands of kilometers and enhance precipitation rates over small areas for short periods of time. The waves tend to occur poleward of a warm frontal boundary, near the inflection point between and upstream upper trough and downstream ridge and downstream of an upper- tropospheric jet streak (Uccellini and Koch 1987). The waves appear to be the result of a combination of shearing instability, geostrophic adjustment, convection, orography, and wave-conditional instability of the second kind (CISK; see, e.g., Uccellini and Koch 1987; Koch and Golus 1988; Koch and Dorian 1988; Koppel et al. 2000).

Gravity waves have been the subject of research on winter storms (Bosart and Cussens 1973; Bosart and Sanders 1986; Uccellini and Koch 1987; Schneider 1990). Bosart and Sanders examined gravity wave conditions during the February 1983 "Megalopolitan" Snowstorm, noting the movement of a region of strong pressure fluctuations and associated enhancements in snowfall and occurrence of convective activity. Schneider (1990), Powers and Reed (1993), and Pokrandt et al. (1996) also examined a gravity wave event during a Midwest snowstorm on 15 December 1987, during which several disturbances created "whiteout" blizzard conditions accompanied by surface pressure falls of as great as -11 hPa per 15 min. In the 15 December 1987 case, the gravity waves induced surface pressure minima that were actually lower than the pressure observations at the cyclone center itself. A winter storm in January 1994 (Bosart et al. 1998) was notable for multiple gravity wave events, which spawned some incredibly heavy snows.

Gravity waves have been noted within a number of Northeast winter storms. According to A. Seimon (2002, personal communication; Seimon is a gravity wave analyst who is well known for his ability to detect and analyze gravity waves and predict subsequent impacts in real time), a gravity wave during a snowstorm from Ohio to New England on 19 January 1987 was associated with heavy snowfall rates in advance of the wave, pressure falls of up to 9 hPa in less than an hour, and a precipitation-free region following wave passage during a time when heavy snow was expected. Gravity wave activity was also observed during the January 2000 snowstorm over the Northeast. More detailed examples of snowstorms within the Northeast urban corridor influenced by gravity waves are summarized below.

a. 11 February 1983

During the Megalopolitan Snowstorm of February 1983, a singular gravity wave (Fig. 6-13a) with a propagation rate derived at 15–25 m s^{-1} moved rapidly

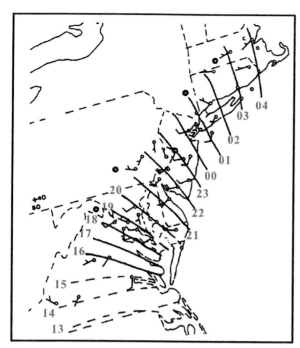

**GRAVITY WAVE MOTION
11 FEBRUARY 1983**

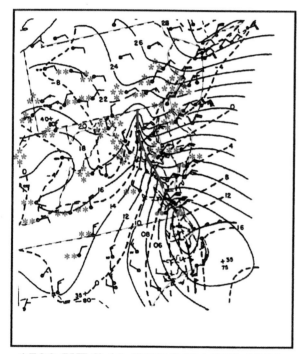

1700 UTC 11 FEBRUARY 1983

FIG. 6-13. (a) Gravity wave isochrone analysis for 1300 UTC 11 Feb–0400 UTC 12 Feb 1983. (b) Manually prepared surface analysis for 1700 UTC 11 Feb 1983. Mean sea level isobars (solid, every 2 hPa). Plotting format of conventional surface observations. [Derived from Bosart and Sanders (1986).]

northeastward through the Northeast urban corridor before accelerating toward northern New England (Bosart and Sanders 1986). Rapid pressure falls and rises accompanied the passage of the wave, increasing to a maximum of 6 hPa h^{-1} pressure falls and 8 hPa h^{-1} pressure rises after 1700 UTC 11 February. A surface analysis (Fig. 7-13b) shows evidence of the wave at 1700 UTC 11 February over northern Virginia with a pronounced pressure trough along the Chesapeake Bay. This gravity wave event was accompanied by an increase in northeasterly winds as pressure fell, followed by their weakening accompanied by some wind shifts to the northwest as the pressure rose. It was also accompanied by very heavy snow and snow pellets at the wave crest. The events were also accompanied by lightning and thunder as the precipitation increased with the approach of the wave crest similar to what has been observed in a midwestern spring convective case (Uccellini 1976). This case is one of the best documented examples of the ability of gravity waves to enhance precipitation rates as a wave *ridge* approaches and to decrease precipitation rates (but enhance wind speeds) as the wave *trough* approaches.

b. 4 January 1994

Another particularly interesting example of a gravity wave influencing the structure of a Northeast snowstorm occurred on 4 January 1994 (Bosart et al. 1998). A high-amplitude gravity wave propagated from North Carolina northeastward through Maine in only 12 h (Fig. 6-14b), propagating at speeds approaching 35–40 m s^{-1}. The gravity wave was accompanied by rapid pressure falls reaching 13 hPa per 30 min, rivaling pressure tendencies observed in landfalling hurricanes. Sea level pressure fell 11 hPa in only 28 min at Providence, Rhode Island, and 9 hPa in only 8 min (!) at Bedford, Massachusetts (Fig. 6-14). The gravity wave was associated with a number of pressure fall–rise couplets that complicated the analysis of surface weather conditions (Fig. 6-14c). As the wave approached, heavy precipitation and increasing wind speeds caused havoc along the Northeast coast. As the wave passed, precipitation ended, producing a "hole" in the overall precipitation shield associated with the cyclone [see satellite imagery in Figs. 12 and 19 of Bosart et al. (1998)]. Numerous reports of rapid rises and falls in sea level pressure along the Northeast coast accompanied this disturbance. Large sea level oscillations (seiches) associated with wave passage occurred in coastal waters off Rhode Island and Maine (Bosart et al. 1998).

4. Mesoscale snowfall related to elevation, thundersnow, and "bay effect" snow squalls

Elevation can also play a key role in the snowfall distribution, exhibited by the late season storm of 9 May 1977 (volume II, chapter 12, Fig. 12.2-9). During some

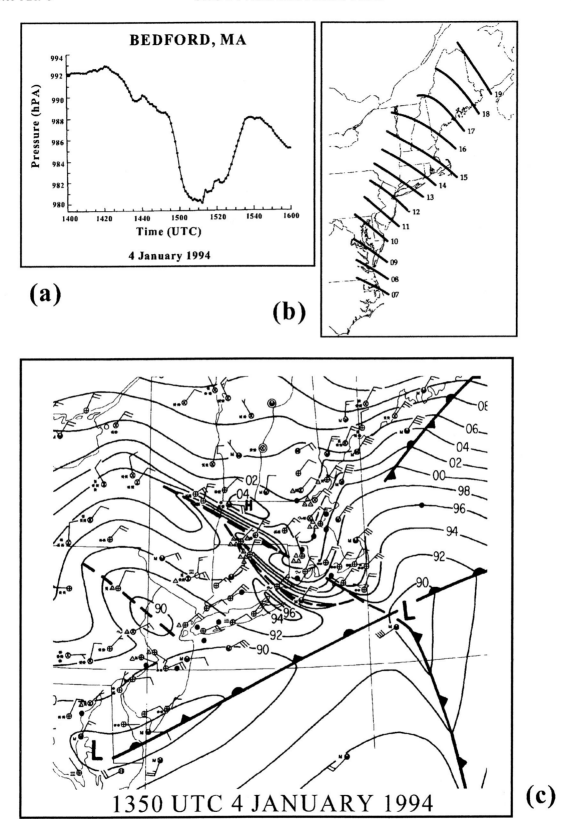

FIG. 6-14. (a) The 1-min pressure vs time (UTC) for Bedford, MA. (b) Gravity wave isochrone analysis for 0700–1900 UTC 4 Jan 1994. (c) Manually prepared surface analysis for 1350 UTC 4 Jan 1994. Mean sea level isobars (solid, every 2 hPa). Plotting format of conventional surface observations. [From Bosart et al. (1998).]

snowfall events, small changes in elevation and local changes in precipitation intensity can spell the difference between a sloppy mixture of rain and snow and rapidly accumulating heavy snowfall. During the snowstorm of 18–21 March 1958, elevation played an important role in enhancing snow amounts across Pennsylvania and Maryland, where 30–50 in. (75–125 cm) fell (volume II, chapter 10.3, Fig. 10-3.1) in the foothills of the Appalachian Mountains, while much smaller amounts were observed at low elevations. During early and late season snowfall, elevation plays a crucial role in snowfall distribution and also plays a role in enhancing (or suppressing) ascent as well.

The occurrence of thunder and lightning is another interesting mesoscale phenomenon that occurs with some heavy snow events. As noted for the February 1983 snowstorm described earlier, some of these events are associated with gravity wave disturbances (Bosart and Sanders 1986; Schneider 1990) but others are not (i.e., Schultz 1999). In Washington 4–6 in. (10–25 cm) of snow fell within a 2-h period on 9 March 1984 with a region of thunderstorms associated with a developing coastal storm [Fig. 6-15; Kocin et al. (1985)]. During the following winter, a summerlike line of thunderstorms on 25 January 1985 was accompanied by heavy snow, leaving 1–2 in. (2.5–5 cm) in the Washington metropolitan area and up to 5 in. (12.5 cm) at Bel Air, Maryland, accompanied by strong wind gusts. The occurrence of lightning and thunder with Northeast snowstorms has been documented for at least 14 of the 30 cases examined in depth, having been reported in at least one of the five major metropolitan airports. (These cases include the snowstorms of March 1956, February 1958, March 1958, March 1960, January 1966, December 1966, February 1972, February 1978, April 1982, February 1983, February 1987, March 1993, January 1996, and December 2000.) The occurrence of thunder and lightning is likely much higher than indicated here. It appears that electrical activity is a common feature in these severe snowstorms, indicating the importance of convection as an integral component of the cyclone that can act to enhance snowfall rates.

The destabilization and moistening of cold air flowing across relatively warm water is an obvious snow mechanism around and downwind of the Great Lakes (i.e., Niziol et al. 1995; Ballentine et al. 1998; Kristovich et al. 2000). An example is the crippling 25-in. (63 cm) snowfall that fell over an 8-h period in Buffalo, New York, on 18 November 2000 and the record 7-ft (210 cm) snowfall over 4 days in late December 2001. The impact of warmer water can also be observed along the East Coast, especially along the Massachusetts coast near Cape Cod, Delaware Bay, and the Chesapeake Bay. On 18 January 1997, a cold, westerly airflow created a narrow snowband just south of the southern coast of New England (Fig. 6-16a). In this case, Nantucket received 8 in. (20 cm) of snow while Edgartown on Martha's Vineyard received only 0.8 in. (2 cm). More typ-

ically, north-northwesterly flow will produce narrow bands of snow developing along the Massachusetts Bay and across portions of Cape Cod with locally heavy snow. Cold air funneling southward along the Chesapeake Bay has also resulted in "bay-effect" snow showers over southeastern Virginia. One such occurrence is shown in radar images presented in Fig. 6-16b on 25 December 1999. In some instances, the snowbands can occur within a larger-scale cyclonic weather system (Fig. 6-17) and result in localized areas of heavy snows, as occurred during the snowstorm of 14 January 1999 over eastern Massachusetts (Figs. 6-17a and 6-17b).

A final example of an even more isolated mesoscale snow event occurred on the night of 6/7 December 1977. A very heavy snow squall produced about 4 in. (10 cm) of snow in less than an hour at the Pennsylvania State University campus in State College, Pennsylvania, an unusually intense snow squall in a location that typically gets no more than an inch (2.5 cm) from "lake effect" snow showers due to its location relatively far removed from the Great Lakes and to the local topography. The following morning, Dover, Delaware, reported 5 in. (12.5 cm) of new snow on the ground. An examination of satellite imagery shows that the same snow squall was responsible for a very narrow snowband less than 20 km in width, spanning these two locations. This snowband is an example of a small mesoscale event, with an areal width bordering on the microscale (see Fig. 6-1) but since it occurred over a period of several hours as it moved from central Pennsylvania to Delaware put this event clearly in the mesoscale. The origin and evolution of this narrow band of snow has not been documented.

5. Other mesoscale considerations

a. The rapid development of snowfall

The accurate prediction of how much and where heavy snowfall will occur is one of the great challenges in operational meteorology today. That challenge is based on the ability to identify, predict, and diagnose the various scales in which snowfall organizes, from snowbands that extend for hundreds of kilometers and can cover thousands of square kilometers, to the snow squall just mentioned, which may be no more than 20 km wide. The Northeast snowstorm of 30 December 2000 (see volume II, chapter 10.30) captures several examples of some of the mesoscale forecast challenges that are associated with Northeast snowstorms. This storm left 12 in. (30 cm) of snow in New York City and more than 25 in. (63 cm) in parts of northern New Jersey. Nearly 10 in. (25 cm) of snow fell in Philadelphia but nearly no snow fell just 20 mi (30 km) to its southwest.

Early numerical weather forecasts of this storm had been pointing to a possible Northeast snowstorm as

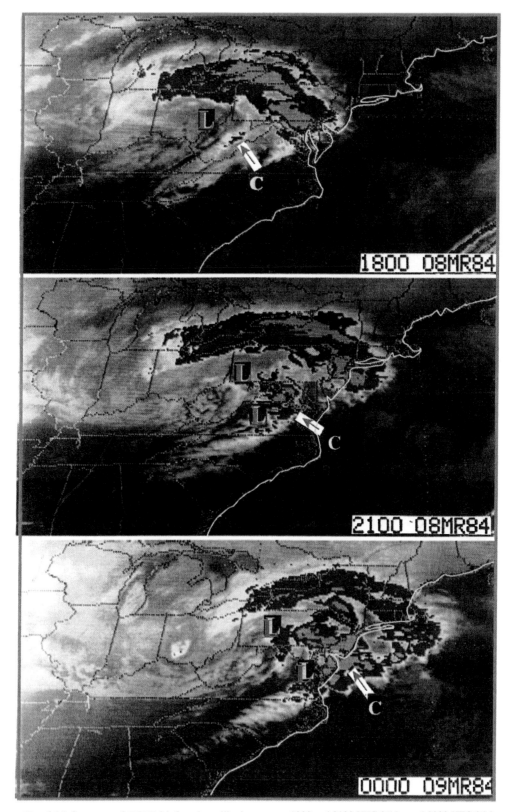

FIG. 6-15. Sequence of *GOES-8* infrared satellite imagery at 1800 and 2100 UTC 8 Mar and 0000 UTC 9 Mar 1984. Arrows point to region of developing convection associated with the secondary development of a surface low pressure system over VA.

18 January 1997

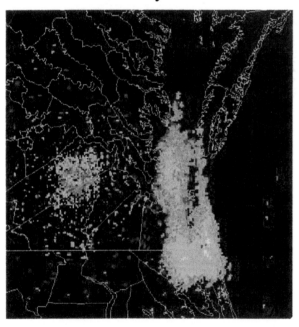

25 December 1999

Fig. 6-16. NEXRAD images of "lake effect"-type snowbands: (a) 18 Jan 1997, clear-air mode reflectivities of snowband off southern New England; and (b) 25 Dec 1999, reflectivities of a Chesapeake "Bay effect" snowband near Norfolk, VA.

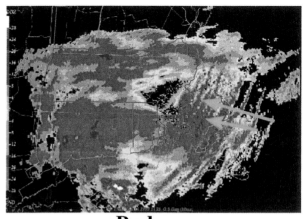

A   Radar

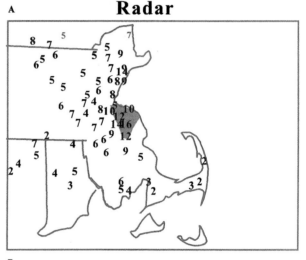

B   Snowfall

Fig. 6-17. Snowfall enhancement due to lake-effect-type snowbands on 14 Jan 1999: (a) clear-air mode reflectivities of snowbands over eastern MA and (b) total snowfall (in.). Shading represents snowfall exceeding 10 in. (25 cm).

much as a week (or more) in advance. As the event drew closer, it appeared more and more likely that a significant low pressure system would develop off the mid-Atlantic coastline. Winter storm watches went up from the middle Atlantic states to New England. A significant Northeast snowstorm did eventually occur but

it was marked by much mesoscale detail that illustrates some of the complexities involved in trying to forecast "synoptic scale" weather systems that always contain mesoscale details that pose significant challenges to the forecast community.

One of the mesoscale considerations with this storm was the sudden onset of cyclogenesis and precipitation. Although most numerical models predicted the onset of cyclogenesis between 0000 and 0600 UTC 30 December 2000, there was little obvious evidence immediately prior to this period of the upcoming cyclogenesis and precipitation development. Between 0000 and 0500 UTC 30 December, cyclogenesis commenced and the surface low deepened rapidly through 1000 UTC 30 December, just southeast of New York City (Fig. 6-18).

A second mesoscale consideration for this case was the rapid development and expansion of the heavy snow. At 0000 UTC 30 December (Fig. 6-18a), some weak radar echoes indicating precipitation aloft are

SURFACE ANALYSES

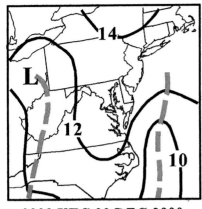 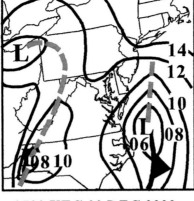 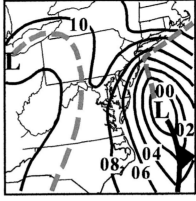

0000 UTC 30 DEC 2000   **0500 UTC 30 DEC 2000**   **1000 UTC 30 DEC 2000**

SURFACE DATA AND RADAR

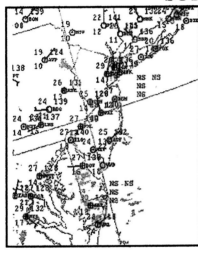 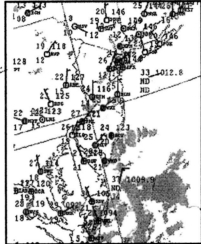 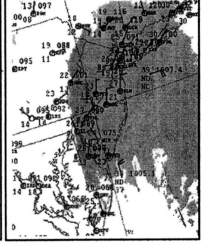

0003 UTC 30 DEC 2000   **0505 UTC 30 DEC 2000**   **1022 UTC 30 DEC 2000**

FIG. 6-18. (top) Surface analyses at 0000, 0500, and 1000 UTC 30 Dec 2000. (bottom) Surface observations and composite radar imagery (gray shading) at 0003, 0505, and 1022 UTC 30 Dec 2000. Times refer to those of the radar imagery.

noted over Virginia and the mid-Atlantic coast but no reports of precipitation reaching the ground. Surface analyses at this time indicated an inverted surface pressure trough off the mid-Atlantic coast but no distinct low pressure center, with lowest pressures around 1009 hPa. Five hours later (Fig. 6-18b), a 1005-hPa low pressure center had now developed east of Elizabeth City, North Carolina, with a broken band of precipitation forming east of the mid-Atlantic coast. At this time, there were a few reports of light freezing rain now reaching the surface in portions of southeastern Virginia and the eastern shore of Maryland. Within the next 5 h (Fig. 6-18c), the precipitation organized and expanded rapidly northward across New Jersey and southeastern New York by 1000 UTC. In this last 5-h period, most reporting stations in New Jersey went

from cloudy skies to heavy snow, with heavy snow also falling in New York City and its northern suburbs, as well. The surface low was now located east of the Delmarva Peninsula coast with an estimated central pressure having fallen to 998 hPa.

In this case, the critical period for the forecast occurred between 0500 and 1000 UTC, when a small, broken area of precipitation quickly organized and intensified over a region home to more than 15 million people. Remarkably, several weather forecast models appeared to predict the timing and location of the cyclogenesis with a considerable amount of skill although the rapid expansion of the snowfall was not forecast nearly as well. The rapidity with which the snowfall expanded and intensified are important mesoscale aspects of this case because in those few hours, crippling snowfall rates quickly impacted millions of lives, fortunately, while most people slept.

A third mesoscale aspect of this storm was the exceptionally sharp transition zone between no snow and heavy snow (Fig. 6-19). This zone was so sharp that residents of Washington, Baltimore, and much of eastern Pennsylvania awoke to a sunny, calm, frosty morning, despite predictions of heavy snow. This tranquil weather stood in stark contrast to the heavy snow falling in nearby Philadelphia and in New York City. The transition zone between heavy snow and little or no snow occurred across the Philadelphia metropolitan area where the western suburbs saw little or no snow while portions of the eastern and northern suburbs reported as much as a foot (30 cm) of snow.

A fourth mesoscale aspect of this storm, and also impacting many other storms, provided an additional challenge for forecasters: the rain–snow line. This feature was observed later in the storm's evolution over much of New England and spelled the difference between more than a foot (30 cm) of snow and much lighter snowfall accumulations. Much of central and eastern New England was spared heavy accumulations because of a changeover from snow to either mixed precipitation or rain.

These critical mesoscale "transition zones" between heavy and no snow, as well as between snow, mixed precipitation, and rain can represent some of the most significant mesoscale forecast considerations during any given winter event. As shown in Fig. 6-20, the incredible variety of precipitation types, changeovers, and combinations during a given winter storm poses tremendous challenges to forecasters. The amount of detail inherent in these storms is complex, and the ability to diagnose and forecast the subtle changes that can have profound effects on whether a given area receives a crippling snowstorm, freezing rain, rain, or no precipitation at all is a task requiring some extremely sophisticated numerical predictions and an enhanced appreciation on the part of forecasters of the mesoscale character of these storms that make or break any forecast. Part of the challenge in predicting where the heaviest snow will occur is often related to predicting the sharp cutoff of snow at the northern and western edge and where the rain–snow transition region will be.

b. Northern and western edge of heavy snow

As noted above, the northern or western edge of the heavy snows often has a sharp cutoff between heavy snowfall and little or no snowfall. During the snowstorm of January 1961 (Fig. 6-21a), several localized regions of snowfall accumulations exceeding 20 in. (50 cm) occurred, mostly toward the northern edge of the larger-scale area of heavy snow. Another example is the interior snowstorm of March 1994 (Fig. 6-21b), which shows a classic synoptic-scale snowband extending from western Virginia northeastward to central New

York. However, the region of heaviest snows (from 20 to more than 30 in.; 50 to 75+ cm) is found very close to the northwestern edge of the heavy snows rather than symmetrically in the center of the band. The 30 December 2000 snowstorm also displayed a band of very heavy snow along the northwestern edge of the precipitation shield, with a sharp gradient between heavy snow and no snow. As mentioned earlier, the Philadelphia metropolitan area was located within this sharp boundary, with heavy snows in the city and little snow in the western suburbs.

This relationship between the locations of the heaviest snows sometimes along the northwestern edge of the precipitation shield lends to some very difficult forecast situations in which forecasters must decide whether there will be heavy snowfall or none at all. For example, a major cyclone may develop well offshore and forecasters can have a mesoscale forecast problem with predicting where and whether the western edge of the heavy snowband will become established onshore or offshore. Suddenly, a forecast of no snow will change to a forecast of heavy snow with a small shift of the precipitation. Portions of eastern Long Island and southeastern New England can be affected, with heaviest snows from Boston to Cape Cod and the islands of Martha's Vineyard and Nantucket, or across eastern Maine.

Two significant New England snowstorms are shown in Figs. 6-22a and 6-22b as examples. The storm of 28 February 1952 has been described as the most severe to strike Nantucket, with hurricane force winds and more than 20 in. (50 cm) of snow. Significant snows fell as far west as Boston and Providence but diminished rapidly just west of these locations. A similar storm brought crippling conditions to Cape Cod on 9 February 1987. The snowstorm of 16-17 December 1981 (Fig. 6-22b) is another example of heavy snow along the northwestern edge of the precipitation band associated with a strong offshore cyclone (see Sanders 1986) that was originally forecast to remain too far offshore to produce heavy snow. Heavy snows backed into eastern New England, producing an unexpectedly heavy snowfall. On 24 February 1989, the heart of the Northeast urban corridor from New York City to Washington was faced with the threat of heavy snow as a slow-moving cyclone moved northeastward just east of the Atlantic coast. Blizzard conditions raged across eastern Virginia, the Delmarva Peninsula and eastern New Jersey, with up to 19 in. (48 cm) in southeastern New Jersey (Fig. 6-22c). However, a tremendous gradient in snowfall existed along the western edge of this snowband, and the metropolitan areas of Washington, Baltimore, Philadelphia, and New York escaped with no snow despite heavy accumulations only 100 km to their east-southeast. The snowstorm of 26 February 1999 (Fig. 6-22d) is a final example of an offshore storm where forecasters were unsure whether it would remain offshore or clip eastern New England. The western edge of the storm left 12 in. (30 cm) of snow or greater from the eastern tip of

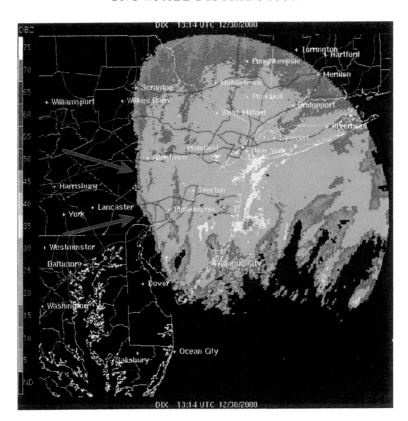

1314 UTC 30 December 2000

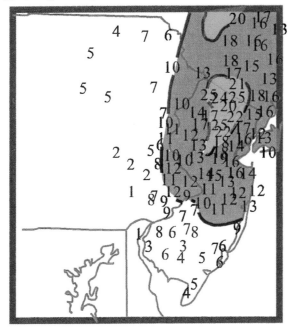

Total Snowfall

FIG. 6-19. (top) NEXRAD base reflectivity at 1314 UTC 30 Dec 2000 showing the sharp western edge of the snowband. (bottom) Total snowfall acccumulation [in.; blue represents 10 in. (25 cm) or greater; green represents 20 in. (50 cm) or greater] following the storm.

PRECIPITATION TYPE
HYPOTHETICAL WINTER STORM

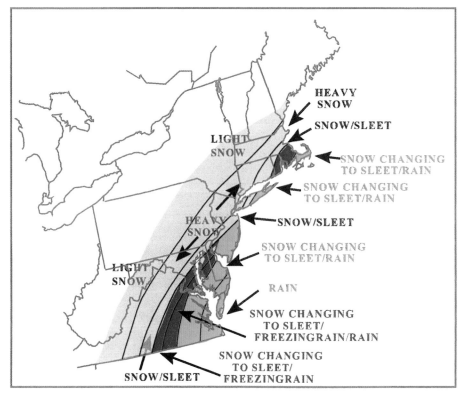

FIG. 6-20. Schematic of precipitation type variability for a hypothetical Northeast winter storm.

Long Island through Cape Cod where up to 20 in. (50 cm) fell. These cases are all examples of intensifying offshore cyclones with the northwestern edge of the snowfall clipping the coastline.

During the New England snowstorm of 6–7 February 1978 (see volume II, chapter 10.17, Fig. 10.17-1), snow expanded westward around an intensifying upper-level cutoff low. The western edge of the snow proved to be a mesoscale forecast problem for the Baltimore–Washington area. In Washington and to the west, accumulations were an inch or two (2.5–5 cm) at most, while the near-eastern suburbs measured 6 in. (15 cm) of wind-blown snow; Baltimore measured 9 in. (23 cm) and up to 18 in. (45 cm) fell near the Chesapeake Bay. The January blizzard of 1996 (volume II, chapter 10.27, Fig. 10.27-1) and the Megalopolitan Snowstorm of 11 February 1983 (volume II, chapter 10.20, Fig. 10.20-1) both exhibited sharp northern edges to the heavy snows. For example, Albany, New York, measured only an inch of snow during the 1996 blizzard while locations only 50 km south measured 20 in. (50 cm). In both cases, numerical model forecasts initially underforecast the northward expansion of the northward edge of the heavy snow, resulting in forecast underpredictions in what were generally two fairly predictable events.

Nicosia and Grumm (1999) noted that bands of heavy snow are often located on the northwestern edge of the synoptic region of snowfall associated with these snowstorms and further contribute to the sharp boundary between no snow and very heavy snowfall concentrated in a narrow band. A combination of frontogenesis found in a region characterized by negative values of equivalent potential vorticity, combined with low values of conditional symmetric stability, and a related frontal circulation regime focus a vertical ascent pattern near the northwest boundary of the larger snow shield associated with the synoptic surface low. In another investigation of heavy precipitation along the northwestern edge of a storm's precipitation field, Martin (1999) used a numerical simulation to examine the role of frontogenesis and quasigeostrophic forcing of heavy precipitation in the northwestern sector of several cyclones, including the 1 April 1997 New England snowstorm (see volume II, chapter 10.28). Martin showed that the precipitation in the northwestern sector of the storms (sometimes called wraparound precipitation) originated in the warm sector of the storm and was associated with a ridge of relatively warm air aloft that had a sharp northwestern edge enhanced by distinct deformation patterns. The sharp northwestern edge may align with the sharply rising western boundaries of the warm and

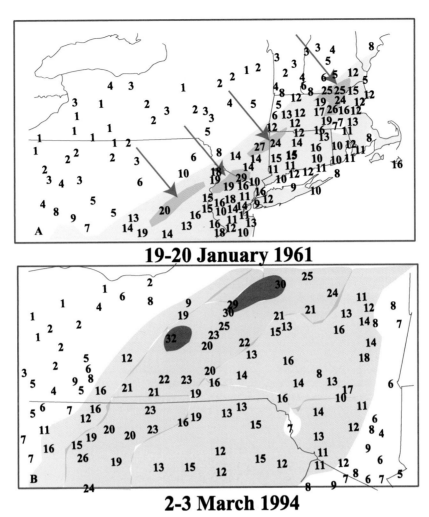

19-20 January 1961

2-3 March 1994

Fig. 6-21. Examples of heaviest snow located along the northwestern edge of the synoptic-scale snowband: (a) 19–20 Jan 1961 and (b) 2–3 Mar 1994.

cold conveyor belts where strong, confluent flow is found.

c. Evolution of the rain–snow line

Forecasters are rarely presented with a straightforward situation in which one form of precipitation will be the only form of precipitation during the storm event (i.e., Keeter and Cline 1991). There often are distinct areas that receive either rain or snow, but in terms of the transitional region it is often difficult to observe, model, and forecast the changing vertical temperature and moisture profiles during a winter storm that can result in small transitional regions of mixed rain and snow, ice pellets, and freezing rain. Other storms may be dominated by large areas of mixed precipitation or precipitation that changes phase over time, as shown in Fig. 6-20.

Precipitation type is related to the melting behavior of snow and the depth of warm layers in the lower troposphere that overlay colder temperatures near the surface (Stewart 1985). The evolution, location, and width of transition zones between rain, mixed precipitation, and snow are critical concerns in forecasting heavy snow. A 20–100-km difference between the forecast and actual location of the rain–snow line has a tremendous impact on the millions of residents of the Northeast urban corridor. Critical changes in temperatures over small distances and over relatively small vertical layers can mean the difference between mixed rain and snow or a crippling snowfall. As an example, the lead author deliberately drove his car through a transition zone of a mixed-precipitation winter storm in December 1996 in Maryland over a period of several hours to observe its spatial and temporal evolution. In this instance the transitional region, which separated mostly rain from accumulating snow, consisted of rain, snow, and a few ice pellets and was no greater than 20 km in width. In addition, this transitional area oscillated slowly northward and southward over a distance of 3–4 km

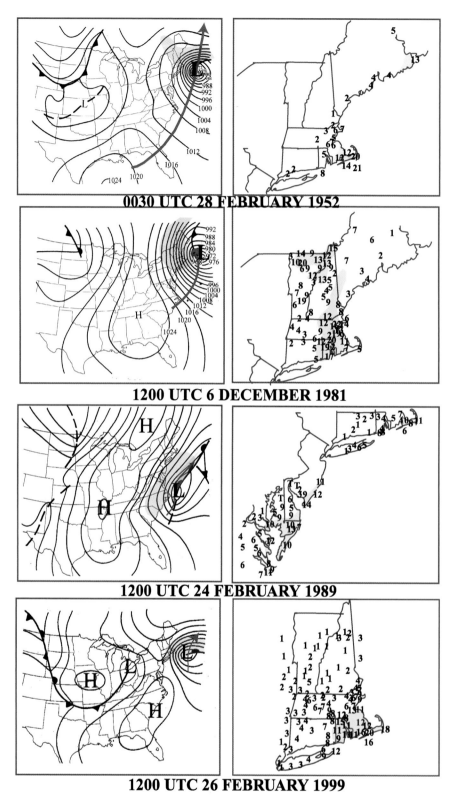

0030 UTC 28 FEBRUARY 1952

1200 UTC 6 DECEMBER 1981

1200 UTC 24 FEBRUARY 1989

1200 UTC 26 FEBRUARY 1999

FIG. 6-22. Examples of snowstorms in which the northern and western edge of the storm presented forecast challenges to the major metropolitan centers. (left) Surface analyses at 0030 28 Feb 1952, 1200 UTC 6 Dec 1981, 1200 UTC 24 Feb 1989, and 1200 UTC 26 Feb 1999. (right) Total snowfall for each event [shading: accumulations exceeding 10 in. (25 cm)].

over a period of a couple hours. An extreme example of a transitional zone was found with the early season coastal storm of 25–26 November 1971, an "interior snowstorm" (see chapter 5 and volume II, chapter 11.1). This storm was characterized by a report (NOAA 1971) of only 7 mi (11 km) separating more than 12 in. (30 cm) of snow from no accumulation on a line from northwestern Rhode Island through eastern Maine.

Snow is associated with below freezing temperatures throughout the entire troposphere, although temperatures may be above freezing for a small depth near the surface. In general, the aggregation of snow crystals is temperature dependent with high aggregation rates near 0°C (Stewart et al. 1984). Therefore, larger flakes are expected as temperatures approach freezing. The depth of above freezing temperatures must be sufficient to melt all the snow before it reaches the surface. Because the latent heat of fusion is extracted from the air during melting, an isothermal layer at 0°C is usually present through a depth of 200–300 m (Stewart et al. 1984). The boundary between rain and snow is characterized by an isothermal 0°C layer about 1 km deep (Stewart 1985).

As discussed in chapter 5, freezing rain occurs when snow melts completely in an upper inversion before falling into a lower subfreezing region. Temperatures in the lower region are not low enough or the layer is not deep enough to allow refreezing to ice pellets to occur. Ice pellets occur when snow partially melts and refreezes before reaching the surface (see Fig. 5-24 for examples of soundings taken during freezing rain and ice pellet events). Often, the regions experiencing a transition between one or another precipitation type occurs over a relatively small domain and is therefore a mesoscale forecast problem. Stewart and King (1987) determined the width of the transition regions for two separate storms, including one whose width varied between approximately 40 and 65 km while the other case exhibited a transition zone width as small as 11 km. Stewart (1992) also reports cases in which the transition region remained over one location for as long as 9–12 h.

Occasionally, rain may unexpectedly change to snow due to evaporative cooling [Fig. 6-23; see Homan and Uccellini (1987)] or when melting of snowflakes becomes a dominant process in atmospheric cooling (Wexler et al. 1954; Bosart and Sanders 1991; Kain et al. 2000). Early on 22 February 1986, forecasters in the Washington area pondered not only whether there would be precipitation but also whether the precipitation would fall as rain or snow. During the afternoon of 22 February, surface temperatures approached 3°C while early morning soundings at Washington Dulles International Airport (IAD) indicated that 850-hPa temperatures were 2°C. With such information, the rain versus snow forecast was a difficult one since forecasters needed to determine if the relatively mild air at 850 hPa could be cooled sufficiently by evaporative cooling to produce snow rather than rain. With rawinsonde observations

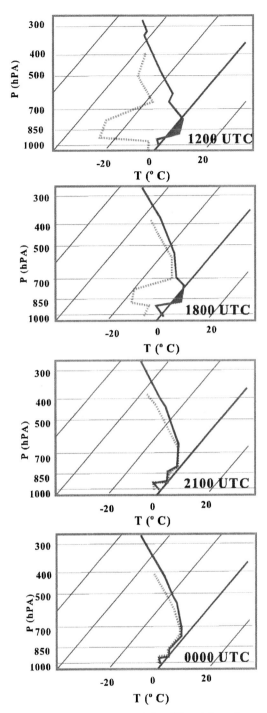

FIG. 6-23. Temperature (solid red line, °C) and dewpoint temperature (dashed green line, °C) for station IAD at (a)1200, (b) 1800, and (c) 2100 UTC 22 Feb 1986, and (d) 0000 UTC 23 Feb 1986. Soundings are plotted on a skew T–log p diagram, which includes pressure (horizontal blue line, hPa) and temperature (angled violet line, °C). [From Homan and Uccellini (1987, their Fig. 17).]

Comparison of old (WSR57) vs new (WSR-88D) Radar

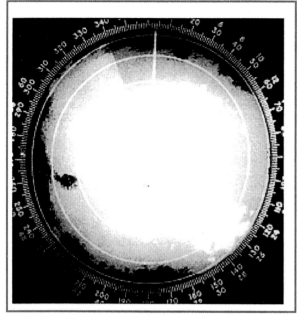

13 March 1993
WSR-57 Radar Patuxent, MD

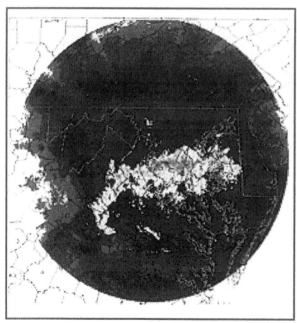

13 March 1993
WSR-88D Radar Sterling, VA

FIG. 6-24. Radar depiction of snowfall on 13 Mar 1993 as seen from (a) a WSR-57 radar at Patuxent, MD, and (b) a WSR-88D radar from the Sterling, VA, National Weather Service office.

available only every 12 h, it would have been impossible to observe vertical temperature changes. However, datasets available from the Genesis of Atlantic Lows Experiment (GALE; Dirks et al. 1988) contained 3-hourly soundings at various operational sounding sites, including Dulles airport near downtown Washington (Fig. 6-23) that show the effect of evaporation on initially dry air below 700 hPa. As the initial precipitation fell through this layer and evaporated, the air cooled from initial readings of 2°–3°C to a temperature just below 0°C as the entire column became saturated. The availability of soundings at 3-h intervals was crucial in monitoring critical temperature and moisture changes that influenced the changeover from mixed rain and ice pellets to moderate and heavy snow.

Cooling due to melting snowflakes also may be sufficient in marginal rain versus snow situations to allow heavy wet snow to fall, rather than rain. This factor was significant during the early season snowstorm of 3–4 October 1987 [see Bosart and Sanders (1991); also see volume II, chapter 12.1), as well as a high-elevation Appalachian snowfall in February 1998 (Kain et al. 2000).

6. Concluding remarks: Impact of NEXRAD radars

The deployment of the Next-Generation Doppler Radar (NEXRAD) in the early 1990s (Crum and Alberty 1993; Friday 1994) has greatly improved the ability to observe the mesoscale structure inherent in snowstorms and has provided the forecasters an important tool to describe the mesoscale character of the snowfall as it is evolving and to make more accurate short-term forecasts. A comparison of reflectivity patterns from the old radars and the NEXRAD radar is shown in Fig. 6-24 and illustrates this advance. As a result, more detailed forecasts and related updates are presently issued to the public that account for short-term changes in snowfall rate based on radar observations. The radars have provided the technology to observe the presence of multiple snowbands that appear to characterize many snowstorms, as well as other structures, including gravity wave phenomena. Recognition of the many ways that snowfall appears on the radars ultimately will provide more insight into the mechanisms that generate them. In any event, the examples provided in this chapter illustrate how the deployment of the NEXRAD Doppler radar system has allowed forecasters (and the general public) to actually see the structure of the snowfall and better anticipate the mesoscale structure of the snowfall and to forecast exactly where the heaviest snowfall is occurring. This new ability to observe the mesoscale structure of Northeast snowstorms will allow the forecast community to better describe and predict these events as well as contribute to our understanding of their formation mechanisms. In the following chapter, the many physical processes that contribute to the mesoscale and synoptic-scale development of Northeast snowstorms are discussed.

Chapter 7

DYNAMICAL AND PHYSICAL PROCESSES INFLUENCING
NORTHEAST SNOWSTORMS

Analyses of the surface pressure, and upper-level geopotential height, wind, temperature, and moisture, plus other selected diagnostic fields described in the preceding chapters, all provide a framework for understanding the many processes that contribute to heavy snowfall in the northeast United States. The complex interactions of cyclonic and anticyclonic weather systems, upper-level troughs, ridges, and jet streaks, influenced by mountains, coastal boundaries, and ocean currents, point to a wide array of dynamical and physical processes that are critical for the development of Northeast snowstorms (see Table 7-1).

In this chapter, brief descriptions of the relevant processes are provided to show how each influences the development of these storms. Although the processes are treated separately to identify their individual contributions, the discussion then focuses on their nonlinear, synergistic interactions, which are crucial for generating the organized regions of heavy snowfall observed in many of these cases. Then, the relationships between upper-level trough–ridge systems and jet streaks with cyclogenesis from a potential vorticity perspective are reviewed. The chapter concludes with a discussion of Sutcliffe's "self-development" concept, a basis for describing the interactions of dynamical and physical processes that contribute to rapid cyclogenesis often associated with heavy snowstorms.

1. Cyclogenesis and upper-level processes

The analyses of the 30 snowstorms highlighted in chapters 3 and 4 emphasize the dominant role of surface cyclogenesis. The development of cyclonic disturbances at sea level requires upper-level divergence to produce a net reduction of mass and an associated reduction in sea level pressures, in a region where the low-level wind field is generally convergent. The net difference between the divergence aloft and the low-level convergence is a measure of the rate of surface cyclonic or anticyclonic development, as deduced by Dines (1925), emphasized by Sutcliffe (1939, 518–519), and later used to formulate a "development equation" (Sutcliffe 1947) that was further refined by Petterssen (1956). Bjerknes and Holmboe (1944) related the vertical distribution of divergence necessary to sustain cyclogenesis to the presence of upper-level trough–ridge patterns, with divergence aloft, convergence near the earth's surface, and a level of nondivergence in the middle troposphere downstream of the wave trough axis (Fig. 7-1a). A reversal of this pattern contributes to anticyclogenesis upstream of the trough axis.

The recognition that upper-level divergence is necessary for surface cyclogenesis was coupled with the requirement for a significant ageostrophic wind component (Sutcliffe 1939; Bjerknes and Holmboe 1944). Bjerknes and Holmboe related the pattern of upper-level divergence associated with upper-level trough–ridge systems in the geopotential height field to "longitudinal" or along-stream ageostrophic wind components resulting from cyclonic and anticyclonic curvature. Subgeostrophic flow at the base of a trough and supergeostrophic flow at the crest of a ridge combine to enhance the divergence (convergence) downstream (upstream) of the trough axis (Fig. 7-1a).

Bjerknes (1951) also discussed the likely contribution from the "transverse" or cross-stream ageostrophic wind components associated with upper-level wind maxima known as jet streaks (see Palmén and Newton 1969, p. 199). Jet streaks embedded within the general jet stream pattern enhance and focus upper-level divergence and their resultant vertical motion fields (Fig. 7-1b) and also contribute to surface high and low pressure cells in the exit and entrance regions of the jets (Bjerknes 1951; Riehl et al. 1952; Reiter 1969; Uccellini and Kocin 1987; and others).

Meteorologists have long recognized that measuring the divergence profile associated with cyclogenesis was impractical (see, e.g., Austin 1951, p. 632; Petterssen 1956, p. 294), since upper-level wind measurements were not sufficiently accurate to diagnose the relatively small difference between the upper- and lower-level divergence that governs surface pressure falls associated with cyclogenesis. Since divergence was difficult to determine due to unreliable and scarce wind measurements, kinematic and dynamical diagnostics were derived to approximate the divergence by applying the vorticity and thermodynamic equations. These approaches related sea level cyclone development to cyclonic vorticity advection in the middle and upper troposphere (see Sutcliffe 1939, 1947; Holton 1979, 119–

TABLE 7-1. Meteorological phenomena involved in the development of Northeast snowstorms. Scales are defined as synoptic (24–48 h; >=2000 km) and large mesoscale or meso-α (3–24 h; 100–2000 km).

Meteorological phenomena and *scale*	Important processes for the development of Northeast snowstorms
Upper-level trough-ridge pattern *Synoptic*	Provides source of vorticity and kinetic energy, upper-level divergence, and associated ascent; amplification of this feature often marked by decreasing wavelength, increasing amplitude, and the development of diffluence downwind of a "negatively tilted" trough axis
Upper-level jet streaks *Meso-α*	Increased upper-level divergence, kinetic energy, and vorticity and advection patterns; combination of jet streaks is an important factor in juxtaposing cold and relatively warmer/moist air masses and enhances and focuses ascent patterns required for heavy snows
Processes in the lower troposphere, interactions with the Appalachians, cold-air damming, coastal frontogenesis [*meso-α* (and smaller)]	Focuses low-level baroclinic zone, ascent, and temperature advections in a narrow region along the coast; coastal front enhances precipitation rates
Low-level jet *Meso-α*	Increases the transport of moisture into region of heavy snow, warm-air advection, and associated ascent pattern; plays a significant initial role in focusing mass flux divergence contributing to the cyclogenesis along the coast
Cold anticyclone extending across the eastern Great Lakes, southern Canada, or New England north of the incipient low *Synoptic*	Provides cold air required to maintain snow along the coast, a process augmented by upper-level confluent flow over the Northwest and cold-air damming east of the Appalachian Mountains
Sensible and latent heat fluxes *Meso-α* (and smaller)	Warms the lower troposphere by reducing the static stability of the planetary boundary layer over the ocean; provides a continuing source of moisture for heavy precipitation over a large area
Latent heat release *Meso-α*	Provides extra heat for intensifying cyclogenesis and also enhances the dynamic processes and associated deepening rates of the surface cyclone through the self-development process

143), which could be readily diagnosed from the geopotential height fields. Using this method, the divergence can be inferred from patterns of vorticity advections in upper troughs and ridges (Palmén and Newton 1969, 316–318), where maxima and minima in the vorticity fields are typically found (Fig. 7-1c). The contribution of jet streaks to the divergence could also be inferred from the vorticity-advection patterns associated with the enhanced wind shears on the cyclonic and anticyclonic sides of the jet (Fig. 7-1d).

Charney (1947) recognized the importance of Bjerknes and Holmboe's work relating upper-level troughs to surface cyclogenesis, but also noted that many upper-level troughs are observed without surface cyclogenesis. Charney (1947) and Eady (1949) introduced the concept of baroclinic instability, which identified preferred wavelengths of trough–ridge systems that were likely to produce cyclonic development and distinguished between troughs that induce surface cyclogenesis from those that do not. The baroclinic instability theory also points to factors other than upper-level troughs that help to explain why, at certain wavelengths, the upper-level trough–ridge systems amplify, with corresponding surface cyclonic development. The thermal structure and related advection patterns, with cold-air advection upstream of a trough and warm-air advection downstream of the trough are noted as important elements in the transformation of eddy potential to eddy kinetic energy required to produce the amplifying wave systems normally associated with surface cyclogenesis.

With an approach based on the development equation,

Sutcliffe and Forsdyke (1950) and Petterssen (1955, 1956) offered the rather simple concept that cyclone formation is favored when an upper-level trough, marked by a cyclonic vorticity maximum and cyclonic vorticity advections, approaches a frontal system or low-level baroclinic zone, where thermal advections are large. Sutcliffe and Forsdyke hypothesized that the low-level thermal advection pattern could act to enhance the upper-level vorticity advections, and thus the divergence, in a manner that they termed self development (also discussed by Palmén and Newton 1969, 324–326). The vorticity and thermal advection concepts have since been combined within the quasigeostrophic framework (see, e.g., Holton 1979, chapter 7) and have been related to the horizontal structure of tropospheric waves and vertical circulation patterns that constitute cyclogenesis. More recently, Keyser et al. (1989) and Loughe et al. (1995) have devised and applied diagnostic techniques that isolate the longitudinal and transverse circulations associated with trough–ridge patterns and jet streaks, respectively, that can be linked directly to the vertical motions associated with cyclogenesis. This "ψ vector" approach has shown that both the longitudinal and transverse components make important contributions to the vertical circulation associated with cyclogenesis.

a. Surface cyclogenesis and upper-level trough–ridge patterns

Surface cyclogenesis associated with Northeast snowstorms involves either a primary low pressure center

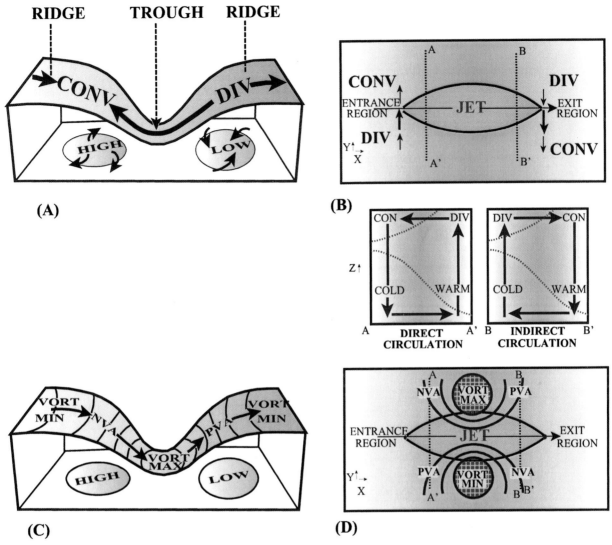

FIG. 7-1. (a) Schematic relating the along-stream ageostrophic wind at upper levels to patterns of divergence associated with an idealized upper-level wave and surface high and low pressure couplets (after Bjerknes and Holmboe 1944). (b) Schematic of transverse ageostrophic wind components and patterns of divergence associated with the entrance and exit regions of a straight jet streak (after Bjerknes 1951). Vertical cross sections illustrate vertical motions and direct and indirect circulations in the entrance region (line A–A') and exit region (line B–B') of a jet streak. Cross sections include two representative isentropes (dotted), relative positions of cold and warm air, upper-level divergence, and horizontal ageostrophic wind components within the plane of each cross section. (c) Schematic of maximum (cyclonic) and minimum (anticyclonic) relative vorticity centers and advections (nva, negative or anticyclonic vorticity advection; pva, positive or cyclonic vorticity advection) associated with an idealized upper-level wave. (d) Schematic of maximum and minimum relative vorticity centers and associated advection patterns associated with a straight jet streak.

that develops near the Gulf of Mexico or a secondary low pressure center that develops along the Southeast or Middle Atlantic coast (Fig. 7-2) and tracks northeastward approximately 100–300 km offshore. These cyclone systems encompass a wide range of processes throughout the depth of the atmosphere that 1) juxtapose warm and cold air, 2) entrain huge amounts of water vapor into the regions of precipitation, and 3) organize, focus, and enhance ascent, all of which are necessary for the production of heavy snow. The snowstorms also require conditions that enhance low-level baroclinicity near or immediately along the coastline and maintain

cold air over the coastal plain so that snow rather than rain falls within the urban areas. Thus, an environment conducive to heavy snowfall is usually not provided solely by the presence of a cyclone, but also requires an interaction with a cold anticyclone of Canadian origin, poised to the north and northwest of the developing storm system. Furthermore, the evolution of the cyclones that produce these snowstorms is linked to upper-level trough–ridge and embedded jet streak patterns that evolve in a manner consistent with the baroclinic instabilities and self-development concepts noted above.

Although the evolution of the trough–ridge config-

CYCLONE EVOLUTION

GULF OF MEXICO ATLANTIC COAST

ATLANTIC COASTAL REDEVELOPMENT

T

T

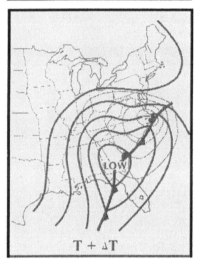

T + ΔT

T + ΔT

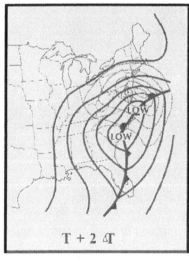

T + 2 ΔT

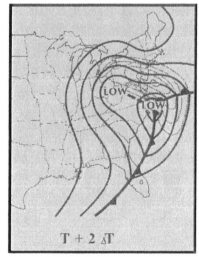

T + 2 ΔT

uration varied from one storm to another (see chapter 4), 28 of the 30 cases displayed an increase of amplitude, and all 30 cases showed a decreasing half-wavelength between the trough and downstream ridge axes during cyclogenesis (Fig. 7-3a). Both of these characteristics are indicative of nonlinear processes that increase upper-level divergence during surface cyclogenesis (see, e.g., Sutcliffe and Forsdyke 1950; Palmén and Newton 1969, 324–326). Furthermore, a marked diffluence of the upper-level height contours downstream of the trough axis (Fig. 7-3b) was displayed in all 30 cases, which is another signature of increasing upper-level divergence over the developing cyclone (Scherhag 1937; Bjerknes 1954; Palmén and Newton 1969, 334–335). This pattern was especially pronounced during the period of most rapid cyclogenesis and the development of heavy snowfall. The diffluence of the height contours increased dramatically as the trough axes assumed a northwest to southeast orientation, or what is commonly referred to as a negative tilt (Fig. 7-3b).

All of the upper troughs contained one or more cyclonic vorticity maxima, producing the vorticity advection patterns that are commonly used to infer upper-level divergence. Although the vorticity advections were not directly evaluated for each case, the increasing amplitudes and decreasing wavelengths of the trough–ridge systems suggest that cyclonic vorticity advection was increasing over the East Coast during cyclogenesis in many, if not all, cases. The vorticity maxima tracked eastward or northeastward across the middle Atlantic states in most of the cases. Heavy snowfall is much less likely when the vorticity maxima propagate to the west or north of the region since the lower-tropospheric warm-air advection associated with these situations erodes the cold air, producing rain, freezing rain, or ice pellets, rather than snow (as is shown in chapter 5).

The merging of multiple vorticity maxima into one cohesive vorticity maximum is known to play an important role in some major cyclone events (see, e.g., Hakim et al. 1995, 1996; Kocin et al. 1995; Bosart, et al. 1996; Bosart 1999, chapter 4). In this sample of Northeast snowstorms, wave mergers are noted in nearly half the cases and act to strengthen the cyclonic vorticity advection and upper-level divergence, thereby increasing cyclone development. The capacity of upper-level troughs to either merge or "fracture" may be linked to the large-scale background flow and evolution (Thorncroft et al. 1993; Bosart and Bartlo 1991; Hakim et al. 1995; Bosart 1999) through a variety of lateral and vertical interactions (Bosart 1999). The dilemma in deter-

mining whether separate troughs and associated vorticity features will merge or "phase together" remains a critical concern to forecasters who must predict surface cyclone development and heavy snowfall using numerical guidance, which displays varying degrees of skill in predicting these interactions (Bosart et al. 1996).

b. Upper-level jet streaks

The relatively straightforward characterization of the upper-level troughs in the 30-storm sample can be extended to the description of the structure and evolution of jet streaks during individual storms. The presence of jetlike, finite-length wind maxima is known to enhance upper-level vorticity advections and divergence, which also contribute to surface cyclogenesis (e.g., see Bjerknes 1951; Reiter 1963, 1969; Palmén and Newton 1969; Uccellini and Kocin 1987; Loughe et al. 1995).

In addition to the increasing amplitudes and decreasing wavelengths of upper trough–ridge systems, "transverse," or cross-stream, ageostrophic components in the entrance and exit regions of upper-level jet streaks also contribute to divergence aloft. Namias and Clapp (1949), Bjerknes (1951), Murray and Daniels (1953), Uccellini and Johnson (1979), and others found that the entrance region of an idealized jet streak is marked by a transverse ageostrophic component directed toward the cyclonic shear side of the jet (see Fig. 7-1). This component represents the upper branch of a direct transverse circulation that converts available potential energy into kinetic energy for air parcels accelerating into the jet. The direct circulation is marked by rising motion on the anticyclonic or warm side of the jet and by sinking motion on the cyclonic or cold side of the jet (Fig. 7-1b). In the exit region, the ageostrophic component is directed toward the anticyclonic shear side of the jet (Fig. 7-1b), representing the upper branch of an indirect transverse circulation (Fig. 7-1b) that converts kinetic energy to available potential energy as air parcels decelerate upon exiting the jet. This circulation features rising motion on the cyclonic or cold side of the jet and sinking motion on the anticyclonic or warm side of the jet. These transverse circulation patterns are consistent with the vorticity advection concepts described by Riehl et al. (1952) and illustrated in Fig. 7-1d.

The influence of curvature effects in masking the contribution of jet streaks to upper-level ageostrophy and divergence, and their associated vertical motion fields, is discussed in studies by Shapiro and Kennedy (1981), Newton and Trevisan (1984), Uccellini et al. (1984),

---

←

FIG. 7-2. Schematic representations of cyclogenesis associated with Northeast snowstorms. (left) "Gulf–Atlantic cyclones," in which a cyclone develops and moves northeastward from the Gulf of Mexico along the Atlantic Coast. Occasionally, new low pressure centers develop farther northeastward along the path of the cyclone along the Atlantic coast and are termed center jumps. (right) "Atlantic coastal redevelopments" occur as a primary cyclone nears the Appalachian Mountains, fills, and a secondary, new low pressure center forms along the Atlantic coast. Solid lines are isobars and thin dotted lines represent isallobars for regions of greatest pressure falls.

Evolution of a diffluent, negatively-tilted
500 hPA trough

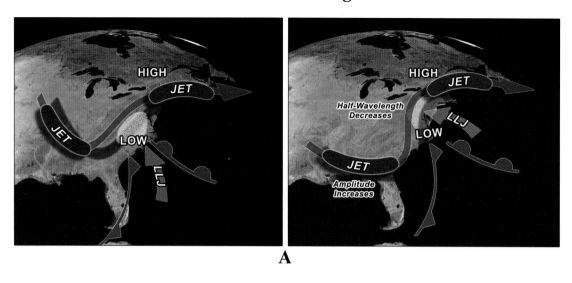

A

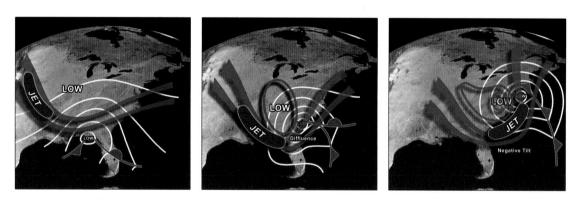

B

FIG. 7-3. (a) Sequence of upper troughs, upper jet streaks, surface frontal and pressure systems, and low-level jet streaks during cyclogenesis. (b) Schematic representation of the evolution of a diffluent, negatively tilted upper-level trough during cyclogenesis.

Keyser and Shapiro (1986), Kocin et al. (1986), Uccellini and Kocin (1987), Keyser et al. (1989), Moore and Vanknowe (1992), and Loughe et al. (1995). These studies indicate that curvature complicates the simple two-dimensional relationships (as shown in Fig. 7-1b) between the ageostrophic wind field associated with jet streaks and divergence. Diabatic processes, especially those related to latent heat release, can also enhance the vertical motions associated with jet streak circulations (see, e.g., Cahir 1971; Uccellini et al. 1987). Despite the complications introduced by curvature and diabatic processes, Brill et al. (1985) and Loughe et al. (1995) demonstrate that these transverse circulations make a significant contribution to the divergence aloft and re-

sultant vertical motion patterns in the entrance and exit region of the jet streak.

The contribution of jet streak circulations to surface cyclogenesis has also been shown in numerous studies. Hovanec and Horn (1975) and Achtor and Horn (1986) illustrate the importance of the left-front exit region of jet streaks in cyclogenesis to the lee of the Rocky Mountains. Mattocks and Bleck (1986) and Sinclair and Ellsberry (1986) emphasize the role of the indirect circulation in the exit region of upper-level jet streaks in the development of cyclones in the lee of the Alps (and the Gulf of Genoa), and the northern Pacific Ocean, respectively. Uccellini et al. (1985, 1987) show how vertical circulations associated with upper-level jet streaks

SCHEMATICS OF JET-RELATED CIRCULATIONS UPPER- AND LOW-LEVEL JETS

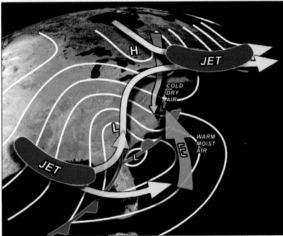

FIG. 7-4. (top) Lateral coupling of two separate upper-level jet streaks, one associated with an upper trough nearing the East Coast and the other associated with confluent upper flow over New England. Low-level cold-air advection is located beneath the confluent entrance region of an upper jet while a low-level jet associated with warm-air advection is located beneath the jet exit region over the Southeast. (bottom) Contribution of the lateral coupling of jet streaks to the low-level temperature and moisture advections during East Coast cyclogenesis.

contributed to two periods of surface cyclonic development during the Presidents' Day Storm. Additional support for the roles of jet streaks in surface cyclogenesis has been provided by Whitaker et al. (1988), Barnes and Colman (1993), and Lackmann et al. (1997, 1999).

As emphasized by Uccellini et al. (1984) and Wash et al. (1988), the contribution to upper-level divergence and its associated ascent pattern by relatively straight jet streaks (as inferred by the magnitude of the absolute vorticity advection) can be as large (or larger) than that computed for trough–ridge configurations. This result is confirmed by Loughe et al. (1995), who utilize a "Ψ vector" analysis of the February 1983 "Megalopolitan"

snowstorm (see volume II, chapter 10.20), and show the transverse circulation associated with the exit region of the upper-level jet produces an ascent pattern equivalent to that linked to the trough–ridge couplet.

For this sample of Northeast snowstorms, a jet streak approaching the base of the upper-level trough nearing the East Coast was observed in every case, with the surface cyclogenesis occurring in the diffluent exit region of the jet (Fig. 7-4). A separate polar jet streak is also evident north and east of the developing surface cyclone in many of the 30 cases. This separate jet can undergo marked amplification as the storm intensifies and, in some cases, attains wind speeds exceeding 80–90 m s^{-1}. These separate jet streaks have well-defined transverse circulation patterns in their exit and entrance regions, which have a marked effect upon the storm's evolution. For many of these cases (i.e., see Uccellini and Kocin 1987), an indirect transverse circulation has been diagnosed within the exit region of the southern jet streak associated with the upper-level trough approaching the East Coast. A direct transverse circulation is located within the entrance region of the jet streak associated with the upper-level confluence zone over Canada.

Uccellini and Kocin's (1987) analyses of the February 1983 case and seven other cases and the resultant schematic representation of the lateral coupling of the transverse circulations around the separate jet streaks (Fig. 7-4) illustrates the following:

i)   As an upper-level trough over the Ohio and Tennessee Valleys approaches the East Coast and a jet streak approaches the base of the trough and propagates toward the diffluent region downwind of the trough axis, a surface low pressure system develops. The cyclogenesis occurs downstream of the trough axis, within the exit region of the jet streak, where upper-level divergence (cyclonic vorticity advection) related to both the trough and the jet favors surface cyclonic development.

ii) A separate upper-level trough over southeastern Canada and a jet streak embedded in the confluence zone over New England are associated with a cold surface high pressure system, which is usually found beneath the confluent entrance region of the jet streak.

iii) Indirect and direct transverse circulations that extend through the depth of the troposphere are located in the exit and entrance regions of the southern and northern jet streaks, respectively.

iv) The rising branches of the transverse vertical circulations associated with the two jet streaks appear to merge, contributing to a large region of ascent along sloped isentropic surfaces, which produces clouds and precipitation between the diffluent exit region near the East Coast and the confluent entrance region over the northeastern United States.

v) The advection of Canadian air southward (Fig. 7-4, bottom) in the lower branch of the direct circulation over the northeastern United States maintains the cold lower-tropospheric temperatures needed for snowfall along the East Coast. In many cases, the lower branch of this circulation pattern is enhanced by the ageostrophic flow associated with cold-air damming east of the Appalachian Mountains (see section 3c).

vi) The northward advection of warm, moist air (Fig. 7-4, bottom) in the lower branch of the indirect circulation over the southeastern United States rises over the colder air north of the surface low. An easterly or southeasterly low-level jet streak (see the next section) typically develops within the lower branch of the indirect circulation beneath the diffluent exit region of the upper-level jet. This low-level flow enhances the moisture transport toward the region of heavy snowfall and focuses the low-level thermal advection patterns along the coast.

vii) The combination of the differential vertical motions in the middle troposphere and the interactions of the lower branches of the direct and indirect circulations appears to be highly frontogenetic [as computed by Sanders and Bosart (1985a) for the February 1983 storm], increasing the thermal gradients in the middle and lower troposphere during the initial cyclogenetic period. This finding is consistent with the increase in available potential energy observed by Palmén (1951, p. 618) during the early stages of cyclogenesis.

As depicted in Fig. 7-4, these separate transverse circulation patterns appear to be "laterally coupled" such that the vertical branches associated with each jet streak merges, enhancing the ascent immediately north of the surface low that focuses the heaviest snowfall within the coastal plain (Uccellini and Kocin 1987; Loughe et al. 1995). The important role of jet streaks and their attending circulation regimes in contributing to the ascent pattern necessary for heavy snowfall is consistent with Bjerknes's (1951) vision of jet streams combining with the trough–ridge patterns to produce the vertical motions associated with surface cyclogenesis.

The "lateral coupling" of the jet streaks southeast and northeast of the developing surface low also makes a significant contribution to the low-level temperature and moisture advection patterns. As depicted in Fig. 7-4, the advection of cold Canadian air southward along the coast is associated with the lower branch of the direct circulation associated with the northern jet streak. Meanwhile, the northward advection of warm, moist air into the region experiencing heavy precipitation occurs in the lower branch of the indirect circulation associated with the southern jet streak. The moisture transport and horizontal temperature advections are further enhanced by the formation of a low-level jet streak (LLJ) in the

lower branch of the indirect circulation. The LLJ is "vertically coupled" to the upper-level jet streak through the mass momentum adjustments within the exit region of the upper-level jet (Uccellini and Johnson 1979), a process that is further enhanced by sensible and latent heat release near the coastal front (Uccellini et al. 1987). Thus, the jet streak–induced transverse circulation patterns not only enhance the ascent necessary for heavy precipitation, but also contribute to the differential temperature and moisture advection patterns, which juxtapose the cold air and the warm, moist air in a manner necessary for heavy snowfall along the coast.

While this characteristic jet streak pattern is rather straightforward, the vertical structure and temporal evolution of the jet streaks during a heavy snow event can complicate the picture. Many cases were marked by large temporal changes in wind speed and in the vertical and horizontal location of the jet core. The evolution of these systems is often difficult to diagnose using the 12-hourly operational radiosonde data and by relying solely on standard isobaric analyses at selected levels without the use of cross-section and isentropic analyses (Uccellini and Kocin 1987) or diagnostic procedures based on numerical model output with 1–3-h temporal resolution (Brill et al. 1991; Manobianco et al. 1992). These problems in assessing the evolution of multiple jet streaks complicate the use of simple conceptual models that relate cyclogenesis to the patterns of upper-level divergence in the left-front exit and right-rear entrance regions of relatively straight jet streaks. However, the availability of the reanalysis datasets at 6-h intervals (see DVD included with volume II) provides a fresh perspective for examining the evolution of multilevel jet streaks (see individual discussions in chapter 10 in volume II).

2. Low-level processes

While upper-level troughs, jet streaks, and their associated vorticity advections contribute to the formation and development of the storms, physical processes in the lower troposphere also play important roles in creating conditions for heavy snowfall. Charney (1947), Sutcliffe (1947), Petterssen (1955), and many others have shown that thermal advection patterns below 700 hPa are important components of cyclogenesis, affecting surface deepening rates, influencing vertical motions, and enhancing the energy conversions associated with developing cyclones. The contribution of thermal advections to the vertical motions associated with cyclogenesis can be assessed from the "development equations" of Sutcliffe (1947) and Petterssen (1955, 320–339), the "ω equation" (see, e.g., Krishnamurti 1968), and from the quasigeostrophic framework (e.g., Holton 1979, chapter 7), which has been used extensively in diagnostic studies of cyclogenesis.

The lower-tropospheric thermal advection patterns associated with Northeast snowstorms are strengthened

EVOLUTION OF THE LLJ, 850 hPA LOW AND LOWER TROPOSPHERIC TEMPERATURES DURING A NORTHEAST SNOWSTORM

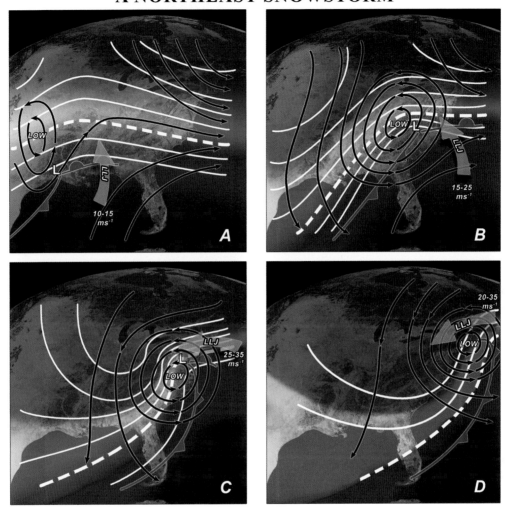

Fig. 7-5. Evolution of the low-level jet, 850-hPa low, and lower-tropospheric temperatures and wind fields during a Northeast snowstorm, including streamlines (solid arrows), isotherms (white lines, thick dashed line represents 0°C isotherm), surface fronts, and low-level jet positions and typical speeds.

during coastal cyclogenesis (Fig. 7-5). As shown in volume II, every storm in the 30-case sample developed in association with a preexisting isotherm gradient or baroclinic region at the 850-hPa level, which intensified and evolved into an "S shaped" configuration during the most intense stage of cyclogenesis. The circulation center at the 850-hPa level developed near the coast, above and slightly to the west of the surface low pressure center, close to the inflection point of the S-shaped isotherm configuration. The warm-air advection pattern is usually marked by 20–35 m s⁻¹ winds directed nearly perpendicular to the isotherms and provide a marked indication of the intense vertical motion patterns associated with these storms.

The lower-tropospheric thermal advections surrounding the 850-hPa low center increased during cyclogenesis. Part of this increase was attributable to the baroclinic instability and associated enhancements of the thermal gradients that are inherent within a cyclogenetic environment. Another part of this increase is due to the effects of topography and sensible heat flux from the ocean on the boundary layer temperature structure.

The Appalachian Mountains, the coastline, and the warm waters of the Atlantic Ocean and the Gulf Stream are particularly important factors. As discussed by O'Handley and Bosart (1996), the Appalachians have a significant impact on the redevelopment of cyclones and the subsequent track of the cyclone center as the upper-level trough system approach the East Coast. Although the Appalachians may facilitate cyclone development through "vortex tube" stretching as air flows over the mountains (i.e., Holton 1979, 87–89; Smith 1979, 163–169), they probably make a more significant contribution to cyclogenesis along the East Coast by entrapping cold air between the mountains and the coastline (O'Handley and Bosart 1996), a process that enhances the baroclinic zone along the coast, as illustrated in Fig. 7-6. This process, termed cold-air damming (Richwien 1980; Forbes et al. 1987; Stauffer and Warner 1987), is marked by a distinctive inverted surface high pressure ridge between the coast and the Appalachian Mountains chain that occurs as the cold air supplied by an anticyclone over New England or Canada is channeled southward along the eastern slopes of the Appalachians. Damming contributes to the southward advection of cold air and is maintained by a combination of ascent, evaporation, orographic, frictional, and isallobaric effects (Forbes et al. 1987; Bell and Bosart 1988; Doyle and Warner 1993a,b).

The cold-air advection associated with damming combines with the fluxes of sensible and latent heat from the ocean surface that warm and moisten the overlying atmosphere (Petterssen et al. 1962; Sanders and Gyakum 1980; Bosart 1981; Gyakum 1983a,b; Kuo and Low-Nam 1990) to intensify the thermal contrast and enhance the baroclinic zone near the coastline. This process, known as coastal frontogenesis (Bosart et al. 1972; Bosart 1975; Ballentine 1980; Bosart 1981; Bosart and Lin 1984; Forbes et al. 1987; Stauffer and Warner 1987), is confined to the lower troposphere, typically precedes cyclogenesis, and focuses low-level convergence, baroclinicity, warm-air advection, and surface vorticity. Coastal frontogenesis was observed in 25 of the 30 cases studied, and is an integral component of the sequence of processes that interact to enhance the temperature gradients associated with the coastal front and the ascent needed to generate heavy snowfall in the urban corridor.

Low-level jet streaks

The intensification of the low-level thermal advections is also related to the development of low-level jet streaks (LLJ) directed at a significant angle to the isotherms (Fig. 7-5). The LLJ, which was observed in all 30 cases, was directed primarily from an east-southeasterly direction, with maximum wind speeds generally between 20 and 35 m s^{-1} at 850 hPa. The LLJ often slopes upward from the oceanic planetary boundary layer toward the coastal plain, where it rises toward 700 hPa over the cold air mass trapped between the Appalachians and the coastline by cold-air damming processes. The LLJ is also an important factor in enhancing and focusing the moisture transport toward the area of heavy precipitation. Given the magnitudes of the wind speeds in the core of the LLJ and the amount of moisture in the oceanic boundary layer, the magnitude of the

EVOLUTION OF COLD AIR DAMMING AND COASTAL FRONTOGENESIS

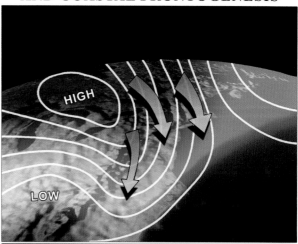

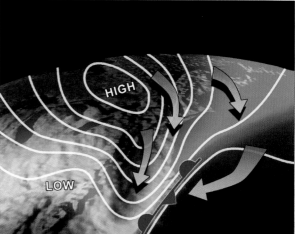

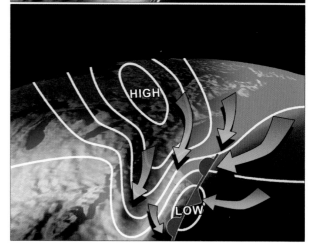

Fig. 7-6. Schematic representation of the evolution of a surface anticyclone, its associated wind and temperature fields, related to the development of "cold-air damming," "coastal frontogenesis," and coastal cyclogenesis.

moisture transport in Northeast snowstorms can approach those computed for spring convective cases in the Midwest (Uccellini et al. 1987). In addition to increasing the warm-air advections and moisture transports during cyclogenesis, the enhanced wind speeds associated with the LLJ also increase the moisture and heat fluxes from the ocean surface into the planetary boundary layer, a process that appears to heat and moisten the airstreams that feed directly into the developing cyclone (Mailhot and Chouinard 1989).

Other interesting aspects of the LLJ include 1) the ageostrophic nature during its intial development (also see discussion for each case in volume II, chapter 10) and 2) its linkage to the rapid pressure falls that often correspond to the onset of secondary cyclogenesis, or the rapid northeastward motion of the primary coastal low, called center jumps (see chapter 3). The rapid formation of the LLJ corresponds to a separate low-level maximum in mass divergence, which has been shown (Uccellini et al. 1987) to make a significant impact on surface pressure tendencies during the rapid development phase of the cyclone near or along the East Coast. The rapid parcel accelerations associated with the formation of the LLJ focuses a low-level mass divergence maximum between 900 and 800 hPa along the coast, giving rise to the initial surface pressure falls that mark the onset of secondary coastal cyclogenesis.

Thus, the LLJ has a significant impact on 1) the moisture transports that fuel the developing cyclone, 2) the temperature advections that illustrate the baroclinic character of the storm, 3) the associated vertical motion pattern that contributes to heavy snowfall, and 4) the mass divergence pattern that accompanies the onset of secondary cyclogenesis near the coast.

3. Cold surface anticyclones

Anticyclones that cross southeastern Canada or surface high pressure ridges that extend eastward from anticyclone centers over the southern Canadian plains or the northern American plains provide the cold low-level air that enables precipitation to reach the ground as snow within the Northeast urban corridor. These anticyclones originate over northern Canada or the polar regions, maintain central pressures often exceeding 1030 hPa, and remain in a position to reinforce the advection of cold air over the coastal plain despite the warm-air advection pattern aloft, as described earlier. The net result is an intensifying low-level baroclinic zone that also acts to focus the vertical motion pattern and related low-level frontogenesis within the regions of heaviest snow (Fig. 7-6). The surface anticyclone remains generally to the north-northwest of the surface low, resulting in a low-level flow of cold air that is minimally influenced by the warming effects of the nearby Atlantic Ocean.

The displacement of an initially cold surface anticyclone too far to the east produces a flow of ocean-warmed air toward the Northeast coast, often resulting

SCHEMATIC OF UPPER CONFLUENCE JET STREAK AND SURFACE ANTICYCLONE

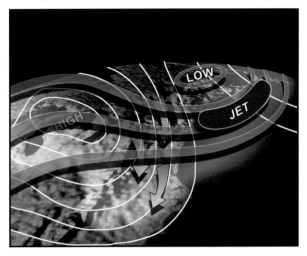

FIG. 7-7. Schematic representation of upper-level confluence, jet streaks, and surface anticyclones.

in rainfall or snow changing to rain along the immediate coastline. While the presence of an anticyclone to the north should not be viewed as a prerequisite for heavy snowfall (e.g., the 9–10 February 1969 storm followed an anticyclone that moved eastward off the coast prior to cyclogenesis), it is a common feature of many of the cases presented here.

The surface anticyclone, like the surface cyclone, is intimately associated with and linked to the upper-level trough–ridge patterns and jet streaks that influence Northeast snowstorms. The maintenance of a cold anticyclone to the north of the developing low is linked to a consistent pattern of upper-level confluence in the northeastern United States and southeastern Canada (Fig. 7-7). In all 30 cases, the anticyclone was located upwind of an upper-level trough crossing eastern Canada, almost always beneath a region of confluent geopotential heights (25 cases), and usually within the entrance region of an upper-level jet streak. The upper confluent flow within the entrance region of the jet streak over southeastern Canada forces a direct transverse circulation pattern that extends throughout the entire troposphere, acting to maintain the surface anticyclone and to enhance the advection of cold air southward along the coastal plain, playing a critical role in the development of most of these snowstorms (Uccellini and Kocin 1987). Thus, the juxtaposition of the surface high and low pressure centers and their evolution during these heavy snow events reflects the complex interaction of multiple upper-level troughs and ridge systems and their associated laterally coupled jet streaks that are important elements required for the development of these storms (Figs. 7-3 and 7-4).

4. Sensible and latent heat release

Traditional descriptions of cyclogenesis during the 1950s and 1960s (e.g., Pettersen et al. 1962) viewed sensible heat and moisture fluxes to be of secondary importance since they tended to be a maximum in the cold, dry air following a cyclone. Furthermore, with the increasing focus on the application of simplified two-layer models to the simulations of cyclones, the relative importance of latent heat release to baroclinic processes was significantly downplayed (Uccellini 1990). Bosart (1981) and Bosart and Lin (1984) showed how these fluxes can be large in the arctic air ahead of a developing cyclone along the East Coast (in this case, the Presidents' Day Storm of February 1979) and are likely to be large in many cases in which cold air over the Northeast precedes the cyclone. These studies and others, especially Sanders and Gyakum (1980) and Sanders (1986), help set the stage for later numerical experiments that showed that the sensible heating off the East Coast can make a significant contribution to the overall development of these storms (e.g., Uccellini et al. 1987; Kuo and Low-Nam 1990).

Danard (1964) recognized that the inability of the first numerical models to accurately predict cyclogenesis was partially due to the neglect of latent heat release. He found it necessary to include latent heat release in the model simulations to account for the distribution and magnitude of the vertical motion pattern associated with cyclogenesis and its observed deepening rates. This finding renewed the interests of the research community in examining the role of latent heat in cyclogenesis and led to a succession of many papers describing the role of diabatic processes on rapid cyclogenesis (Uccellini 1990). Several examples include Krishnamurti's (1968) application of the quasigeostrophic "ω equation" to study the life cycle of a rapidly intensifying cyclone demonstrates that latent heat release is a major contributor to the enhancement of vertical motions. An isentropic-based budget study of a cyclone over the Midwest by Johnson and Downey (1976) shows that the mass circulations and the associated angular momentum transports required for cyclogenesis are greatly influenced by the release of latent heat. In addition, a modeling study by Chang et al. (1982) clearly illustrates the impact of latent heat release on the deepening rate of the surface cyclone and on the structure of the accompanying upper-level trough–ridge system and its related wind fields.

→

FIG. 7-8. Schematic illustration of the impact of sensible heat fluxes in the boundary layer on the temperature advection pattern east of the cyclone center (marked by an L): (a) at time 0000, prior to sensible heat influence; (b) at time T +ΔT, when sensible heat flux has acted to increase temperature gradients and associated warm-air advection; and (c) at time T +2ΔT, when increased wind speed associated with deepening surface low further enhances temperature advection patterns. [Derived from Uccellini (1990).] Blue lines represent upper-level height contours and orange lines represent low-level isotherms.

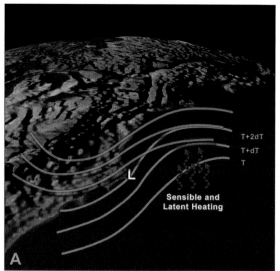

Initial Time

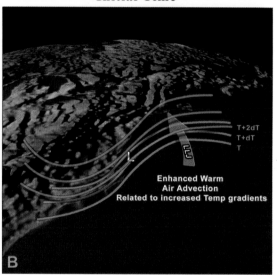

Initial Time + ΔT

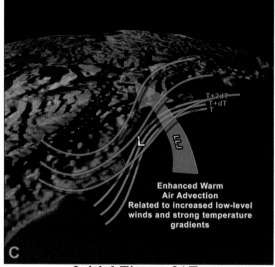

Initial Time + 2ΔT

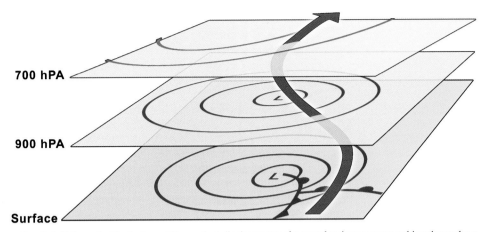

FIG. 7-9. Schematic illustration of the vertical displacement of a parcel trajectory approaching the cyclone from the south-southeast and passing through the area of precipitation associated with the cyclone and crossing the geopotential contours at the 700-hPa level (marked by an open wave) at a significant angle (from Uccellini 1990).

The asymmetrical distribution of clouds, precipitation, and associated latent heat release to the north and west of a developing cyclone (while a dry slot works toward the cyclone center from the southwest) is a characteristic feature of nearly all observed extratropical cyclones, ranging from small-scale, oceanic polar lows to the major coastal storms that affect continents around the world (Uccellini 1990). Twenty-six of 30 cases exhibited a period of rapid sea level development along the East Coast or over the western Atlantic Ocean, which coincided with the occurrence of heavy snowfall to the north and west of the deepening cyclone. The simultaneous observations of rapid deepening rates and the occurrence of heavy precipitation indicate that latent heat release has a significant impact on the development of these storms. Nevertheless, assessing the relative importance of sensible and latent heat release compared to the contribution of dynamical processes, described earlier, is difficult if not counterproductive, as is shown in the next section.

5. The positive feedback between physical and dynamical processes during East Coast cyclogenesis

There are an increasing number of model-based sensitivity experiments which indicate that individual dynamical and diabatic processes, discussed in this chapter and summarized in Table 7-1, make important contributions to cyclogenesis. Furthermore, there is increasing evidence that the rapid deepening phase of extratropical cyclogenesis is dependent on the nonlinear, synergistic interactions among all these physical processes, and not simply on their individual contributions (Uccellini et al. 1987; Uccellini 1990; Stein and Alpert 1993; Alpert et al. 1995).

The difficulty in assessing the relative contributions of latent heat, sensible heat fluxes, and baroclinic instability to rapid cyclogenesis is linked to the feedbacks that occur among the processes within a relatively small domain and over a very short period of time. For example, if sensible heat fluxes over the Atlantic Ocean east of a surface low act to increase the temperature gradients immediately along the coastline (as is shown in Figs. 7-8a and 7-8b), the developing surface low is affected not only by the diabatic additions of heat and moisture in the planetary boundary layer, but also by the baroclinic processes associated with the increased temperature gradients and their resultant increased thermal advections. The feedback does not stop there. A more intense low-level baroclinic zone depicted in Fig. 7-8b could also be expected to contribute to stronger low-level wind speeds that could further enhance the warm-air advection patterns and deepening rate of the storm (Fig. 7-8c).

The increased wind speeds depicted in Fig. 7-8c represent the development of the LLJ [see section 7b (1)]. This increase in wind speed is related to two factors: 1) an isallobaric contribution, due to the change of the pressure gradient force associated with the deepening cyclone (Naistat and Young 1973) and propagating jet streaks (Uccellini and Johnson 1979), and 2) the vertical parcel displacement in a region where the pressure gradient force changes with height (Uccellini et al. 1987). In Fig. 7-9 the vertical parcel displacement east and northeast of a developing surface cyclone, also enhanced by sensible and latent heat fluxes and latent heat release, can lead to a situation in which an air parcel ascends from one level, possibly marked by a closed circulation (e.g., 900 hPa) in which the parcel trajectory is directed parallel along the geopotential heights and ascends to a level marked by an open wave trough (e.g., 700 hPa), in which the parcel is now directed toward lower geopotential heights. As a result, as the parcel rises, it accelerates rapidly toward lower geopotential and turns to the right as it attempts to remain in a balanced state. Model-based trajectory results described by

Uccellini et al. (1987) and Whitaker et al. (1988) illustrate that this process can operate over a very short period of time (as little as 2–4 h) and contributes to ageostrophic wind speeds exceeding 35 m s^{-1}, and a subsequent rapid increase in parcel velocities from 18 to 30 m s^{-1}, in 3 h.

The acceleration of the east to southeasterly flow through the precipitation region near the 850–600-hPa layer is generally located beneath the more southwesterly diffluent flow aloft associated with an upper-level trough–ridge system approaching the East Coast from the west. Whitaker et al. (1988) demonstrate in a simulation of the 1979 Presidents' Day Storm that the transition layer between the accelerating southeasterly airstream and the southwesterly flow farther aloft characterized a region in which the mass divergence increased to a maximum during the storm's period of rapid development.

These results indicate that the deepening rates of extratropical cyclones are related to complex interactions between thermodynamic and dynamic processes, which are dependent on the horizontal and vertical distributions of the pressure gradient force especially as it relates to the transition from a closed circulation to an open wave between 900 and 600 hPa and latent heat release. Within the transition layer between a closed circulation in the lower troposphere and an open wave or trough aloft, the release of latent heat poleward and east of the developing cyclone would be especially important for enhancing the parcel accelerations, divergent airflow, surface pressure tendency, and associated rapid development rate of the cyclone.

Model sensitivity study for the Presidents' Day Snowstorm of February 1979

The feedbacks of sensible and latent heating on temperature gradients, parcel accelerations, energy transformations, and patterns of thermal and vorticity advections are not confined to the lower and middle troposphere. A model sensitivity study by Chang et al. (1982) shows the dramatic impact of latent heat release upon the temperature, geopotential height, and wind fields throughout the entire troposphere. Similar results were obtained for a series of numerical simulations conducted for the 18–19 February 1979 Presidents' Day cyclone (Uccellini 1990), which are presented below.

Four 24-h model simulations of the President's Day Storm, initialized at 1200 UTC 18 February, were run to illustrate the sensitivity of the numerical simulation of the storm to the incorporation of boundary layer physics (sensible and latent heating) and latent heat release. The numerical model, the initialization procedure, and a "full physics" simulation, incorporating both diabatic and adiabatic processes, are described by Whitaker et al. (1988). The other three model simulations include an adiabatic simulation, a simulation that includes the effect of latent heat release but no boundary layer fluxes,

and a simulation with the boundary layer fluxes included but with no latent heat release, following the experimental design described by Uccellini et al. (1987).

The sea level pressure (SLP) maps derived from the four simulations and verifying at 1200 UTC 19 February 1979 are shown in Fig. 7-10. The adiabatic simulation produces an inverted trough (Fig 7-10a) along the East Coast, rather than a well-developed cyclone. By including either latent heat release (Fig. 7-10b) or boundary layer fluxes (Fig. 7-10c), a modest improvement in the forecast of cyclogenesis is achieved over the adiabatic simulation. However, when both boundary layer fluxes and latent heat release are included (Fig. 7-10d), the cyclogenesis is more accurately simulated. The 32-hPa difference between the full-physics simulation and the adiabatic simulation (Fig. 7-10e) of the sea level pressure illustrates the dramatic influence of diabatic processes when sensible and latent heating and latent heat release act synergistically with the upper-level trough–ridge pattern and jet streaks (Uccellini et al. 1987) to produce more realistic and more explosive cyclogenesis.

The impact of the sensible heating within the planetary boundary layer and latent heat release within an expanding region of precipitation occurs not only at the surface but throughout the entire troposphere. For example, the large geopotential height difference of greater than 160 gpm at 850 hPa (Fig. 7-11a) reflects the deeper cyclonic circulation in the full-physics simulation. The impact of incorporating both adiabatic and diabatic processes on the 850-hPa wind field (Fig. 7-11b) is equally dramatic, with 35 m s^{-1} differences in the wind field computed east and northwest of the developing storm system.

At 300 hPa, the height difference between the full-physics and adiabatic simulations reverses sign and exceeds 120 gpm north-northeast of the surface cyclone (Fig. 7-11c). This increase in geopotential height reflects a tendency for the full-physics simulation to retard the eastward movement of the downstream upper-level ridge as the trough approaches from the west. This contrasts with the adiabatic simulation, in which the upper ridge continues to move steadily eastward.

Associated with the slower eastward movement of the upper ridge axis is the maintenance and enhancement of the 300-hPa jet streak as depicted by the 40 m s^{-1} difference between the full-physics and adiabatic simulations (Fig. 7-11d). The wind enhancement takes the form of an anticyclonic "outflow jet" similar to that isolated for a Midwest cyclone by Chang et al. (1982) and for a spring convective complex by Maddox et al. (1981) and may help explain the dramatic development of similar outflow jets in such cases as January 1966 (see volume II, chapters 10.9) and March 1993 [see Bosart et al. (1996), Dickinson et al. (1997), and Bosart (1999); also see discussion of the diabatic development of such jet systems in the next section through the use of potential vorticity concepts).

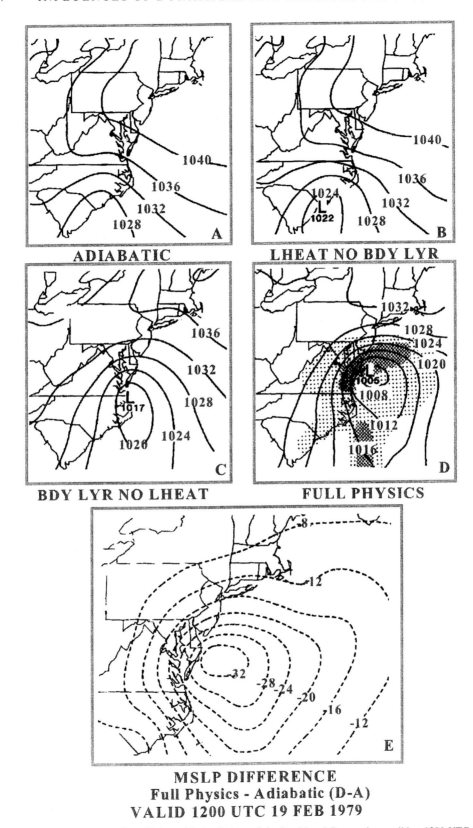

FIG. 7-10. Sea level pressure analyses from 24-h model simulations of the Presidents' Day cyclone valid at 1200 UTC 19 Feb 1979: (a) adiabatic simulation, (b) simulation with latent heat included but no boundary layer fluxes, (c) simulation with boundary layer fluxes but no latent heat release, and (d) full-physics simulation. Total precipitation greater than 2 cm for full-physics simulation indicated by light shading; greater than 4 cm, darker shading. (e) Difference in the sea level pressure (hPa) between the 24-h full-physics and adiabatic model simulations of the Presidents' Day Storm valid at 1200 UTC 19 Feb (from Uccellini 1990).

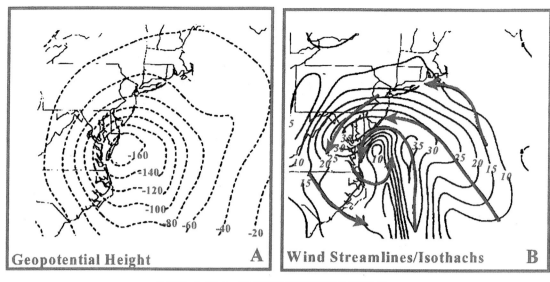

850-hPA DIFFERENCES

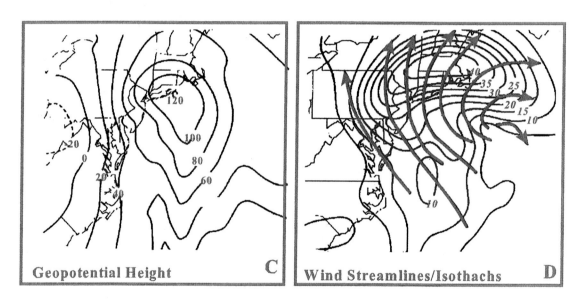

300-hPA DIFFERENCES

FIG. 7-11. (a) Difference in the 850-hPa geopotential height field (gpm) and (b) isotachs (m s⁻¹) and streamlines of the wind field difference between the 24-h full-physics and adiabatic model simulations of the Presidents' Day Storm valid at 1200 UTC 19 Feb 1979. (c), (d) Same as in (a), (b) except for the 300-hPa level (from Uccellini 1990).

The wind speed increase reflects a tendency for diabatic processes to maintain the upper-level ridge downstream of the developing cyclone, as well as to slow its eastward progression, acting to intensify the upper-level jet. Comparisons between the 850- and 300-hPa differences (Fig. 7-11) with the SLP differences (Fig. 7-10) reveal that the entrance region of the outflow jet (Fig. 7-11d) lies directly above the area of the maximum SLP difference between the full physics and adiabatic simulations (Fig. 7-10). Therefore, it appears that the acceleration of the flow into this upper-level jet occurs over the SLP difference maxima located north and east of the surface cyclone.

One might expect that the impact of latent heat release on an extratropical cyclone would be to collocate the ridge aloft with the area of maximum latent heat release and maximum surface pressure falls. However, the region of heaviest precipitation, latent heat release (Fig.

7-10d), and the largest SLP differences exhibited between the adiabatic and full-physics simulation (Fig. 7-10e) is located 350 km upstream of the region marked by the largest positive geopotential height differences at 300 hPa (Fig. 7-11c). Therefore, the upper ridging does not occur over the region of maximum latent heat release, but is located directly over the region where the lower-tropospheric warm-air advections are enhanced due to the increasing wind speeds and cyclonic circulation at 850 hPa (Fig. 7-11b). Thus, the comparisons of models incorporating different combinations of dynamical and diabatic processes strongly suggest how these combinations evident in the full-physics simulation are interlinked and work "synergistically" in the development of the Presidents' Day Storm; for example, the net effect of combining physical processes on the rate of cyclogenesis is greater than the sum of the individual parts (first described by Uccellini et al. 1987).

The 300- and 850-hPa difference fields in Fig. 7-11 indicate that the enhanced sensible and latent heat release and the feedback on the lower-tropospheric warm-air advections and upper-level trough and jet streaks retard the eastward movement of the upper-level ridge and enhance the upper-level jet northeast of the surface low.[1] The presence of an amplifying upper-level jet streak, especially as demonstrated by such cases as March 1993 (volume II, chapter 10.24) and the Presidents' Day Snowstorm of February 2003 (volume II, chapter 10.31), and the shortening wavelength between the advancing upper trough and its decelerating downstream ridge are key factors that have been identified for many of the Northeast snowstorms. The feedback mechanisms described above seem to be an indication of processes that enhance upper-level divergence and support the surface cyclogenesis immediately along the coastline, a key component of many of the snowstorms examined in this monograph.

6. Stratospheric extrusions, tropopause folds, and related potential vorticity contributions to East Coast cyclones

Upper-level trough–ridge systems and jet streaks not only provide the divergence aloft needed for surface cyclogenesis, but also contribute to the intensification of upper-level fronts and the related vertical distributions of potential vorticity (see, e.g., Danielsen 1968).

Attempts have been made to link surface cyclogenesis to the downward extrusion of stratospheric air into the upper and middle troposphere along intensifying upper-level frontal zones through the principle of conservation of isentropic potential vorticity (IPV), where IPV is a function of the static stability and wind shear. As stratospheric air, marked by high IPV values, descends into the troposphere, the air mass is stretched and the static stability decreases significantly. Consequently, the absolute vorticity increases with respect to parcel trajectories as long as the stratospheric values of IPV are preserved, promoting cyclogenesis. Kleinschmidt (1950) relates the advection of a stratospheric reservoir of high IPV associated with a low tropopause to cyclogenesis, going so far as to state that the stratospheric reservoir "is essentially the producing mass of the cyclones" (Kleinschmidt 1957, p. 125). Hoskins et al. (1985; see chapter 5d) discuss the impact of stratospheric IPV anomalies on surface cyclogenesis. Through an "invertibility principle" expressed by Kleinschmidt (1950), Hoskins et al. show that a positive IPV anomaly that extends downward from the stratosphere into the middle troposphere provides an optimal situation for enhancing the IPV advection in the middle to upper troposphere. This acts to induce a cyclonic circulation that extends throughout the entire troposphere to the earth's surface (Fig. 7-12), a hypothesis demonstrated for a Midwest cyclone by Davis (1992), who shows the close relationship of the surface cyclonic development to a potential vorticity anomaly.

Kleinschmidt (1957) also explores the means by which a stratospheric air mass is detached from its main reservoir and is subsequently displaced equatorward, hypothesizing that a disturbance associated with the jet stream is likely responsible for the equatorward displacement of the stratospheric air mass and its descent toward the middle troposphere during the period of cyclogenesis. As the stratospheric air descends, so does the tropopause in a process termed tropopause folding. A tropopause fold is defined by Reed (1955) and Reed and Danielsen (1959) as an extrusion of stratospheric air within upper-tropospheric baroclinic zones downward from the tropopause to the middle and lower troposphere. The concept of a tropopause fold complemented studies pointing to the importance of subsidence in the upper and middle troposphere as a mechanism contributing to upper-level frontogenesis (Reed and Sanders 1953; Newton 1954) and to surface cyclogenesis (Staley 1960; Bleck 1973, 1974; Boyle and Bosart 1983, 1986; Bleck and Mattocks 1984). It now appears that the transverse circulations associated with tropopause folding along the axis of an upper-level jet–frontal system is the mechanism that acts to displace the stratospheric air down toward the 500–700-hPa layer (Danielsen 1968; Uccellini et al. 1985; Whitaker et al. 1988; Bosart et al. 1996; Lackmann et al. 1997; Bosart et al. 2003).

For the February 1979 Presidents' Day cyclone,

---

[1] The downwind displacement of enhanced ridging aloft from the maximum in the SLP difference between the full-physics and adiabatic simulations is also in agreement with Palmén's (1951, 610–611) discussion on the asymmetric nature of extratropical cyclones. Palmén emphasized that the asymmetrics involved with extratropical cyclogenesis is critical for describing the vertical distribution of vorticity and divergence required for the simultaneous decrease in the sea level pressure and convergent cyclonic airflow in the lower troposphere, and the divergent anticyclonic airflow in the middle troposphere immediately downstream of the surface low pressure center during its development phase.

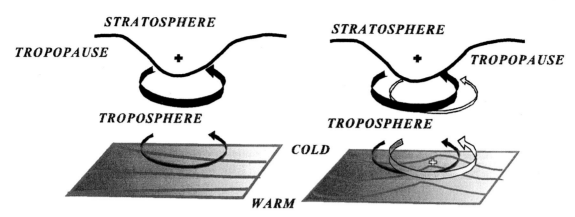

FIG. 7-12. A schematic picture of cyclogenesis associated with the arrival on an upper-level IPV anomaly over a low-level baroclinic region. (left) The upper-level IPV anomaly, indicated by a solid plus sign (+) and associated with the low tropopause shown, has just arrived over a region of significant low-level baroclinicity. The circulation induced by the anomaly is indicated by solid arrows, and potential contours are shown on the ground (Hoskins et al. 1985). (right) A low-level IPV anomaly can also induce a cyclonic circulation indicated by the open arrows that act to reinforce the circulation pattern induced by the upper-level IPV anomaly.

Uccellini et al. (1985) examine an intensifying polar front jet and deepening trough over the central United States (see Fig 7-13). A well-defined tropopause fold marked by the downward extrusion of stratospheric IPV values to 700 hPa was analyzed 1500 km upstream of the cyclogenetic region and 12–24 h prior to rapid surface cyclogenesis along the East Coast (Fig. 7-14). Uccellini et al. also show that subsidence along the axis of the polar front jet contributed to the tropopause fold, and they trace the stratospheric air mass to a position just upstream of the cyclogenetic region during the storm's rapid development.

Whitaker et al. (1988) utilized a numerical model to analyze the tropopause fold and stratospheric extrusion over the central United States and its subsequent eastward displacement toward the coastal region where rapid cyclogenesis occurred. A three-dimensional perspective of the model-simulated stratospheric extrusion and its eastward displacement was generated on the University of Wisconsin's Man computer Interactive Data Access System (McIDAS; Hibbard et al. 1989; Uccellini 1990) and is shown from a southern perspective in Fig. 7-15, with selected trajectories in Fig. 7-16, and from an eastern perspective in Fig. 7-17 at 6-h intervals between 0000 and 1800 UTC 19 February 1979.

The tropopause fold and stratospheric extrusion (as indicated by the descent and horizontal advection of the 2 PVU IPV surface, where 2 PVU = 2×10^{-6} m^2 s^{-2} K kg^{-2} represents stratospheric values of IPV) that preceded and accompanied the intensification of the surface cyclone off the East Coast are depicted in Fig. 7-15. The detailed diagnostic computations that relate the stretching and downward displacement of the IPV anomaly as it propagates toward the East Coast, and its associated increase in absolute vorticity following the parcel trajectories shown in Figs. 7-16 and 7-17 can be found in Whitaker et al. (1988). These results point to a more active role for the subsynoptic-scale processes associated with jet streaks and the associated upper-level fronts in the extrusion of stratospheric air during the precyclogenetic period and subsequent spinup of a surface cyclone, in contrast to the concept that these upper-level processes are a passive consequence of cyclogenesis, an issue discussed by Palmén and Newton (1969, 256–258; Keyser and Shapiro 1986, p. 593; Uccellini 1990; Lackmann et al. 1997).

The three-dimensional simulations depicted for the Presidents' Day Storm (Figs. 7-15–17) also point to a separate region of high IPV confined to the lower troposphere. Diabatic processes primarily associated with the vertical and horizontal distribution of latent heat release within a low-level baroclinic zone contribute to the development of low-level IPV anomalies (Kleinschmidt 1957, 134–136; Gyakum 1983b; Bosart and Lin 1984; Boyle and Bosart 1986; Whitaker et al. 1988). The low-level trajectories in Figs. 7-16 and 7-17 that 1) approach the Presidents' Day Storm from the east and wrap around the center and 2) approach the storm center from the south and rise rapidly before turning anticyclonically are influenced by the latent heat simulated by the model (Whitaker et al. 1988) and illustrate the nonconservative aspects of the low-level IPV maximum in a region of heavy precipitation. The low-level IPV maximum, and the associated thermal advection pattern, also induce a cyclonic circulation extending upward throughout the entire troposphere, as emphasized by Hoskins et al. (1985; see Fig. 7-12). The low-level maximum adds to the circulation induced by the upper-level IPV maximum, as long as the low-level anomaly remains downwind of the upper-level anomaly, maintaining a positive feedback between the two.

The relationships between extrusions of stratospheric air and other Northeast snowstorms are depicted for the March 1993 Superstorm (Fig. 7-18) and for the snowstorm of 24–25 January 2000 (Fig. 7-19). A three-dimensional depiction of the 2-PVU IPV surface for the

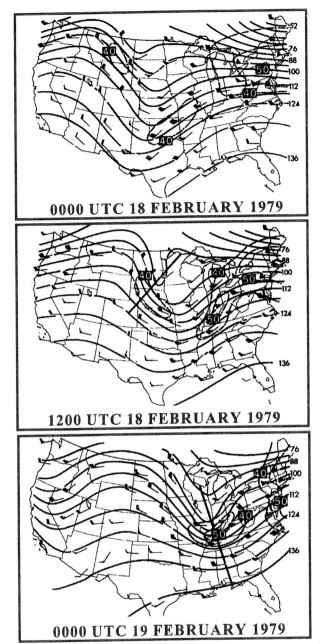

FIG. 7-13. Isentropic analyses for the 312-K surface for (top) 0000 UTC 18 Feb 1979, (middle) 1200 UTC 18 Feb 1979, and (bottom) 0000 UTC 19 Feb 1979, prior to the rapid development of the Presidents' Day Storm (see Fig. 10.18-2 in volume II). Includes Montgomery streamfunction analysis (solid, $100 = 3.100 \times 10^5$ m^2 s^{-2}) and isotachs (dotted–dashed, m s^{-1}). Wind barbs represent observed speeds (whole barbs, 10 m s^{-1}; half barbs, 5 m s^{-1}). Solid lines represent cross sections shown in Fig. 7-14 (from Uccellini 1990).

in the planetary boundary layer ascend through this low-level IPV maximum in much the same manner as is depicted for the 1979 Presidents' Day Storm (see Figs. 7-16 and 7-17). A distinct IPV maximum also can be seen descending from the stratosphere immediately upwind of the developing low. The IPV maximum is nearly collocated with the orange area in the water vapor distribution (Fig. 7-18b) and represents the stratospheric extrusion of dry air into the rapidly developing cyclone in much the same manner diagnosed for the 1979 Presidents' Day Storm. Bosart et al. (1996) describe in detail the sequence of trough mergers and related evolution of the IPV distribution that ultimately leads to this dramatic extrusion of stratospheric air upwind of the developing cyclone depicted in Fig. 7-18a.

An analysis of IPV for the 24–26 January 2000 snowstorm is depicted in Fig. 7-19a, along with the water vapor imagery in Fig. 7-19b. This IPV analysis is derived from the operational NCEP Eta numerical model initial fields at 1200 UTC 24 January and is combined with analyses of sea level pressure and equivalent potential temperature (θ_E) in the planetary boundary layer. The analysis depicts a very narrow tongue of stratospheric IPV located just upwind of the developing surface low, collocated with the distinct dry slot in the water vapor imagery along the Gulf coast. Again, the collocation of the high IPV with a distinct dry slot in the water vapor imagery provides supporting evidence for the role of stratospheric extrusions in the rapid spinup of East Coast cyclones, which characterize many of the Northeast snowstorms described in this monograph.

The application of potential vorticity concepts to rapid cyclogenesis has also been accomplished through the use of "dynamic tropopause maps," which combine the distribution of IPV and potential temperature to depict the tropopause and related stratospheric distribution of potential vorticity onto one chart (Morgan and Neilsen-Gammon 1998; Bosart 1999; Nielsen-Gammon 2001). An advantage of this relatively new mode of analysis is that it explicitly accounts for diabatic heating through nonconservation of potential vorticity. For example, the rapid development of an upper-level jet streak within the downstream ridge of a cyclone-producing trough can be linked to diabatic processes also through use of a potential vorticity perspective, as done by Bosart et al. (1996), Dickinson et al. (1997), and Bosart (1999). Their analyses of the March 1993 Superstorm link the rapid development of an 85 m s^{-1} jet streak near 250 hPa (see volume II, chapter 10.24) to diabatic heating since potential vorticity is not conserved along the portion of the dynamical troopopause near the core of the amplifying jet streak.

The model results from Whitaker et al. (1988) for the 1979 Presidents' Day cyclone (Figs. 7-15–17) and the supporting evidence provided by the IPV diagnostics for the March 1993 Superstorm (Fig. 7-18) and the January 2000 snowstorm (Fig. 7-19) all point to the im-

March 1993 Superstorm reveals the existence of two distinct IPV maxima as the surface cyclone was developing along the southeast coast of the United States (Fig. 7-18a). The low-level IPV maximum is collocated with the rapidly developing surface low pressure system and heavy precipitation. Low-level trajectories originating

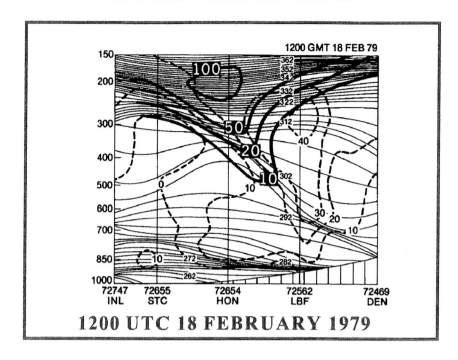

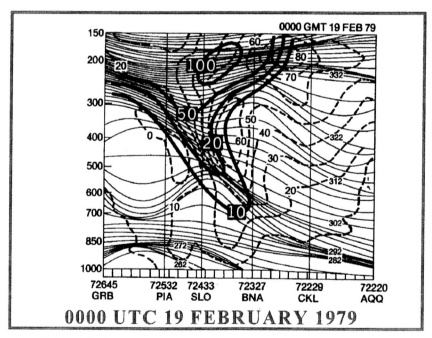

FIG. 7-14. (top) Vertical cross section from International Falls, MN (INL), to Denver, CO (DEN), for 1200 UTC 18 Feb 1979. (bottom) Vertical cross section from Green Bay, WI (GRB), to Apalachicola, FL (AQQ), for 0000 UTC 19 Feb 1979. Shown are isentropes (solid, K), geostrophic wind (dashed, m s⁻¹) computed from the horizontal thermal gradient in the plane of the cross section, and potential vorticity, heavy solid where 10 = 1 PVU. Potential vorticity analysis only shown for upper portion of frontal zone and stratosphere (from Uccellini et al. 1985).

portance of stratospheric extrusions of high IPV for the rapid intensification of the Northeast snowstorms. These results, combined with recent diagnostic studies by Gyakum (1983b), Davis (1992), Bosart et al. (1996), Lackmann et al. (1997), and many others lend support to the hypothesis that the spinup of a cyclone occurs as the upper-level IPV anomaly approaches the East Coast from the west and extends down toward the separate,

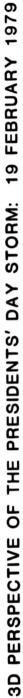

3D PERSPECTIVE OF THE PRESIDENTS' DAY STORM: 19 FEBRUARY 1979

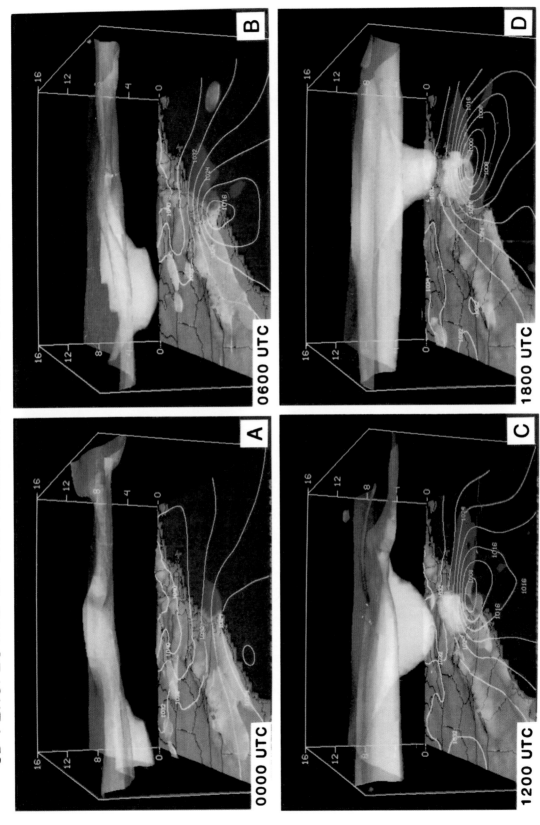

Fɪɢ. 7-15. Three-dimensional perspectives, as viewed from the south, of the 2-PVU IPV surface, and sea level pressure isobar pattern (hPa) derived from the numerical simulation of the Presidents' Day cyclone of Whitaker et al. (1988) for (a) 0000, (b) 0600, (c) 1200, and (d) 1800 UTC 19 Feb 1979. The three-dimensional perspectives in Figs. 7-15–7-17 were derived from Hibbard et al. (1989) using the University of Wisconsin's three-dimensional McIDAS system.

3D PERSPECTIVE OF THE PRESIDENTS' DAY STORM: 19 FEBRUARY 1979

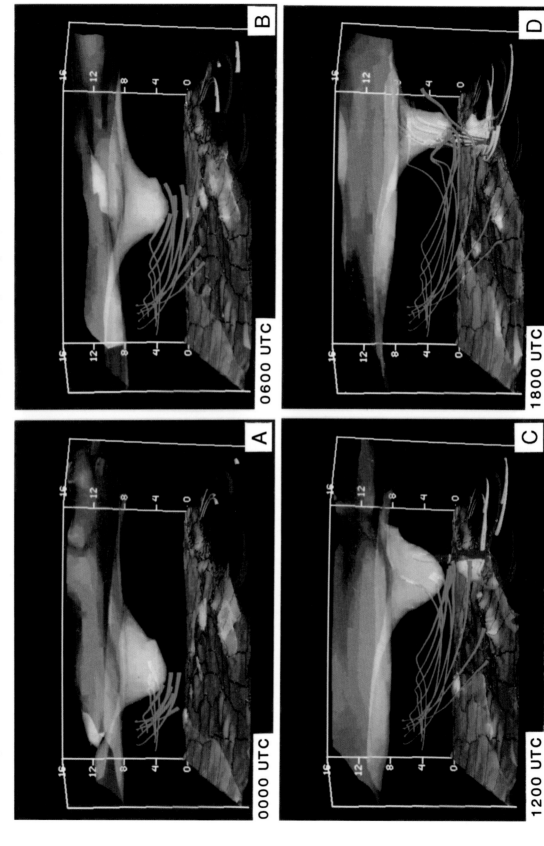

FIG. 7-16. Three-dimensional potential vorticity perspective as in Fig. 7-15, but with sea level pressure analyses removed and trajectories included, derived from 15-min model output as described by Whitaker et al. (1988). Blue trajectories originate within the stratospheric extrusion west and north of the cyclone; yellow and red trajectories originate in the low levels within the ocean-influenced planetary boundary layer.

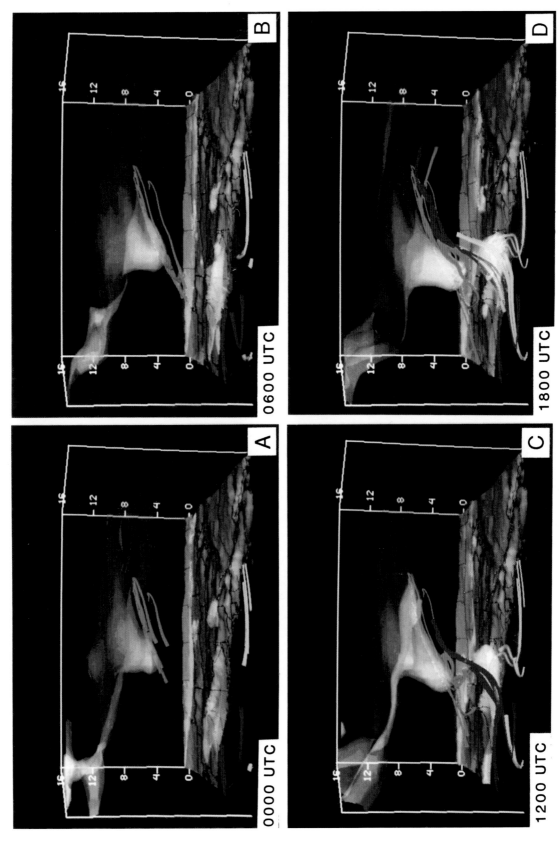

FIG. 7-17. Three-dimensional, model-generated potential vorticity perspective and trajectories as in Figs. 7-15 and 7-16, but as viewed from an eastern perspective, illustrating the sloped nature of both the potential vorticity surface and the trajectories approaching the storm from the south.

EVIDENCE OF THE TROPOPAUSE FOLD
PRIOR TO
THE MARCH 1993 SUPERSTORM

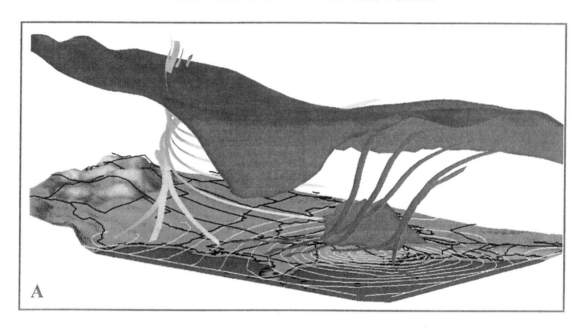

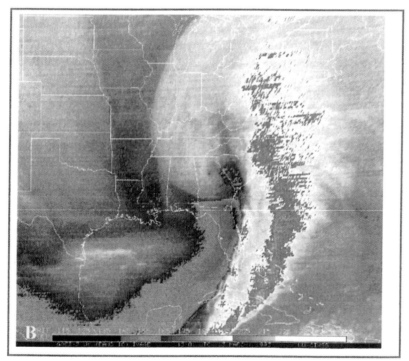

FIG. 7-18. (a) Three-dimensional perspective of the 2-PVU IPV surface, and sea level pressure isobar pattern (hPa) derived from the numerical simulation of the Mar 1993 Superstorm. The 6.7-μm GOES water vapor image on 13 Mar 1993 was provided by C. Velden, University of Wisconsin.

TROPOPAUSE FOLD
PRIOR TO THE
25 JANUARY 2000 SNOWSTORM

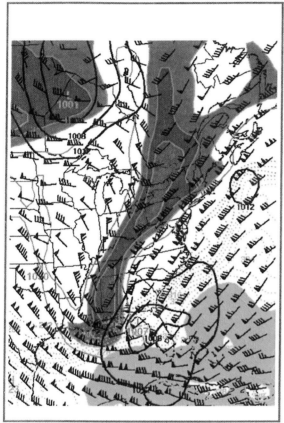

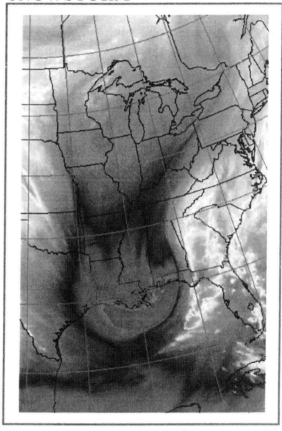

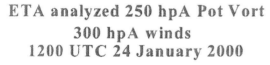

ETA analyzed 250 hpA Pot Vort
300 hpA winds
1200 UTC 24 January 2000

Water Vapor Imagery
1215 UTC 24 January 2000

FIG. 7-19. (left) Eta Model initial analysis of 250-hPa potential vorticity, 300-hPa winds, mean sea level pressure, and equivalent potential temperature at 1200 UTC 24 Jan 2000. (right) The 6.7-μm GOES water vapor image at 1215 UTC 24 Jan 2000.

low-level IPV anomaly extending up from the boundary layer toward 700 hPa. Furthermore, the existence and implied importance of two separate IPV anomalies support the important roles, and related synergistic interactions, of upper-level dynamical processes and low-level diabatic processes in the development of Northeast snowstorms, as discussed in the previous section.

7. Sutcliffe's self-development concept

The model sensitivity study and related diagnostic computations (section 5a) and the potential vorticity framework (section 6) provide a framework to link the interaction of various dynamic and physical processes to the rapid development of cyclones and related snowstorms along the East Coast. The "self-development"

concept, relating common conditions leading to rapidly intensifying cyclones, defined by Sutcliffe and Forsdyke (1950), discussed by Palmén and Newton (1969, 324–326), and applied to two East Coast cyclones (Roebber 1993), also provides a basis for describing the interactions of dynamical and diabatic processes that contribute to rapid cyclogenesis often associated with heavy snowstorms.

The approach of an upper-level trough and/or jet streak toward a low-level baroclinic zone and a preexisting and typically weak surface cyclone form the basis for self-development (Fig. 7-20). These features act to focus and enhance the effects of warm-air advection, and sensible heat and moisture fluxes in the planetary boundary layer, in combination with latent heat release above 850 hPa north and east of the surface

SUTCLIFFE'S "SELF-DEVELOPMENT"

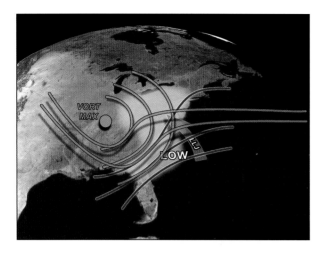

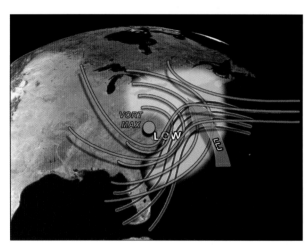

FIG. 7-20. Schematic of the "self-development" concept that describes how the temperature advections, sensible heat fluxes, and moisture fluxes in the planetary boundary layer, and latent heat release associated with cyclogenesis, enhance the amplitude of upper-level trough–ridge systems, and decrease the wavelength between trough and ridge axes, which contributes to an increase in upper-level divergence that further enhances cyclogenesis. Orange lines represent vorticity contours, blue lines are upper-level height contours, and red lines are low-level temperature contours.

low. The net effect is to warm the lower to middle troposphere near the axis of the upper-level ridge and to increase the divergence aloft between the ridge and the approaching trough axes, a process discussed in de-

tail by Bjerknes (1951, p. 599). The warming affects the divergence aloft by slowing the eastward progression of the upper-level ridge and increasing its amplitude. Upstream, a trough moves eastward and amplifies, partly in response to an amplification in the cold-air advection pattern to the west of the surface low pressure center. The decrease in the wavelength between the advancing trough and the slowing ridge, an increase in the diffluence corresponding to the spread of geopotential height lines downstream of the trough axis, and the increase in the maximum wind speeds of the upper-level jet streaks all combine to enhance the divergence in the middle to upper troposphere above the surface low. The increased upper-level divergence, represented by an enhancement of the cyclonic vorticity advections in Fig. 7-20, acts to deepen the cyclone even further. This is the basic concept of self-development.

As the surface low deepens, the lower-tropospheric wind field surrounding the storm increases in intensity, especially to the north and east of the surface low, where the contribution of isallobaric effects and vertical motions to air parcel accelerations is large. An LLJ develops to the north and/or east of the storm center, enhancing 1) the moisture and heat fluxes within the oceanic planetary boundary layer, 2) the moisture transports toward the region of heavy precipitation, and 3) the warm-air advections east of the developing cyclone (Fig. 7-20). Thus, the development of the LLJ further contributes to the self-development process that continues until the cyclone occludes and the heating is effectively cut off.

Self-development depends on the following conditions: 1) the existence of upper-level trough–ridge systems and jet streaks that focus the divergence aloft, a necessary condition for maximizing mass divergence and ascent immediately downstream of the developing surface low; 2) an asymmetrical distribution of clouds and precipitation that focuses the latent heat release and associated dynamic feedbacks on the downstream ridge and polar jet streak, both factors that can enhance the middle-tropospheric divergence north and east of the storm center; and 3) warming due to an enhanced low-level jet and warm-air advection pattern immediately north and east of the developing coastal cyclone. The concept accounts for the adiabatic, quasigeostrophic framework that has been applied to cyclogenesis (Holton 1979, chapter 7) and also for the various interactions among the dynamical and diabatic processes shown in Fig. 7-8. As such, Sutcliffe's self-development concept represents a basis upon which the contribution of the various physical and dynamical processes to rapid cyclogenesis along the East Coast, as summarized in Table 7.1, can be described.

Chapter 8

SUMMARY, FORECAST ADVANCES, AND A NORTHEAST SNOWFALL IMPACT SCALE (NESIS)

The spectacle of heavy snow and high winds in the northeast United States has been documented by the earliest settlers dating to the 17th century (Ludlum 1966). These storms, often referred to as nor'easters, have drawn the attention of some of this country's most notable statesmen, including Benjamin Franklin, George Washington, and Thomas Jefferson; poets such as Robert Frost and John Greenleaf Whittier; and the hundreds of millions of inhabitants who have called the region home from Virginia northeastward to Maine during the past four centuries. Some of the most legendary events, such as the "great snow" of 1717, the Washington–Jefferson Snowstorm of 1772, the blizzards of 1888 and 1899, the 1922 "Knickerbocker" storm, and the great New England snowstorm of 1978 can be recalled for generations by those who have either lived through these events or learned about them through local lore. Legends and fascination aside, these winter storms have also posed enormous challenges to the meteorological research and operational communities who have attempted to understand and predict them, often with mixed results. This chapter will summarize some of the key issues raised in this monograph, focusing on the climatological, synoptic, and mesoscale aspects of these storms, and will also address the advances that have been made in forecasting Northeast winter storms, as well as the forecast challenges that remain. Finally, a Northeast Snowfall Impact Scale (NESIS) is presented to address the need to communicate to the public the impact of these winter storms and to provide a quantitative measure of the impact of heavy snowfall on the heavily populated Northeast urban corridor.

1. Climatological–synoptic–mesoscale review

The climatological and synoptic overviews of many Northeast snowstorms that have occurred throughout the 20th century are focused on the Northeast urban corridor that includes five major metropolitan areas: Washington, D.C.; Baltimore, Maryland; Philadelphia, Pennsylvania; New York City, New York; and Boston, Massachusetts. Follwing are several important aspects of the snowstorms that affect the urban corridor:

- Mean seasonal snowfall increases from 15 in. (38 cm) in Washington to 25 in. (62 cm) as one moves north-

eastward through Baltimore, Philadelphia, and New York City, to approximately 40 in. (100 cm) in Boston.
- The seasonal snowfall can be quite variable with totals ranging from minima of 0.1 in. (0.3 cm) in Washington, 0.7 in. (2 cm) in Baltimore, a trace in Philadelphia, 2.8 in. (7 cm) in New York City, and 9.0 in. (23 cm) in Boston to maxima of 55.4 in. (126 cm) in Washington, 62.1 in. (158 cm) in Baltimore, 63.1 in. (161 cm) in Philadelphia, 75.6 in. (192 cm) in New York City, and 107.6 in. (273 cm) in Boston.
- During a given season, total snowfall includes a variety of light, moderate, and heavy snow events, which vary considerably from one season to the next. For moderate snow events (snowfall of 4–10 in.; 10–25 cm), there is a gradual increase in the frequency of these storms as one moves northward from Washington and Baltimore (where 1 moderate event per season can be expected), through Philadelphia and New York City (where 1.5 events per season can be expected), and toward Boston (where up to 3 moderate events can be expected on a seasonal basis). This general increase in the number of moderate snow events between Washington and Boston should be expected since there is a greater likelihood that more precipitation will fall as snow in the colder latitudes of New England. The distribution also reflects the impact of the eastward-moving storms that track just south of New England, intensify off the coast of Maine, and yield significant snows in that region while producing little or no snowfall from New York City and points south.
- The heavy snow events [snowfall greater than 10 in. (25 cm)] occur with much less frequency than the light-to-moderate snows and can be considered rare events in the Northeast urban corridor. Washington, Baltimore, and Philadelphia average one such storm every 3–4 yr, New York City averages one every 2–3 yr, while Boston records one storm every 1.5 yr. The nearly uniform distribution of heavy snow events between Washington and New York City is linked to cyclones that typically originate farther south, which then move northeastward along the coast producing heavy snow from Virginia toward New York and New England.
- A maximum of these heavy snowstorms is recorded

in the month of February for all the cities but Boston, lending validity to the notion that February is the month for the "big snows."

Snowfall records that extend up to and over a 125-yr period for the five major cities and other locations in the Northeast urban corridor were subjected to a spectral analysis to determine whether seasonal snowfall exhibits any periodic behavior. This analysis revealed that spectral peaks can be identified for the seasonal snowfall for each of the five cities, which appear to have non-random fluctuations on the order of 5–7, 10–12, and 20 yr. The clarity of this signal is muddled by the fact that the spectral bands are not necessarily dominant and there is considerable variability diagnosed from city to city.

Given the city-to-city variability in the spectral analysis and the multiple peaks, it is difficult to identify a specific, dominant, periodic cycle that can be used as a basis for describing the seasonal snowfall record. A more dominant "signal" in the seasonal snowfall history is the existence of distinct episodes of abnormally high snow totals, which appear as spikes in the annual records. These abnormally high snowfall totals tend to be clustered in 2- or 3-yr periods and are often related to the occurrence of one or more major snowstorms within a season, and can follow or precede longer periods in which below normal snowfall is observed. This characteristic of the annual snow record may point to a more "episodic" character of the annual snowfall distribution than a periodic nature. The net result is a long period where seasonal snowfall may be of little consequence (e.g., the 1980s into the early 1990s), followed by a relative onslaught of major snowstorms, accompanied by high winds and cold temperatures.

Three periods of excessive snows and related major snowstorms stand out during the second half of the 20th century. The first period extends from the late 1950s through the 1960s, which was characterized by many snowy winters and major snowstorms and was the snowiest such period of the 20th century. The late 1970s saw several memorable cold waves and snowstorms, especially from 1977 through 1979, which were preceded by the tranquil winters of the early and middle 1970s and followed by the relatively tranquil 1980s (with a few exceptions). The last period, extending from 1993 through 1996, included the winter of 1995/96, by far the snowiest winter of the century, and two other seasons with numerous snow and ice storms. During the March 1993 "Superstorm," approximately 100 million Americans saw snow from this storm, perhaps the most widespread Eastern snowstorm in recent history. During the January 1996 "Blizzard of '96," 30 million people are estimated to have experienced a snowfall exceeding 20 in. (50 cm), a paralyzing event. This snowy period followed the relatively quiet period of the 1980s and early 1990s and was followed by the quiet late 1990s. Despite all the activity between 1993 and 1996, there were still two winters surrounding the snowy winter of

1995/96 that saw minimal seasonal snowfall, including the least snowy season of the century for selected sites throughout the Northeast.

To summarize, an examination of winter seasons within the Northeast urban corridor shows the following:

- the heavy seasonal snowfall tends to be episodic in nature, and
- the snowy winters often occur with distinct major snowstorms, which are relatively rare.

Federal, state, and municipal agencies face enormous challenges in planning, budgeting, and responding to such events. The rare and episodic character of Northeast snowstorms also adds to the challenge for meteorologists who have to forecast these storms with the confidence and reliability required so that proper precautions can be taken in advance of their occurrence.

Two modes of surface cyclonic development characterize the 30-sample study of heavy snowstorms within the Northeast urban corridor. One mode is the "Gulf of Mexico–Atlantic coast" development, where a surface low forms near the Gulf of Mexico and moves northeastward along the Atlantic coast, sometimes with the surface low pressure center "leapfrogging" up the coast. The other mode is the "Atlantic coastal redevelopment," which involves a primary cyclone diminishing in intensity as it approaches the Appalachians while a secondary low develops along the Carolina coast and heavy snow falls across the Northeast urban corridor. Many of these coastal cyclones deepen rapidly, with the cyclones exhibiting an average deepening period of 36 h and a mean decrease in pressure of 30 hPa.

Two modes of anticyclone configurations also characterize the sample. In general, the location of the anticyclone, typically over eastern Canada, north to northwest of the developing cyclone, plays a crucial role in providing cold, near-surface air (without modification from the Atlantic Ocean) that allows the precipitation to reach the ground as snow.

At upper levels, very distinctive patterns are observed with the snowstorm sample. Prior to heavy snowfall, a "split flow" regime with an upper ridge over western Canada and a strong shortwave disturbance and associated or jet stream across the southwestern United States are common signatures. The surface cyclones are associated with upper troughs with increasing amplitudes, decreasing wavelengths, diffluence downwind of the trough axis, and three modes of evolution. The most common mode involves the evolution of an open-wave trough into a closed low, or "cutoff" low. Another common signature is the presence of a deep trough and jet stream over eastern Canada, sometimes associated with a high-amplitude ridge over the North Atlantic. Within these jet streams, localized jet streaks are also common signatures, with a dual-jet pattern and a related "lateral" coupling of transverse circulation patterns identified as important signatures that characterize many of the snowstorms.

Thirty snowstorms are selected to represent heavy snow situations in the heavily populated Northeast corridor. An additional 37 cases classified as "near miss" snowstorms are also examined to distinguish what conditions lead to heavy snowfall and what conditions reduce the likelihood of snowfall in the urban corridors. Three categories of near-miss cases are examined, including so-called interior snowstorms, moderate snowstorms, and ice storms. For interior snowstorms, a number of factors favor heavy interior Northeast snowfall amounts with snow changing to rain along the coast. These factors include a more pronounced amplification of the upper trough–ridge that allows 1) the surface low to track farther west than their heavy snow counterparts and 2) the surface anticyclone to drift farther east. These factors combine to force warmer air toward the coastline, favoring rain versus snow in the urban corridor while the heaviest snows occur inland.

With moderate snowstorms, snowfall in the Northeast urban corridor is diminished with weaker upper-level troughs, less well-defined jet streaks, weaker surface cyclogenesis, and faster-moving storm systems. However, the same mechanisms (marked by confluent flow over the northeast United States) that allow cold, low-level air to remain near the coast in the major snowstorm sample are also present here. With ice storms, heavy snow does not materialize in a number of the scenarios that allow cold air to remain at the surface and warm air to flow aloft. One significant scenario involves a very strong anticyclone, similar to the heavy snow situations, but with an upper-level trough that moves northwest of the urban corridor, rather than just to the south, allowing warmer air to flow northeastward above a shallow dome of cold air remaining at the surface. The resulting vertical temperature profiles support freezing rain and/or sleet with little or no snow.

The mesoscale aspects of Northeast snowstorms, both in terms of mesoscale detail found within synoptic-scale snowbands, as well as relatively small, isolated areas of significant snowfall, are also described and related to the presence of jet streaks, slantwise instabilities, frontogenesis, inverted sea level troughs, gravity waves, and "bay effect"–type enhancements. The mesoscale nature of Northeast snowstorms is often the most difficult to diagnose and predict. Rapid changes in the development and decay of heavy snowbands as well as boundaries (such as the rain–snow line and the observation of heavy snowbands located toward the northwest region of the overall snowfall distribution) are mesoscale aspects that represent profound challenges to forecasters.

Recent advances, especially the Next-Generation Doppler Radars (NEXRAD), have improved the ability to observe the mesoscale structure of snow events and greatly facilitate short-term predictions. Improvements in forecasting mesoscale snowfall are likely to rely on continued advancements in numerical weather prediction models and observing systems such as satellites and radar. Advancements in computer technology and associated forecast systems that allow forecasters ready access to the data from models and observing systems, and provide for the rapid dissemination of short-term forecasts and warnings, will also be crucial for improving the entire suite of forecast products associated with winter storms.

2. Processes contributing to the episodic character of Northeast snowstorms

What factors account for the episodes of heavy snow along the Northeast coast? Perhaps part of the answer lies with the nature of the global circulation within which these storms develop and propagate. Several aspects of the general circulation pattern appear to play important roles in the occurrence of snowstorms that affect the Northeast urban corridor and are listed below:

- The establishment of an upper trough in eastern Canada in association with a negative height anomaly in the middle and upper troposphere is a very important factor. On many occasions, this negative height anomaly is also coupled with a positive height anomaly associated with a blocking ridge across northeastern Canada, Greenland, the Davis Strait, the Labrador Sea, Iceland, and the surrounding North Atlantic. These features are correlated to the negative phase of the North Atlantic Oscillation (NAO), in which a positive height anomaly over the North Atlantic near Greenland and Iceland correlates with a negative height anomaly across the northeastern United States and eastern Canada. This pattern acts to maintain an upper-level trough–jet streak across eastern Canada that can result in reinforcing cold-air outbreaks in the northeast United States.

- The negative phase of the NAO is particularly favorable for Northeast snowstorms since the confluent upper-level height pattern typical with a deep eastern Canadian trough and its associated jet streaks are important components for sustaining a strong surface anticyclone as a source for the low-level cold air along the Northeast coast through a direct circulation within the entrance region of an upper-level jet streak. An intriguing finding is the occurrence of several snowstorms following the nadir of a negative NAO period. This finding implies that the negative phase of the NAO may establish conditions for heavy snowfall but those conditions must "relax" for the snowstorms to affect the Northeast urban corridor.

- Another global circulation pattern that can contribute to Northeast snowstorms is the appearance of an upper ridge across the north Pacific that can extend into Alaska. When this ridge is located north of a separate westerly flow regime, a split-flow pattern develops. This flow regime provides the southward advection of cold air from northern Canada to the United States,

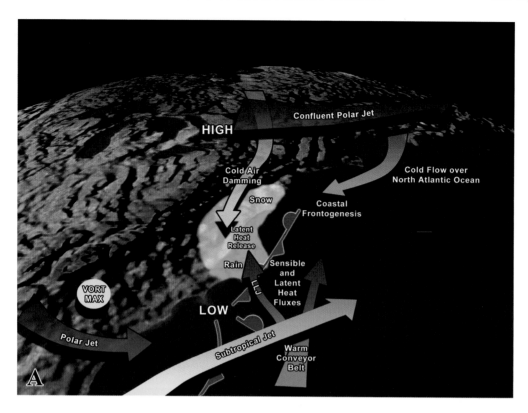

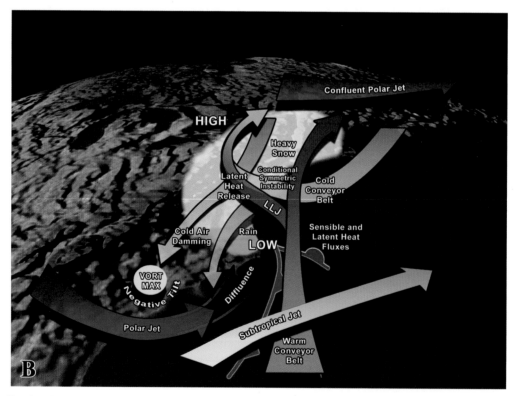

FIG. 8-1. Schematics of factors that contribute to heavy snowfall along the Northeast urban corridor (a) during the initial development of the snowstorm and (b) when the snowstorm is fully developed.

New York's Big Snow of 1947

FORECASTS

December 25, 9:30 P.M.   Tonight cloudy with some snow possible toward morning.
Friday - Cloudy with occasional snow ending during the afternoon
Followed by partial clearing

December 26, 10:30 A.M.   About 7 inches of snow has fallen up to this time.
Indications are for snow to continue to late afternoon or evening
And accumulate 8 to 10 inches.

FIG. 8-2. (top) U.S. Weather Bureau forecasts made prior to the 26 Dec 1947 snowstorm. (bottom) Front page of *New York Times* on 27 Dec 1947, the day following the snowstorm.

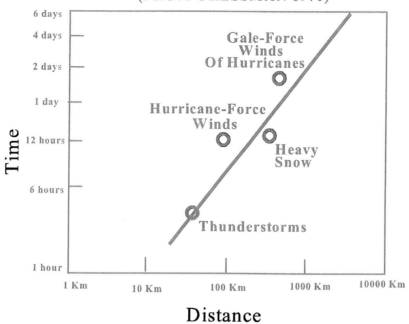

FIG. 8-3. Limits of predictability of public weather forecasts (from Cressman 1970).

while also supporting weather disturbances that move from west to east in the southern branch that could lead to cyclogenesis in the southern third of the United States. These cyclones can then develop into major winter storms that draw in both the moisture from the Gulf of Mexico and Atlantic Ocean, and the colder air associated with the northern stream.

• The episodic nature of the snow events may also be associated with other global circulation regimes such as the Southern Oscillation (SO; or El Niño or ENSO) that have been linked (with the warm phase termed El Niño) to enhanced cyclonic activity that tracks across the southern United States and then along the Northeast coast. On occasion, the increased cyclone activity may be associated with enhanced snowfall along the East Coast when cold air combines with the enhanced storminess. However, in other years, snowfall may be low since cold air is prevented from entering the northeast United States when an El Niño episode is very strong, resulting in the heavy precipitation falling as rain rather than snow along the East Coast. During the opposite cold phase, or La Niña, snowfall is typically below normal in the Northeast urban corridor.

While the large-scale circulation pattern provides an important framework within which these storms evolve, the development of the snowstorm itself represents a manifestation of a complex interaction among many physical processes that occur on the synoptic to me-

soscales (Fig. 8-1; Table 7-1). Upper-level trough–ridge systems provide divergence and ascent required for cyclogenesis. Jet streaks embedded within this trough–ridge system help focus the ascent patterns, transport potential vorticity toward the developing cyclonic circulation, and enhance low-level temperature gradients and moisture transports that are required for heavy snowfall. Cold anticyclones generally have to be positioned to the north of the developing cyclone to sustain the source of low-level cold air required for snow. Mesoscale processes such as cold-air damming, coastal frontogenesis, low-level jet streaks, and conditional symmetric instability all contribute to enhancing the baroclinic environment for cyclogenesis and focus the moisture transports and ascent that enhance the snowfall rates within the urban corridor. Sensible and latent heat fluxes over the ocean and latent heat release within the developing cyclone all act to contribute to the cyclone's rapid intensification. When all of these processes are combined during cyclogenesis in such a way to maximize the thermal advections and moisture transports toward the coast, enhance ascent, and still maintain a deep enough layer of cold air along the coast (Fig. 8-1b), heavy snow is the outcome in the Northeast urban corridor.

3. Advances in the prediction of Northeast snowstorms

Sixty years ago, weather forecasting was based on examining surface weather observations, applying basic

NUMERICAL FORECAST EXPERIMENT FROM 1953: THE NOVEMBER 1950 APPALACHIAN STORM

12 HOUR FORECAST
500 hPa
0300 UTC 25 NOV 1950

(Charney and Phillips, 1953)
Jour. Meteorology

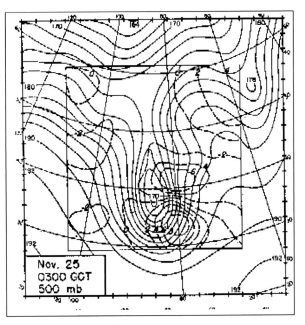

ANALYSES

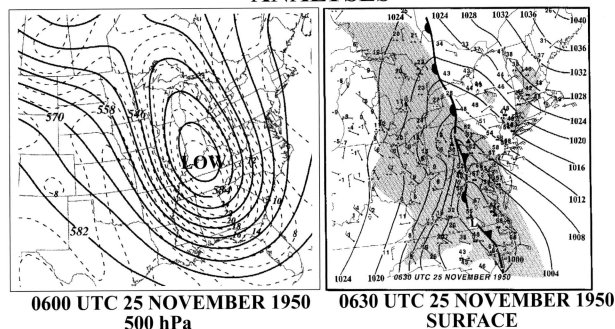

0600 UTC 25 NOVEMBER 1950
500 hPa

0630 UTC 25 NOVEMBER 1950
SURFACE

FIG. 8-4. (top) A 12-h forecast of the 500-hPa geopotential height field at 0300 UTC 25 Nov 1950 (from Charney and Phillips 1953). (bottom left) Analyzed 500-hPa geopotential heights (solid) and vorticity (dashed) at 0600 UTC 25 Nov 1950 (from NCEP–NCAR reanalysis dataset). (bottom right) Surface analysis at 0630 UTC 25 Nov 1950.

physical and thermodynamic principles, and then making a subjective forecast based mainly on the experience level of the forecasters. [For a description of how forecasts were made in the 1940s, the reader is referred to Stewart (1941).] Weather forecasting in the early and middle 20th century was considered more of a subjective art than a science, an art that unfortunately became perhaps better known for spectacular failures than successes. For example, New York City's greatest snowfall, on 26 December 1947, was severely underpredicted, even as the heavy snow enveloped the city (Fig. 8-2). Similar situations occurred during the major snowstorms of February 1969 and 1979. Because of such highly publicized forecast failures, the occasionally correct forecast was not accepted with much confidence. Public officials and agencies responsible for public safety were slow to react since forecasts did not have the credibility required for people to take necessary action to mitigate the potential impact of an impending storm. Cressman's (1970) review of forecast skill in the 1960s shows that the time interval for a useful snow prediction was on the order of 12 h, which was lower than the 1-day skill limit placed on hurricane forecasts at that time (Fig. 8-3).

As discussed by Shuman (1989), Platzman (1979, 1990), and N. A. Phillips (2001, unpublished manuscript), the first numerical prediction experiments, during January and February 1949, were applied to winter cyclones. These experiments were based on a one-level barotropic model run on the Army's Electronic Numerical Integrator and Computer (ENIAC) in Aberdeen, Maryland. Charney and Phillips (1953) utilized these first pioneering experiments to develop a two-level geostrophic model, and then expanded that model to three levels (N. A. Phillips 2001, unpublished manuscript). The first "test case" for these models was the November 1950 "Appalachian Storm," which is discussed in volume II, chapter 9. Some results from the two-level model are reproduced from Charney and Phillips (1953) as Fig. 8-4 and show a 12-h prediction for 500-hPa heights associated with a major storm development over the eastern third of the country.

While the forecasts for the November 1950 cyclone may be considered crude by today's standards, they provided the foundation for one of the great advances for the science of meteorology: the development of complex numerical model systems applied to weather prediction problems and used in real-time weather forecasts. There were a number of early successes, but the failures were memorable and provided the impetus for improving the models through the 1960s and into the 1970s. The major breakthrough during this period was the introduction of hemispheric, and then global, models in the 1970s and 1980s that incorporated physical processes such as latent and sensible heating and that make better use of global datasets to initialize the models. Following the successful early forecasts of the great New England snowstorm of February 1978, even 3 days in advance (Fig.

8-5, top), the operational numerical predictions of the Presidents' Day Snowstorm of February 1979 (Fig. 8-6, bottom) failed to forecast the rapid cyclogenesis or the heavy snow even 12 h in advance. Much research followed to help resolve limitations of the operational models (Bosart 1981; Bosart and Lin 1984; Uccellini et al. 1984, 1985, 1987), especially with regard to the importance of increased model resolution and improved boundary layer parameterizations. Research led to improved forecasts of oceanic sensible and latent heat fluxes, as well as cold-air damming and coastal frontogenesis. Meanwhile, operational forecasts of a late season snowstorm in April 1982 (see volume II, chapter 10.19) and the February 1983 "Megalopolitan" (Fig. 8-7; volume II, chapter 10.20) demonstrated a high degree of success, even though some details about the northern spread of snowfall in February 1983 were not well predicted.

Despite a steady improvement in the forecasts for the middle and upper-tropospheric circulation regimes in the 1960s and 1970s [as measured by skill scores for the 500-hPa height prediction (Bonner 1989)], forecasts of surface cyclones and their associated weather did not improve until the middle to late 1980s (Sanders 1987; Uccellini et al. 1999). The introduction of sophisticated global data assimilation systems, higher-resolution global and regional numerical models, and the application of these models on state-of-the-art supercomputers around 1985 paved the way for improved predictions of cyclogenesis as far as 5 days in advance (Uccellini et al. 1999). Numerical model forecasts of winter storms even captured mesoscale detail in some instances (see Kocin et al. 1985). In 1987, two major January snowstorms were relatively well forecast (see volume II, chapters 10.21 and 10.22) while some of the early forecasts of the February 1987 snowstorm (see volume II, chapter 10.23) failed to merge separate troughs and therefore lacked any significant cyclogenesis, a process also noted by Reed et al. (1992). However, within 1–2 days of the storm's greatest impact, the forecast models began to resolve the upper trough interaction and better simulated the cyclogenesis. In November 1987 (see chapter 6), operational models suggested an early season snowstorm across the Northeast urban corridor, but failed to differentiate the large differences in snowfall accumulations between heavy snow in Washington and Boston, and light snowfalls in Philadelphia and New York City. By the early 1990s, major primary and secondary cyclone events were often being forecast with "useful skill" 3–5 days in advance, even for "data void" regions of the North Atlantic and Pacific Oceans

$\rightarrow$

FIG. 8-5. Medium- and short-range forecasts of the Feb 1978 New England snowstorm. (top left) An 84-h forecast from the primitive equation (PE) baroclinic model valid at 1200 UTC 7 Feb 1978, (top middle) 72-h manual forecast from the National Meteorological Center (NMC, now known as the National Centers for Environmental Prediction; from Brown and Olson 1978), (top right) surface analysis at 1200 UTC 7 Feb 1978. (middle) Short-range (12–48-h forecasts) of 500-hPa geopotential heights (vorticities) and surface pressure (1000–500 hPa geopotential thickness) from NMC LFM model initialized at 1200 UTC 5 Feb 1978. (bottom) Verifying 500-hPa heights (winds) and surface pressure (fronts, precipitation type: green, rain; blue, snow) between 0000 UTC 6 Feb 1978 and 1200 UTC 7 Feb 1978.

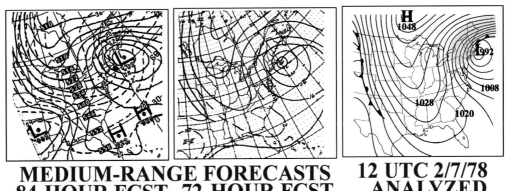

MEDIUM-RANGE FORECASTS
84-HOUR FCST　72-HOUR FCST　　**12 UTC 2/7/78 ANALYZED**

SHORT-RANGE FORECASTS

FORECAST

500 hPa

SURFACE

12 HOUR FCST　24 HOUR FCST　36 HOUR FCST　48 HOUR FCST

ANALYZED

500 hPa

SURFACE

00 UTC 2/6/78　12 UTC 2/6/78　00 UTC 2/7/78　12 UTC 2/7/78

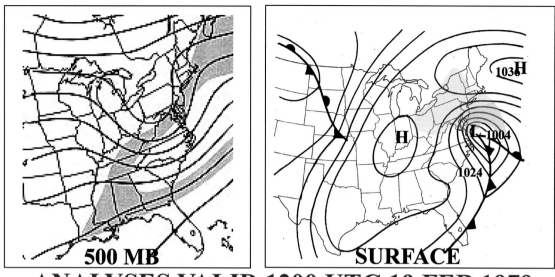

ANALYSES VALID 1200 UTC 19 FEB 1979

500 hPa

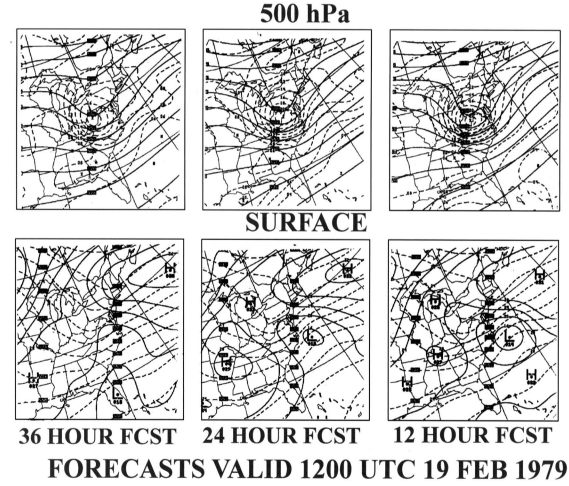

SURFACE

36 HOUR FCST   **24 HOUR FCST**   **12 HOUR FCST**

FORECASTS VALID 1200 UTC 19 FEB 1979

FIG. 8-6. Analyses and short-range forecasts of the February 1979 Presidents' Day Snowstorm. (top) The 500-hPa geopotential heights and winds and surface analysis valid at 1200 UTC 19 Feb 1979. (middle) LFM 36-, 24-, and 12-h forecast 500-hPa geopotential heights (and vorticity; dashed) valid at 1200 UTC 19 Feb 1979. (bottom) LFM 36-, 24-, and 12-h forecast sea level pressure (1000–500 hPa geopotential thickness) valid at 1200 UTC 19 Feb 1979.

(see Grumm 1993; Stokols et al. 1991; Tracton 1993; Uccellini et al. 1999).

From the summary presented above, one can conclude that the accuracy of weather forecasting, even for a rare event such as a major Northeast snowstorm, can be attributed to the introduction of numerical models into the forecast process beginning in the 1950s, the continued improvements made to the numerical models and global data, and the overall professional development of forecasters whose training and education are based heavily on understanding the strengths and weaknesses of the models. Today, forecasts made within the National Weather Service are based on a modernized "end to end" forecast process (Wernly and Uccellini 2000) that involves the following:

- global observations based heavily on remote observations such as automated surface observations, rawinsondes, satellite and radar data, and wind profiler and aircraft data;
- model-based data assimilation systems that take advantage of the strengths of various observing systems (e.g., horizontal coverage of satellite temperature and wind data) while minimizing their weaknesses (e.g., poor vertical resolution of satellite data);
- 16-day global model simulations run once per day and 84-h mesoscale model simulations run four times each day;
- statistical models used to forecast weather elements directly from the numerical model output;
- production of national guidance products by central and regional hydrological forecast offices for hydrometeorological, hydrologic, aviation, marine, tropical, and severe storm predictions;
- production of local forecast and warnings for the general public, and aviation, marine, and fire weather personnel by local forecast offices; and
- coordination and outreach among forecasters and the national/local emergency management community.

This process has served as a basis for several remarkable forecasts of East Coast cyclones in the early and middle 1990s (e.g., the October 1991 "Halloween Gale" or "Perfect Storm" along the East Coast, and the December 1992 storm that produced record snows in the Appalachian Mountains; see volume II, chapter 11.1).

The forecasts for the March 1993 Superstorm (Fig. 8-8; see volume II, chapter 10.24) represent perhaps one of the best efforts at predicting a Northeast snowstorm (Uccellini et al. 1995). The superstorm was predicted 5 days in advance with winter storm and blizzard warnings issued with unprecedented lead times up and down the East Coast (Fig. 8-9). Forecasts for 1–3 ft (30–90 cm) of snow were issued a day prior to the storm and verified over a large area of the Appalachian Mountains and the Northeast. The vast distribution of the snow and the evolution of the rain–snow line near the Northeast urban corridor were both predicted with an accuracy

and a level of confidence that led to the public recognition that followed. The front page of the *Boston Herald* issued prior to the storm (Fig. 8-10) serves as a reminder of the impact of a successful forecast for a major winter storm and the reaction of the public to the forecasts. The newspaper's emphasis on the likelihood of a major winter storm, expected impacts, and the preparedness efforts of the emergency management community in anticipation of the storm saved lives and contributed to mitigation efforts to minimize public danger. The contrast of this front page treatment of the March 1993 Superstorm to the weather forecast of rain followed by "fair" weather, which preceded the "Blizzard of '88" a century before (see Fig. 9-1, volume II) serves as a dramatic illustration of the advances made in forecasting major nor'easters and represents a defining moment in the public's confidence in those forecasts to take necessary steps to mitigate its impact.

The forecasts for the January Blizzard of 1996 (see volume II, chapter 10.27) that crippled the East Coast represented a repeat performance for the entire forecast community. While numerical forecasts 3–5 days in advance were not as consistent in forecasting an East Coast storm, operational models later converged on the development of a major snowstorm 1–2 days prior to the heavy snow (Fig. 8-11). This resulted in forecasts issued for snowfall "measured in feet," rather than in inches, across the Northeast urban corridor a day prior to the snowfall. Even with the early indications of a major event for the mid-Atlantic states, early forecasts did not adequately foretell the heavy snow falling from New York City to Boston, similar to the forecast of the 1983 Megalopolitan Snowstorm (see volume II, chapter 10.20). However, those areas were alerted to the potential for another historic snowstorm a day in advance. Warnings for blizzard conditions and heavy snowfall dominated the media a day before, allowing governors of several states to declare states of emergency and major airlines to cancel and reroute flights before the first flakes fell.

The forecasts for the March 1993 and January 1996 storms are remarkable not only for their accuracy and public response, but especially because the magnitudes of these two storms place them well within the realm of the "rare event" that has, in the past, made such storms unforecastable and unforgettable. Model simulations of the record-setting New England snowstorm of 31 March–1 April 1997 (see volume II, chapter 10.28) had mixed success 3 days in advance, but within 24 h of the major cyclogenesis and heavy snows, model forecasts of the location and intensity of the surface low were remarkably accurate (Fig. 8-12).

Even with the forecast successes noted above, there are no guarantees that all major snowstorms will be predicted with similar accuracy. The forecasts of the March 1993 Superstorm, the January Blizzard of 1996, and the New England snowstorm of 31 March–1 April 1997 led many to believe that new forecast standards

36-hour Forecast

48-hour Forecast

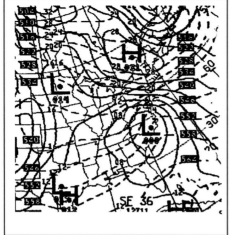

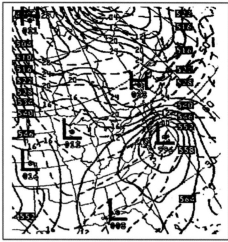

Surface, 1000-500 hPA thickness

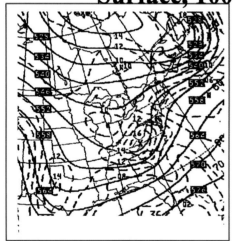

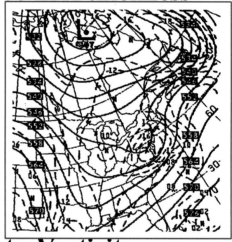

500 hPA Heights, Vorticity

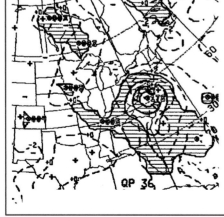

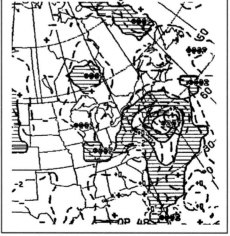

700 hPA VV, Precipitation

Valid 12 UTC 2/11/83　　Valid 00 UTC 2/12/83

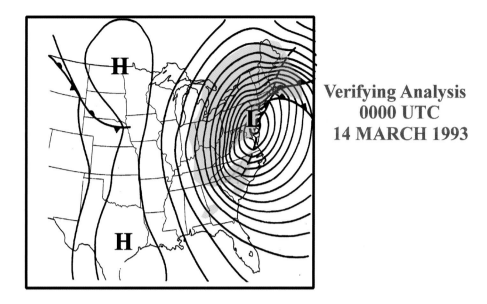

Forecasts valid at 0000 UTC 14 MARCH 1993

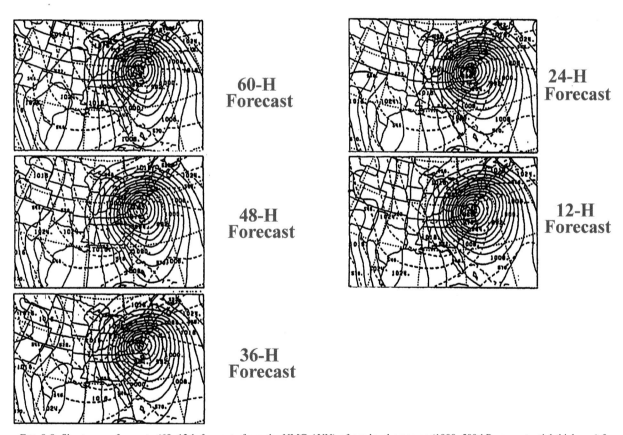

FIG. 8-8. Short-range forecasts (60–12-h forecasts from the NMC AVN) of sea level pressure (1000–500 hPa geopotential thickness) for the Mar 1993 Superstorm, valid at 0000 UTC 14 Mar 1993 (from Caplan 1995), including a surface analysis valid at 0000 UTC 14 Mar 1993.

←

FIG. 8-7. Operational 36- and 48-h forecasts of the Feb 1983 "Megalopolitan" Snowstorm. (top) Sea level pressure (solid) and 1000–500 hPa geopotential thickness (dashed); (middle) 500-hPa geopotential heights (vorticities) and (bottom) 700-hPa vertical motions and precipitation amounts valid at (left) 1200 UTC 11 Feb 1983 and (right) 0000 UTC 12 Feb 1983.

MARCH SUPERSTORM
WATCH AND WARNING LEAD TIME

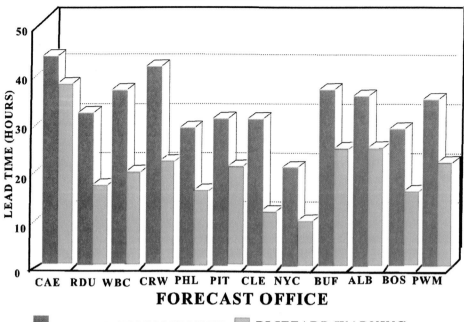

FIG. 8-9. Comparison of lead times for winter storm watches and blizzard warnings [heavy snow warnings were issued at Columbia, SC (CAE)]; forecasts were issued by the local NWS Forecast Offices in the Eastern Region of the National Weather Service: RDU, Raleigh–Durham, NC; WBC, Washington, DC–Sterling, VA; PHL, Philadelphia, PA; CRW, Charleston, WV; PIT, Pittsburgh, PA; CLE, Cleveland, OH; NYC, New York City, NY; BUF, Buffalo, NY; ALB, Albany, NY; BOS, Boston, MA; and PWM, Portland, ME. Lead times are in h. [Adapted from Uccellini et al. (1995).]

had been attained that could be applied to future events. In fact, there were still significant errors contained within these good forecasts, which lends credence to an old aphorism: "the devil lies in the details." In 1993, the observed cyclogenesis in the Gulf of Mexico was vastly underpredicted although most models were correct in predicting a very deep cyclone, but the errors in location and intensity of the initial cyclogenesis likely resulted in poor forecasts of coastal flooding in Florida that led to the loss of several lives. Also, the forecast of heavy snow was not perfect, in that observed snowfall was heavier farther west than indicated, resulting in heavy snowfall in places such as eastern Ohio, which did not expect much snow. Prior to the January 1996 blizzard, some early forecast models indicated that the storm would move out to sea with little effect on New York City and southern New England. As the event drew closer, the models shifted the snows farther and farther north. In the April 1997 storm, the deep cyclone was relatively well forecast but the very heavy band of snow [20–30 in. (50–75 cm)] that affected eastern New England from late on 31 March through the late morning of 1 April was a surprise in its duration and intensity.

Therefore, it appears that while the forecast and timing of the rapid cyclogenesis that accompanies many of these snowstorms have significantly improved since the days of the 1979 Presidents' Day Snowstorm, the details of the resulting precipitation amounts and their locations continue to be a major source of forecast error.

It also appears that the precipitation distribution of some storms appears to be more predictable than that of others. With the dawn of the 21st century, the snowstorms of 24–25 January 2000, 30 December 2000 (see volume II, sections 10.29 and 10.30) and 4–6 March 2001 (see section 8.4) clearly showed the limitations of snowfall prediction without an appreciation of predictability and uncertainties.

The 30 December 2000 and 4–6 March 2001 snowstorms are both cases in which the cyclogenesis was predicted 5 days or more in advance of each storm. However, just 24–48 h prior to the 30 December 2000 snowstorm, some forecast models pointed to a distinct possibility of significant snow from Washington to New York City. While that forecast verified in Philadelphia and New York City, residents from Washington to eastern Pennsylvania woke to sunshine rather than the forecast heavy snow (see chapter 6 and Fig. 6-19). As on 30 December 2000, the 4–6 March 2001 cyclone and snowstorm potential were foreseen 5–7 days in advance. However, with each successive forecast cycle, the area

FIG. 8-10. Headline from the *Boston Herald* for 13 Mar 1993. Reprinted with permission
(from Uccellini et al. 1995).

of predicted heavy snowfall kept creeping northward. As the event drew closer, the snowfall bypassed a substantial portion of the forewarned urban corridor, creating a public furor over "hyping" the storm. Considering that forecasters at one point were looking for the heaviest snow accumulations possibly as far south as North Carolina, while they ultimately occurred from central New York to southern and central New England, points to the conclusion that there is still much that needs to be learned before we can consistently and accurately predict the area and timing of the heavy snow associated with these storms.

The 24–25 January 2000 snowstorm is an example of not only failures in forecasting precipitation amounts and location, but also represents failures in forecasting the onset of cyclogenesis, even 24 h in advance. In this case, medium-range (3–5 day) forecasts missed the event completely. Short-term models (1–2 day) erroneously forecast the cyclone and its associated heavy precipitation farther east than was actually observed, resulting in a major unforecast snowstorm in the Washington, Baltimore, and Philadelphia metropolitan areas. As the event approached, the model forecasts still had the storm track 100 km too far east even 24 h before

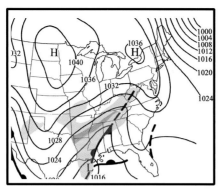

INITIAL ANALYSIS

1200 UTC 6 JANUARY 1996

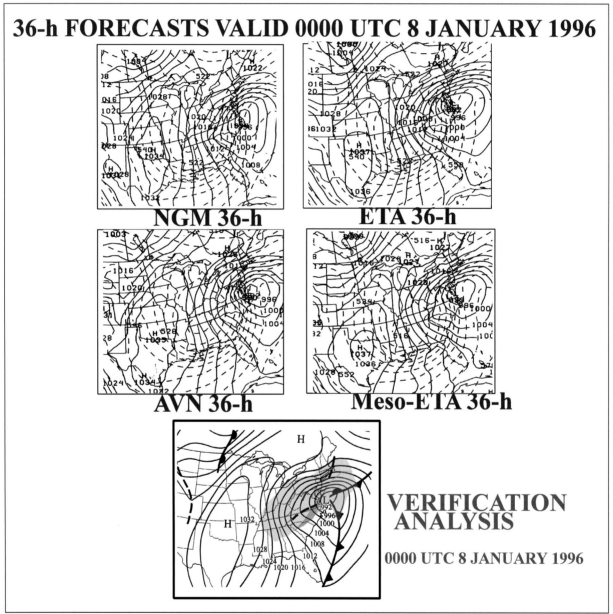

FIG. 8-11. The 36-h forecasts of the January 1996 blizzard. (top) Surface analysis at 1200 UTC 6 Jan 1996, the time of model initialization for the subsequent forecasts. The middle four panels show 36-h forecasts of sea level pressure and 1000–500-hPa thickness valid at 0000 UTC 8 Jan 1996 from the Nested Grid Model (NGM), Eta, AVN, and Meso-Eta numerical models. (bottom) Surface analysis valid at 0000 UTC 8 Jan 1996.

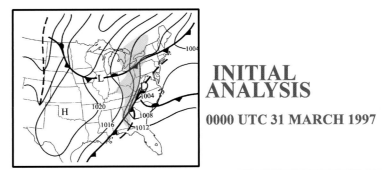

INITIAL ANALYSIS

0000 UTC 31 MARCH 1997

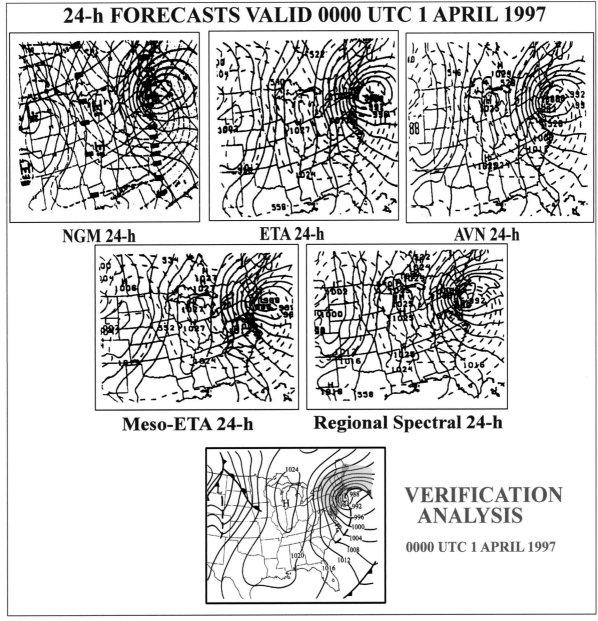

24-h FORECASTS VALID 0000 UTC 1 APRIL 1997

NGM 24-h

ETA 24-h

AVN 24-h

Meso-ETA 24-h

Regional Spectral 24-h

VERIFICATION ANALYSIS

0000 UTC 1 APRIL 1997

FIG. 8-12. The 24-h forecasts of the 31 Mar–1 Apr 1997 snowstorm. (top) Surface analysis at 0000 UTC 31 Mar 1997, the time of model initialization for the subsequent forecasts. The middle five panels show 24-h forecasts of sea level pressure and 1000–500-hPa thickness valid at 0000 UTC 1 Apr 1997 from the NGM, Eta, AVN, Meso-Eta, and regional spectral models. (bottom) Surface analysis valid at 0000 UTC 1 Apr 1997.

the event and kept the heaviest precipitation east of the Washington-to-Philadelphia metropolitan areas (Fig. 8-13). Later model runs did bring the cyclone and precipitation farther west, but it was too little and too late for a successful forecast, even 12 h before the event from North Carolina to southeastern New York (Fig. 8-13c).

Challenges include improving the initialization of numerical models and addressing related observation issues, especially in the regions upstream (Pacific Ocean and polar regions), a step that would add more consistency to the extended-range forecasts of East Coast storms. Another challenge involves improving the mesoscale details of the short-range forecasts, by enhancing the ability to predict the exact timing, position, and movements of the rain–snow line, the area of mixed precipitation, and the position of heavy snow bands and the associated mesoscale distribution of heavy snowfall. This forecast challenge is being addressed by the combined use of Doppler radar and digital geostationary satellite data and the improvements in mesoscale models with resolutions approaching 10 km. There is still much room for improvement to ensure a reliable, consistent forecast ability that can be sustained from one storm to the next, which thus can meet the raised expectations of the public who expect forecasts to be more accurate and consistent.

Another challenge involves extending the range of predictability of Northeast snowstorms and the means to assess properly the degree of uncertainty of these types of systems. Forecasters' observations that some storms appear more reliable to current forecast techniques begs for an understanding of why some storms seem to be inherently predictable several days in advance while others are problematic right up to the event. The continued expansion of ensemble model predictions (Tracton and Kalnay 1993), based on numerical predictions from many perturbed initial states, could serve as the basis for meeting this challenge. Today, medium- to long-range forecasts (3 days and beyond), and even short-range (1–2 days) forecasts, are impacted by a run-to-run model variability that detracts from the accuracy and reliability of the forecast and the confidence with which this forecast can be issued. The ensemble model approach should provide for more stable and more likely forecast solutions out to a week in advance, allowing the forecaster to interact with the public and the emergency management community with a measured level of confidence, or certainty, when forecasting these events.

Another major challenge facing the forecast community remains the forecast of light to moderate snow events that affect the Northeast urban corridor more frequently than the major storms that grab the headlines. Even today, forecasters seem to have greater difficulty with predicting storms that can come and go quickly and affect relatively small domains. Indeed, there is a growing awareness that some of the cyclones associated with the "big storms" may be more predictable in advance than the smaller snow events. The inability to make consistent predictions of the light to moderate events also erodes public confidence. Improvements in mesoscale modeling and more sophisticated data assimilation schemes, based on more detailed mesoscale datasets, will lead to a better understanding of these snow events and our ability to forecast them.

Finally, a continuing effort is envisioned to develop an even more comprehensive group of cases, covering a larger variety of storms over a longer period of time. The compilation of major snowstorm cases will be extended forward in time as new storms occur. It will also be expanded back in time to include all the major Northeast snowstorms for which data are available. Of course, much of the original data prior to the middle 1940s consist only of surface measurements. However, the surface patterns still contribute to an understanding of many of the physical processes described throughout this monograph. Additional research, ranging from climatological reviews to model-based case studies, together with improved observations, especially over the data-sparse oceans, are all needed. It is only through a combination of research efforts that our understanding of these storms will be enriched and our ability to forecast them with acceptable consistency and increasing lead times may be realized.

4. Northeast Snowfall Impact Scale (NESIS) derived from Northeast storm snowfall distributions

A Northeast Snowfall Impact Scale (NESIS), a scale that can be used to make a relative comparison of the snowstorms presented in this monograph and to provide a quantitative means for making such an assessment is presented in this section. As discussed by Kocin and Uccellini (2004), 30 snowstorms described in chapters 3 and 4 form the basis for NESIS. These 30 snowstorms were all characterized by snowfall distributions *with large areas of 10-in. (25 cm) accumulations and greater* and affected millions of people (Table 3-1; Figs. 3-2a,b). These snowfall distributions, mapped onto the population of the northeast United States, provide a "calibrated" basis for NESIS. NESIS makes use of Geographic Information Systems (GIS) technology, which facilitates the digital mapping of snowfall distribution and population density and takes the following form:

Northeast Snowfall Impact Scale (NESIS)

$$= \sum_{n}^{x} [n * (A_n/A_{mean} + P_n/P_{mean})]. \tag{8.1}$$

In equation (8.1), n represents selected values of snowfall (in in.) divided by 10, n = 1 is used for the area of snowfall 10 in. (25 cm) and greater, n = 2 is used for the area 20 in. (50 cm) and greater, and so on.

36-Hour, 24-Hour and 12-Hour forecasts
1200 UTC 25 JANUARY 2000
Eta MODEL

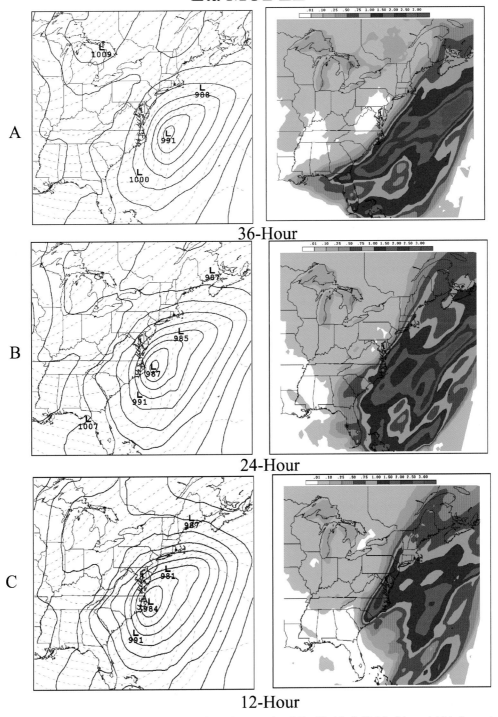

FIG. 8-13. Forecasts of the 24–25 Jan 2000 snowstorm from the operational Eta Model: (left) 36-, 24-, and 12-h forecasts of sea level pressure (solid) and thickness (dashed) valid at 1200 UTC 25 Jan 2000 and (right) 48-, 36-, and 24-h forecasts of total precipitation valid at 0000 UTC 26 Jan 2000.

TOTAL SNOWFALL DISTRIBUTION

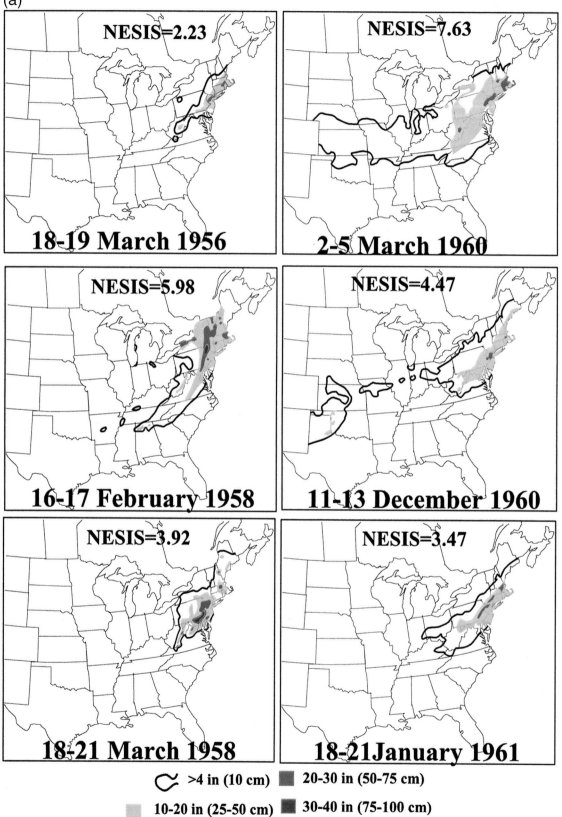

FIG. 8-14a–e. Storm snowfall in excess of 10 (solid line), >25 (light blue), >50 (dark blue), and >75 cm (red) for all 30 snowstorms used to generate the NESIS scale, with corresponding NESIS value.

TOTAL SNOWFALL DISTRIBUTION

(b)

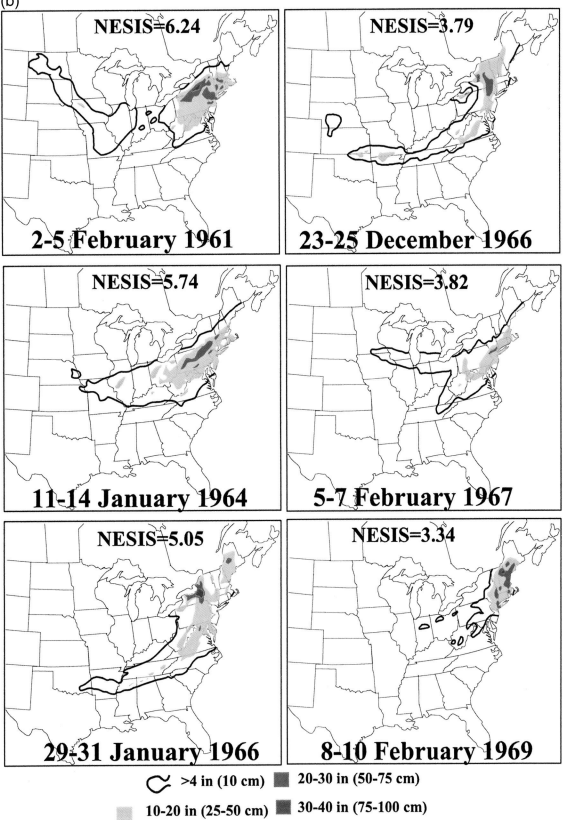

FIG. 8-14b. (*Continued*)

TOTAL SNOWFALL DISTRIBUTION

(c)

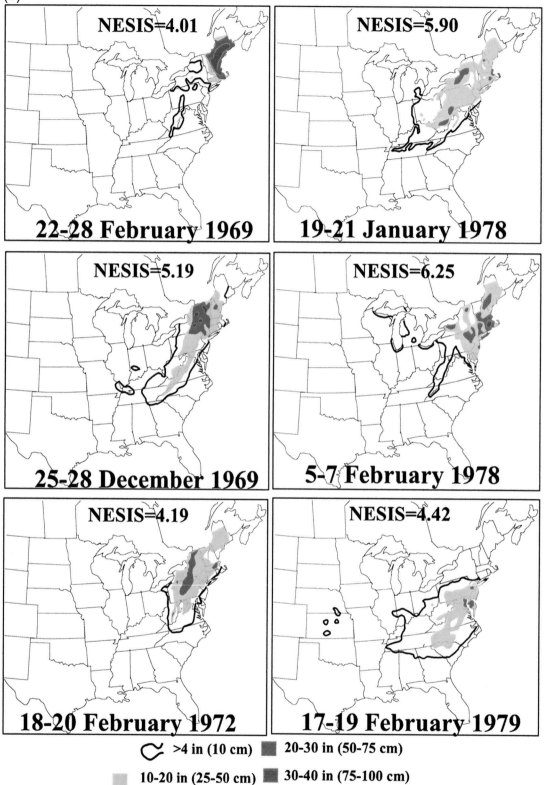

FIG. 8-14c. (*Continued*)

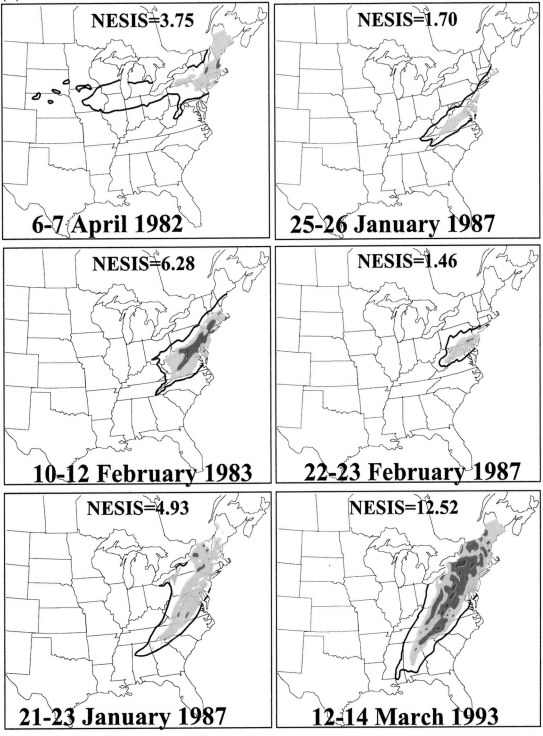

FIG. 8-14d. (*Continued*)

TOTAL SNOWFALL DISTRIBUTION

(e)

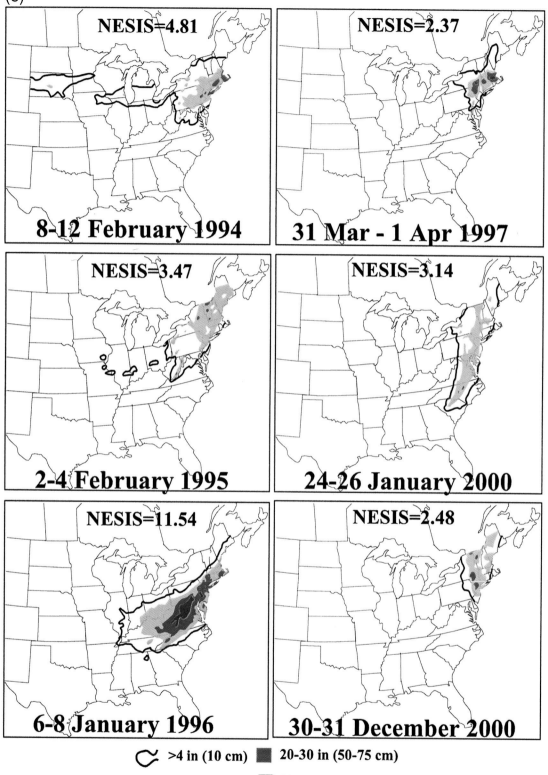

FIG. 8-14e. (*Continued*)

TABLE 8-1. Estimated area ($\times 10^3$ mi^2) and population [(pop.) in millions, 1999 census], affected by snowfall accumulations of 4 (10), 10 (25), 20 (50), and 30 in. (75 cm) during 30 Northeast snowstorms (Fig. 8-14), Ranked (from highest to lowest) NESIS values (using 1999 census data) are computed from the total snowfall distribution of 30 cases used to calibrate the NESIS equation.

Rank	Date	4-in.		10-in.		20-in.		30-in.		NESIS
		Area	Pop.	Area	Pop.	Area	Pop.	Area	Pop.	
1	12–14 Mar 1993	386.0	89.2	283.5	66.8	142.4	19.6	12.9	1.8	12.52
2	6–8 Jan 1996	313.8	82.3	200.1	66.1	90.2	39.8	15.1	5.1	11.54
3	2–5 Mar 1960	590.4	108.6	140.8	57.0	7.6	8.5			7.63
4	10–12 Feb 1983	157.1	58.5	112.6	51.6	33.7	25.7	0.9	0.2	6.28
5	5–7 Feb 1978	220.2	67.4	132.3	48.0	30.7	16.0	0.9	1.2	6.25
6	2–5 Feb 1961	369.3	85.0	114.0	50.7	19.4	8.7	1.4	0.2	6.24
7	14–17 Feb 1958	282.6	72.0	129.2	54.6	20.2	6.0	3.4	0.8	5.98
8	19–21 Jan 1978	295.2	79.5	167.7	53.1	8.3	3.2			5.90
9	11–14 Jan 1964	356.5	87.6	129.6	51.2	10.3	1.5			5.74
10	25–28 Dec 1969	250.6	61.2	138.7	25.9	37.6	4.0			5.19
11	29–31 Jan 1966	371.4	83.1	111.7	22.2	12.3	2.4	1.5	0.5	5.05
12	21–23 Jan 1987	286.9	79.1	153.7	38.4	2.0	0.1			4.93
13	8–12 Feb 1994	280.0	86.5	57.7	39.3	4.4	13.4			4.81
14	11–13 Dec 1960	302.9	68.0	78.5	48.0	0.6	2.5			4.47
15	17–19 Feb 1979	304.0	72.1	88.2	36.9	4.3	3.0			4.42
16	18–20 Feb 1972	206.3	59.5	140.9	24.5	13.5	1.4			4.19
17	22–28 Feb 1969	101.7	20.6	48.4	10.3	40.8	8.2	24.2	4.2	4.01
18	18–21 Mar 1958	146.7	53.7	62.1	40.7	13.8	7.5	3.5	0.7	3.92
19	5–7 Feb 1967	246.0	81.1	50.9	44.8					3.82
20	23–25 Dec 1966	292.2	63.0	89.8	18.1	9.9	1.4			3.79
21	6–7 Apr 1982	258.3	75.5	79.3	28.7	2.1	0.6			3.75
22	2–4 Feb 1995	200.1	62.6	98.0	29.9					3.51
23	18–21 Jan 1961	144.9	57.4	62.3	43.0	5.7	2.9			3.47
24	8–10 Feb 1969	107.5	40.4	66.4	31.2	11.6	9.6			3.34
25	24–26 Jan 2000	205.6	64.9	74.2	23.6	0.3	0.2			3.14
26	30–31 Dec 2000	103.8	40.0	56.5	28.0	3.7	1.4			2.48
27	31 Mar–1 Apr 1997	76.4	31.9	32.0	13.0	13.1	7.0	3.1	2.2	2.37
28	18–19 Mar 1956	64.9	44.8	28.6	32.8	2.6	2.6			2.23
29	25–26 Jan 1987	74.3	34.8	38.0	11.5					1.70
30	22–23 Feb 1987	61.3	35.6	28.3	16.6	0.3	0.1			1.46

To account for snowfall exceeding 4 in., n = 0.4 is also included in (8.1).

Here, A_n is the estimated area of snowfall exceeding n ($\times 10$) in. for any given snowstorm, and A_{mean} is the mean area of snowfall *greater than 10 in. (25 cm)* derived from the 30 major snowstorms described in chapters 3 and 4 for the 50 yr from 1950 to 2000 (Table 3-1). In addition, P_n is the population (in 1999 census figures) estimated to live within the snowfall area A_n (Table 8-1), and P_{mean} is the mean population for the 30 cases (also computed with 1999 census figures) within the area of snowfall greater than 10 in. (25 cm). Area and population are estimated in the GIS system by utilizing a county database and selecting all counties in which *at least half* of the county is analyzed to lie within a given snowfall interval (i.e., 4 in. or greater, n = 0.4; 10 in. or greater, n = 1; 20 in. or greater, n = 2; etc.).

The scale is calibrated by first computing A_{mean} and P_{mean} for 10-in. (25 cm) snowfall accumulations within the 13-state area from West Virginia–Virginia northeastward to Maine for each of 30 cases (Fig. 3-2; Table 3-1). Final values of NESIS are then computed for the *total* snowfall distribution east of the Rocky Mountains (Figs. 8-14a–e; Table 8-1). These steps recognize that the basis for the application of NESIS is the impact of heavy snowfall on the Northeast urban corridor, while

also accounting for the total snow history associated with these storms as they track across the United States.

For example, heavy snowfall from the New England snowstorm of late February 1969 (Fig. 8-14c) was confined solely to eastern New England, while many other snowstorms, such as those in March 1960, January 1964, and February 1979 (Figs. 8-14a–c), were part of more widespread storm systems affecting larger portions of the nation. Thus, NESIS represents a measure of the integrated, or total, impact of a snowfall within and outside the northeast United States, calibrated by the 30 storms from 1950 to 2000 that had the largest apparent impact upon the Northeast urban corridor. Furthermore, the scale provides added weight to the higher snowfall increments (n = 2, 3, . . .), which are generally maximized in the northeastern part of the United States for these selected storms, reflecting the greater potential disruption when very heavy snow falls in the most densely populated areas.

Values of A_n and P_n (area and population) are derived using the Arcview 3.0 GIS software for total areas of snowfall exceeding 10 in. (25 cm) for each case. The areal coverage of 10-in. (25 cm) accumulations from all 30 storms (Table 3–1) ranges from 28 to 212.6 $\times 10^3$ mi^2, with a mean area (A_{mean}) of 91.03 $\times 10^3$ mi^2. The population, derived from 1999 census figures, affected by snowfall accumulations greater than 10 in. (25 cm), ranges from 10.3 to 59.9 million in the 30-case sample, with a mean population (P_{mean}) of 35.4 million.

The NESIS values computed for the 30 cases are ranked from highest to lowest in Table 8-1 and range from 1.46 to 12.52 (Fig. 8-14), with an average value of 4.80. The March 1993 storm has the largest NESIS value of 12.52, given the large areal extent of 4-, 10-, and 20-in. (10, 25, and 50 cm) snows extending west into the Ohio Valley and south across the southeast United States (Fig. 8-14d). The March 1993 storm has, by far, the largest areas of greater than 10- and 20-in. (25 and 50 cm) snowfall among the 30 storms (Table 8-1). The blizzard of January 1996 scores second highest of the 30 cases, and scores closest to the March 1993 event, in part, because it has the largest population affected by 20-in. (50 cm) snows (Table 8-1). Both March 1993 and January 1996 are the only storms in which the area of 10-in. (25 cm) snowfall exceeds 200 × 10³ mi². March 1960 scores third highest with 7.63, reflecting the largest area and population affected by greater than 4-in. (10 cm) snowfall of all 30 cases, much of which occurred west of the Appalachians (Fig. 8-14a). The three snowstorms of February 1983, February 1978, and February 1961 all score above 6. Each case produced large areas of 10- and 20-in. (25 and 50 cm) snowfalls over highly populated regions (Table 8-1), although impacting noticeably smaller areas than the top 3 storms in the 30-case sample.

Five cases scored between 5 and 6 (Table 8-1) and were also widespread storms over populated areas, but the smaller numbers reflect very heavy snowfall greater than 20 in. (50 cm) falling either over smaller areas or less populated areas than those cases that scored 6 or greater (Table 8-1). Six cases score between 4 and 5, and with one exception, are either widespread snowfalls exceeding both 4 in. (10 cm) and 10 in. (25 cm), but with small areas of 20 in. (50 cm) and greater. The one exception is the snowstorm of late February 1969, which scores greater than 4 despite relatively small areas of 4- and 10-in. (10 and 25 cm) snows, but which did have an unusually large area of 20-, 30-, and 40-in. (50, 75, and 100 cm) snows (Table 8-1). Eight storms score between 3 and 4 and these storms have 10-in. (25 cm) snowfalls that tend to cover smaller areas than higher-scoring cases (ranging between 50 and 100 × 10³ mi²; roughly covering an between the size of the state of New York and the combined states of New York, Pennsylvania, and New Jersey; Table 8-1). The snowstorm of March 1958 scores close to 4 because of the relatively higher area and populations affected by greater than 20- and 30-in. (50 and 75 cm) snowfalls. Five cases score less than 3 primarily due to relatively small areas of greater than 10-in. snowfalls (38 × 10³ mi² and less, roughly an area smaller in size than the state of Virginia; Table 8-1).

A summary of the range of NESIS values for the 30 Northeast snowstorms is shown in Fig. 8-15. A majority

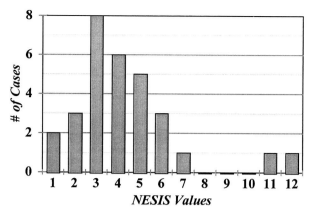

FIG. 8-15. Distribution of NESIS values for 30 cases of Northeast snowstorms used to calibrate the NESIS scale.

of cases attain values in the 3–5 range (19 of 30 cases) while the most significant storms (e.g., March 1993 and January 1996) with the heaviest snows in the most populous areas have NESIS values much greater than the rest of the cases.

a. Application of the NESIS to other cases

In this section, NESIS is applied to an independent sample of 30 "near miss" snowstorms (see chapter 5), four historic cases (see volume II, chapter 9), a major storm system on 4–6 March 2001 that received extensive media coverage for a perceived lack of impact on the major metropolitan areas in the Northeast, and five cases during the winter of 2002/03, including the heavy snowfall associated with the Presidents' Day Snowstorm of 15–18 February 2003 (hereafter Presidents' Day II, to differentiate it from the Presidents' Day storm of 1979).

1) "NEAR MISS" SNOWSTORMS FOR THE NORTHEAST URBAN CORRIDOR

The "near miss" cases described in chapter 5 either produced 1) the heaviest snowfall in the interior locations, west and north of the Northeast urban corridor; 2) moderate snowfall amounts related to storms whose snowfall is generally limited by a number of factors (i.e., rapidly moving storms); or 3) precipitation that fell more as sleet or freezing rain rather than snow. Total snowfall distributions for 30 near-miss cases are analyzed (only interior and moderate snowstorms are included), and areas and populations for 4-, 10-, and 20-in. (10, 25, and 50 cm) total snowfall distributions are shown in Table 8-2, as are the resultant NESIS values.

NESIS values for the interior snowstorms range from 1.86 to only 4.45, and average 3.0, well below the average of 4.8 for the 30 original cases. NESIS values for the moderate snowstorms range from 1.20 to 4.85 and average 2.1. Smaller NESIS values tend to occur with storms in which snowfall is limited to areas of the in-

TABLE 8-2. Dates of 30 interior (INT) and moderate (MOD) cases; area ($\times 10^3$ mi^2) and population [(pop.) millions, 1999 census] affected by greater than 4-, 10-, and 20-in. (10-, 25-, and 50-cm) snowfall; and NESIS values.

Date	Type	4-in. Area	4-in. Pop.	10-in. Area	10-in. Pop.	20-in. Area	20-in. Pop.	NESIS
16–17 Feb 1952	INT	125.2	23.2	66.7	10.2	12.4	1.1	2.17
16–17 Mar 1956	INT	195.5	56.8	92.3	14.6			2.93
12–13 Mar 1959	INT	215.3	56.6	121.1	19.2	7.7	0.2	3.64
14–15 Feb 1960	INT	353.9	47.4	142.1	15.1	23.3	0.6	4.17
6–7 Mar 1962	INT	148.6	37.6	70.0	10.8	19.3	1.3	2.76
19–20 Feb 1964	INT	169.7	54.3	53.4	12.2	3.5	0.4	2.39
22–23 Jan 1966	INT	296.4	60.9	145.1	22.6	6.6	1.5	4.45
3–5 Mar 1971	INT	195.7	37.6	101.6	10.5	23.3	1.9	3.73
25–27 Nov 1971	INT	163.4	26.9	73.4	10.6	6.6	1.1	2.33
16–18 Jan 1978	INT	364.4	62.0	122.1	16.2			4.10
28–29 Mar 1984	INT	124.6	31.0	53.3	11.2	2.1	0.2	1.86
1–2 Jan 1987	INT	164.6	34.0	76.6	11.1			2.26
10–12 Dec 1992	INT	118.7	28.6	61.6	15.1	21.5	5.5	3.10
3–5 Jan 1994	INT	222.3	41.3	76.4	9.3	10.5	1.7	2.87
2–4 Mar 1994	INT	165.4	47.9	109.1	12.3			3.46
3–5 Dec 1957	MOD	87.2	44.8	9.4	11.7			1.32
23–25 Dec 1961	MOD	105.5	44.0	14.8	8.6			1.37
14–15 Feb 1962	MOD	101.4	33.0	33.8	12.7	0.4	0.6	1.59
22–23 Dec 1963	MOD	374.2	75.9	51.3	21.0			3.17
16–17 Jan 1965	MOD	214.5	69.0	15.3	10.3			1.95
21–22 Mar 1967	MOD	62.3	35.9	7.0	15.6			1.20
31 Dec 1970–1 Jan 1971	MOD	151.0	55.3	46.4	6.8	4.4	0.2	2.10
13–15 Jan 1982	MOD	382.2	83.6	133.9	13.4			3.08
8–9 Mar 1984	MOD	120.9	53.0	54.6	5.5			1.29
7–8 Jan 1988	MOD	488.5	80.5	129.7	9.5			4.85
26–27 Dec 1990	MOD	166.0	58.4	12.7	8.8			1.56
19–21 Dec 1995	MOD	260.3	62.0	85.4	19.0			3.32
2–4 Feb 1996	MOD	157.3	56.4	44.1	7.1	0.9	0.1	2.03
16–17 Feb 1996	MOD	136.7	52.6	12.2	11.4			1.65
14–15 Mar 1999	MOD	180.3	49.5	58.8	5.8	1.4	0.1	2.20

terior Northeast while the storms with higher values tend to be more widespread snowfalls covering larger portions of the nation.

Three examples of representative "interior" snowstorms are shown in Fig. 8-16 (left-hand side). These three cases are selected from 15 interior snowstorms described in chapter 5 and produced the heaviest snow inland away from the major metropolitan areas of the Northeast urban corridor. The three cases, 10–11 December 1992, 3–5 January 1994, and 2–4 March 1994 (Figs. 8-16a–c), each produced a large region of greater than 10 in. (25 cm) of snow, as well as significant areas of greater than 20 in. (50 cm) of snow. NESIS values for these interior cases are 3.10, 2.87, and 3.46 (Table 8-2), respectively, reflecting the relatively large areas of heavy snowfall, although much of the snow fell over less populated areas compared to the "major" storms discussed in the previous section.

Three examples of representative "moderate" snowstorms are also shown in Fig. 8-16 (right-hand side) and represent snowfalls in the Northeast urban corridor that are dominated by snowfalls of 4–10 in. [10–25 cm; with some areas of greater than 10 in. (25 cm)]. The each of these three cases, 8-9 March 1984, 26–27 December 1990, and 16–17 February 1996 (Figs. 8-16d–f), affected much of the Northeast urban corridor with 4–10

in. (10–25 cm) of snow, with smaller areas of greater than 10 in. (25 cm). NESIS values for these cases are 1.29, 1.56, and 1.65, respectively (Table 8-2).

Four snowstorms that affected the Northeast as either moderate or interior snowstorms *but* were also widespread snowfalls across other areas of the United States are shown in Fig. 8-17. Three cases, 14–15 February 1960, 22–23 January 1966, and 16–18 January 1978 (Figs. 8-17a–c), were selected as interior snowstorms since the heaviest snows fell across interior portions of the Northeast. However, these three storms also produced heavy snows greater than 10 in. (25 cm) across widespread areas of the Ohio Valley, the Southeast, or the Tennessee Valley. These three storms have NESIS values of 4.17, 4.45, and 4.10 (Table 8-2). These values are higher than other interior snowstorms shown in Fig. 8-16 because the heavy snow fell in areas well beyond the Northeast, impacting other major metropolitan areas, thus elevating the NESIS values. While the moderate snowstorms represented by the three cases in Fig. 8-16 score between 1 and 2, one moderate case in the Northeast on 7–8 January 1988 was also part of a heavy snowstorm across the southern plains and the Southeast. As a result of the widespread snowfall, this moderate snowfall scores 4.85 (Fig. 8-17d; Table 8-2).

Therefore, while the 30 major snowstorms of the pre-

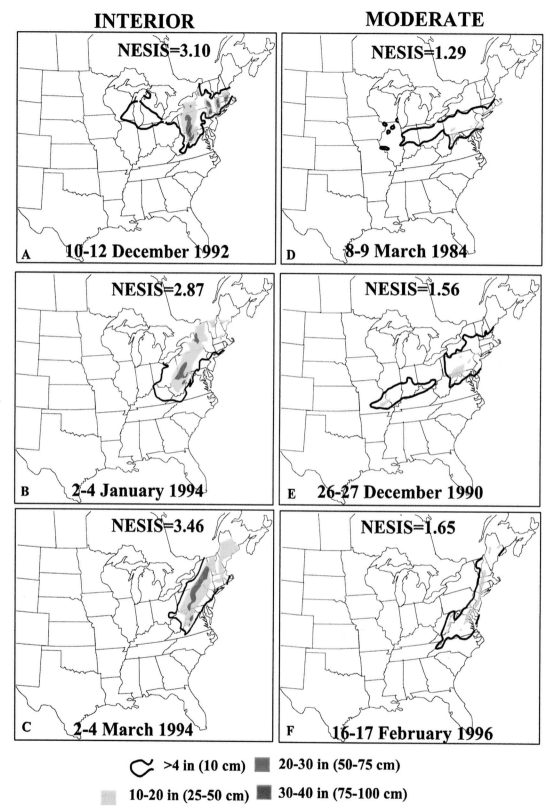

FIG. 8-16. Storm snowfall in excess of 10 (solid line), >25 (light blue), >50 (dark blue), and >75 cm (red) for selection of representative interior and moderate snowstorms, with corresponding NESIS value.

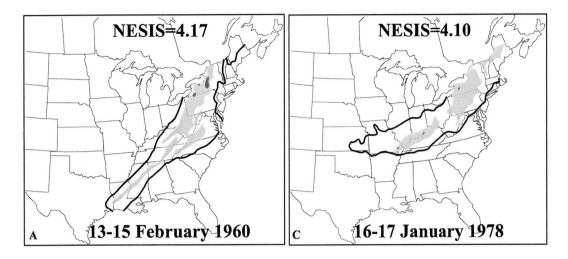

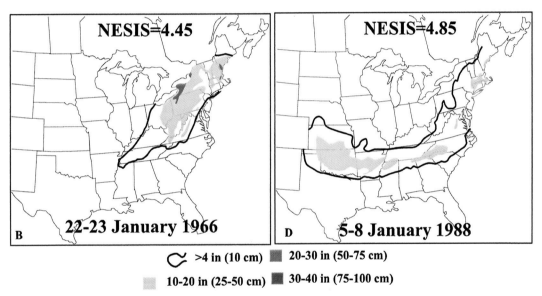

FIG. 8-17. Storm snowfall in excess of 10 (solid line), >25 (light blue), >50 (dark blue), and >75 cm (red) for several widespread interior and moderate snowstorms in which significant snowfall extends outside the Northeast, including NESIS values.

ceding section have an average NESIS value near 5, interior snowstorms that that do not extend far outside the Northeast have NESIS values generally between 2 and 4, while moderate snowstorms within the same bounds have NESIS values generally between 1 and 2. More widespread snowfalls that move across the United States and ultimately affect the Northeast score higher and can raise the NESIS value by as much as 1–2 or more points.

2) HISTORIC CASES

NESIS is also applied to four historical Northeast snowstorms (Table 8-3), which are described in volume II, chapter 9, and shown in Fig. 8-18. The Blizzard of

'88, perhaps the most infamous of all Northeast snowstorms (see Kocin 1983), has an NESIS value (8.34), using the 1999 census, that is lower than either the March 1993 Superstorm (12.52) or the January Blizzard of 1996 (11.54). An examination of the snowfall distribution of the Blizzard of '88 (Fig. 8-18a) shows that the total areal coverage of the snowfall is relatively small, especially when compared with the 1993 and 1996 cases (Figs. 8-14d,e). However, the NESIS value computed for this storm is larger than 28 of the 30 cases examined in Table 8-1, due to the unusually large regions of snowfall greater than 20, 30, and 40 in. (50, 75, and 100 cm, respectively) over populated areas (using 1999 census values).

The February 1899 blizzard [Fig. 8-18b; see Kocin

TABLE 8-3. Dates of four historical snowfall cases and six recent snowfall cases, for area and population [(pop.) millions, 1999 census] affected by greater than 4-, 10-, 20-, and 30-in. (10, 25, 50, and 75 cm) snowfall ($\times$ 10³ mi²), and for NESIS values.

Date	4 in.		10 in.		20 in.		30 in.		NESIS
	Area	Pop.	Area	Pop.	Area	Pop.	Area	Pop.	
Historical									
11–14 Mar 1888*	144.9	52.7	87.9	37.9	48.2	26.1	24.8	12.8	8.34
11–14 Feb 1899	362.1	81.7	181.8	61.7	33.0	20.0			8.11
27–29 Jan 1922	107.1	46.3	62.3	26.0	22.4	1.0	10.5	1.4	3.63
26–27 Dec 1947	114.0	46.9	35.4	31.1	5.3	16.5	0.5	1.7	3.50
2001–03									
4–6 Mar 2001	161.1	40.2	105.1	21.6	30.4	5.6	1.8	0.1	3.53
4–5 Dec 2002	269.7	64.7	6.1	0.4					1.99
25 Dec 2002	345.3	72.8	91.3	18.5	13.8	1.5	4.4	0.2	4.42
3 Jan 2003	211.1	35.6	77.4	10.9	11.0	1.5			2.65
6–7 Feb 2003	88.4	50.2	6.1	5.5					1.18
15–18 Feb 2003	303.5	78.2	142.0	59.2	51.9	40.9	2.7	0.2	8.91

* Mar 1888 also has areas and populations for snowfall exceeding 40 in. (100 cm; area = 7.8 $\times$ 10³ mi²; population = 1.9 million) and 50 in. (125 cm; area = 0.8 $\times$ 10³ mi²; population = 0.2 million; 1999 census).

et al. (1988)] was a widespread snowstorm from the southeastern United States northeastward to New England. This storm culminated one of the coldest periods ever recorded in the eastern United States and paralyzed the eastern third of the country. The NESIS value of 8.11 computed with the 1999 census population data is also higher than 28 of the 30 cases and is slightly lower than the 1888 blizzard. The relatively high value is due mainly to the large distribution of greater than 10- (25 cm) and 20-in. (50 cm) snowfall amounts along the entire coast from Virginia to Maine, home to a large proportion of Northeast residents, as well as significant snow throughout the Southeast and the Tennessee Valley (Fig. 8-18b).

Both the 1922 "Knickerbocker" storm (Fig. 8-18c) and the December 1947 New York City snowstorms (Fig. 8-18d) produced record snowfall for Washington and New York City, respectively, but have NESIS values of only 3.63 and 3.50, respectively. These values are less than the average for the 30-case sample, due to the comparatively small areal extent of heavy snowfall, even though the heavy amounts are both focused locally in densely populated regions. Therefore, while the greatest snowstorms in New York City's and Washington's histories factor into the NESIS computations, the storms have a lower overall impact given the small areal extent of the snowfall associated with these storms. These two examples illustrate the role of NESIS in providing an integrated measure of the regional snowfall, rather than focusing on *local* snowfall measurements as a major impact agent.

3) RECENT CASES

(i) The 4–6 March 2001 snowstorm

As another illustration of how the scale may be applied to assess the impact of a storm, NESIS is applied

to a snowstorm that occurred on 4–6 March 2001. For a week to 3 days prior to the snowstorm, numerical weather prediction models and forecasters saw the potential for this storm to affect the Northeast urban corridor with heavy snow from Washington to New York City. Forecasters in several local media outlets warned about the potential impact of this storm up to 4 days in advance. The National Weather Service issued numerous winter storm watches, affecting possibly tens of millions of people, 1–2 days in advance and subsequently issued warnings from Philadelphia to Boston. National Weather Service medium-range forecast discussions, combined with the emphasis placed on the storm's potential by the media, raised the expectations within the general population that this would be a major snow event. However, the forecast heavy snowfall continued to shift farther and farther north with each succeeding series of numerical weather forecasts after 2 March 2002, resulting in the suspension of winter storm watches and warnings across much of the area from Pennsylvania southward. The storm was viewed by many as a disappointment because the dire forecasts (and resulting "hype") for a major snowstorm did not materialize from New York City southward.

A major cyclone did develop along the East Coast on Monday, 5 March, as predicted, but the snowstorm occurred farther north and inland. The storm left an area of heavy snowfall (greater than 25 cm; Fig. 8-19a) larger than the mean area for the 30 major snowstorms shown in Table 8-1 (105 $\times$ 10³ mi² vs 91 $\times$ 10³ mi²). This area covered much of the Boston metropolitan area but missed the urban corridor between Washington and New York City, although eastern suburbs of New York City received up to 16 in. (40 cm) of snow. Using 1999 census data, an estimated 21 million people were affected by greater than 10-in. (25+ cm) amounts, considerably less than the 35.4 million mean for the 30

HISTORIC SNOWSTORMS

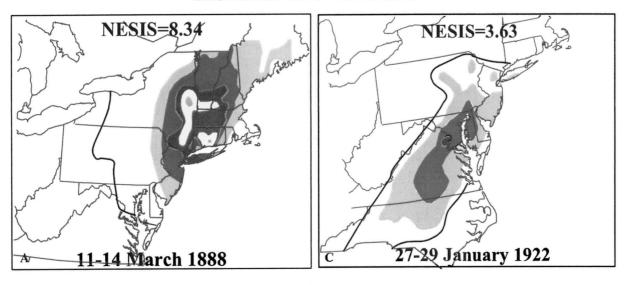

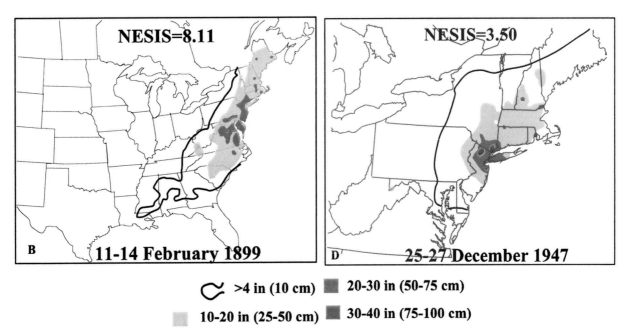

FIG. 8-18. Storm snowfall in excess of 10 (solid line), >25 (light blue), >50 (dark blue), and >75 cm (red) for four historic snowstorm cases including (a) the 11–14 Mar 1888 "Blizzard of '88," (b) the 11–14 Feb 1899 "Blizzard of 1899," (c) the "Knickerbocker" snowstorm of Jan 1922, and (d) New York City's "big snow" of Dec 1947.

cases shown in Table 8-1. In addition, more than 5 million people are estimated to have been affected by greater than 20-in. (50 cm) amounts.

This case yields an NESIS value of 3.53 (Table 8-3), which is considerably lower than the average of 4.80 for the 30 major snowstorm cases but slightly higher than the representative examples for interior snowstorms shown in Fig. 8-16. Had the earlier forecasts verified and the heaviest snowfall occurred approximately 300 km farther south, affecting the New York to Washington corridor, more than 45 million people would have wit-

nessed snowfall accumulations exceeding 10 in. (25 cm), resulting in an NESIS value estimated between 4.5 and 6. The NESIS demonstrates that this storm was comparable, in terms of area and population affected, to some of the lower-scoring major snowstorms documented in Table 8-1 and scored similarly to the many interior snowstorms also discussed in chapter 5. However, the media and public reaction that this snowstorm did not "measure up" to expectations related to forecasts prior to its development is also confirmed by the use of the NESIS scale.

RECENT SNOWSTORMS

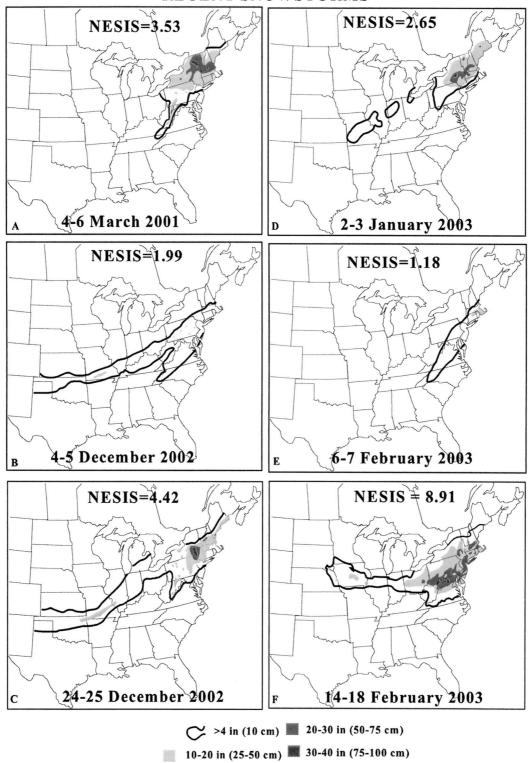

FIG. 8-19. Storm snowfall in excess of 10 (solid line), >25 (light blue), >50 (dark blue), and >75 cm (red) for five recent snowstorms, including (a) 4–6 Mar 2001, (b) 4–5 Dec 2002, (c) 24–25 Dec 2002, (d) 2–3 Jan 2003, (e) 6–7 Feb 2003, and (f) the Presidents' Day II storm of 14–18 Feb 2003.

(ii) The snowstorms of 4–5 December 2002, 24–25 December 2002, 3–4 January 2003, and 6–7 February 2003

The winter of 2002/03 produced many significant snowfalls for the Northeast urban corridor and culminated in the Presidents' Day II snowstorm of 15–18 February, one of the heaviest snowstorms to affect the Washington–Boston corridor. The first significant snowfall threat in two winters greeted much of the Northeast urban corridor on 4–5 December 2002 as a general snowfall of 5–9 in. (13–23 cm) occurred from Virginia to southern New England. The area of heaviest snowfall in the Northeast (Fig. 8-19b) is similar in location to the snowstorms of February 1983 and January 1996 (see Fig. 3-2b), but maximum amounts were much less. Few sites reported 10 in. (25 cm) or greater but the region exceeding 4 in. (10 cm) was widespread, including the Ohio Valley and the southern plains. The early December snowfall scores an NESIS value of 1.99 from the widespread occurrence of greater than 4-in. (10 cm) snows from the southern plains, to the Missouri and Tennessee Valleys, and into the Northeast, but scored less than the values exhibited by many of the 30 cases used to derive the scale since few areas received 10 in. (25 cm) or greater. This snowfall was a classic moderate snowstorm, as described in chapter 5, that had an NESIS value scoring near 2 with snowfall widespread within and outside the Northeast.

The Christmas 2002 snowstorm was associated with a rapidly developing cyclone that left a large band of snowfall exceeding 20–30 in. (50–75 cm; Fig. 8-19c) across the Mohawk and Hudson Valleys of New York and the Catskill Mountains. Snow changed to rain and then back to snow in the metropolitan areas from Washington to Boston, and near-blizzard conditions occurred in the New York City metropolitan area as 4–12 in. (10–30 cm) fell as the surface low deepened to 970 hPa. This storm also affected a large area outside the northeast United States, including the southern plains and Great Lakes. The Christmas storm rates a 4.42, scoring heavily because of widespread areas of greater than 4 in. (10 cm) of snow from the southern plains through the Ohio Valley and the Northeast (where much heavier snow fell). This storm represents a classic interior snowstorm for the Northeast in which heaviest snows fell mostly outside the major metropolitan areas. However, it scored higher than a representative interior snow because significant snows extended well outside the Northeast.

Only a week later, on 3–4 January 2003, another cyclone developed and was associated with the heaviest snows again occurring over the Mohawk and Hudson Valleys of New York and the Catskills (Fig. 8-19d), as well as central New England. For the most part, however, snowfall did not materialize from Washington to New York City while the Boston metropolitan area exhibited large variations in snowfall as rain fell in the immediate city while many suburbs saw significant snow. The January 2003 storm scores 2.65, another typical interior snowstorm with similarities to the Christmas snowstorm, but the snowfall was much less widespread.

On 6–7 February 2003, an area of 4–10 in. (10–25 cm) of snow affected the entire Northeast corridor again from Virginia to Massachusetts with a region from northeastern Connecticut, northern Rhode Island, and eastern Massachusetts experiencing greater than 10 in. of snow (25 cm; see Fig. 8-19e). The snowstorm of 6–7 February 2003 scores a 1.18, representative of a moderate snowstorm that affected a similar area of the Northeast as did the snowfall of 4–5 December 2002, but this snowstorm was more limited in scope, with little significant snow outside the Northeast.

(iii) Presidents' Day II Snowstorm (15–18 February 2003)

The most significant and widespread snowstorm since the January Blizzard of '96 affected many major metropolitan areas of the middle Atlantic states and New England on 16–18 February 2003 (it affected the Midwest on 14–15 February). For the Northeast urban corridor from Washington to Boston, the snowfall was one of the heaviest on record and paralyzed a wide area during a long holiday weekend. An area of 4–10 in. (10–25 cm) of snow affected much of the Northeast with a large area of greater than 10-in. (25 cm) accumulations from northern Virginia through central New England (Fig. 8-19f). Snowfall exceeding 20 in. (50 cm) was unusually widespread within the area from Washington to Boston and some notable totals include 27.5 in. (70 cm) at Boston (the greatest 24-h snowfall on record) and a storm total of 28.2 in. (72 cm) at Baltimore (the greatest snowfall storm total on record). The population of the area (using the 1999 census) affected by snowfall exceeding 20 in. (50 cm) was comparable to, if not slightly larger than, the Blizzard of 1996 (41 million, compared to 39.8 million; Table 8-3 vs Table 8-1). Scattered areas of 30-in. (75 cm) totals were reported from West Virginia to New York with a few reports of greater than 40 in. (100 cm) near the western panhandle of Maryland. The heavy snow extended into the Ohio Valley and the Midwest (Fig. 8-19), resulting in an NESIS value of 8.91 (Table 8-3), the third highest value of 70 cases examined to date. This high value reflects the number of highly populated areas affected by snowfall of near and greater than 20 in. (50 cm). Nearly 80 million people are estimated to have been affected by greater than 4 in. (10 cm) of snow from the Midwest to the Northeast while nearly 60 million people experienced greater than 10 in. (25 cm) of snow.

b. Categorizing NESIS values

The NESIS values calculated for the 70 snowfall cases, including the ranges and means of the snowfall dis-

TABLE 8-4. NESIS categories (1–5), their corresponding NESIS values, number of the 70 total cases within each category, and a descriptive "adjective."

Category	NESIS values	No. of cases	Description
1	1–2.49	23	Notable
2	2.5–3.99	22	Significant
3	4–5.99	16	Major
4	6–9.99	7	Crippling
5	10.0+	2	Extreme

tribution and populations affected, allow a quantitative means to partition the NESIS values into several categories. A categorical ranking of 1–5 is proposed (Table 8-4; similar to the Saffir–Simpson scale) that utilizes the divisions inherent in the NESIS values (all 70 cases scored between 1 and 12) to separate the snowfalls into similar categories based on area, population, and the occurrence of very heavy snowfall. These categories, their corresponding NESIS values, and total number of cases within each category are shown in Table 8-5.

1) CATEGORY 1 (NESIS = 1.0–2.49): NOTABLE

All storm snowfall distributions examined in this paper score NESIS values of 1 and greater. A total of 23 out of 70 cases occur in this category and include 5 of the 30 cases used to calibrate the scale, 5 of the interior snowstorms, and 11 of the moderate snowstorms. Of the recent cases, the 3–5 December 2002 snowstorm and the 6–7 February 2003 snowstorm fall into this category. These storms are "notable" for their large areas of 4-in. (10 cm) accumulations and small areas of 10-in. (25 cm) snowfall. An example of a category 1 snowfall is shown in Fig. 8-20a.

2) CATEGORY 2 (NESIS = 2.5–3.99): SIGNIFICANT

A total of 22 cases out of 70 occur in this category and include 8 of the 30 original cases, 7 of the interior snowstorms, and 3 of the moderate snowstorms. The historical snowstorms of January 1922 and December 1947, the heaviest snowfalls in Washington and New York City, fall into this category. Of the recent cases examined, two cases (March 2001 and January 2003) also fit into this category. This category includes storms that produce "significant" areas of greater than 10-in. (25 cm) snows while some include small areas of 20-in. (50 cm) snowfalls. A few cases may even include relatively small areas of very heavy snowfall accumulations [greater than 30 in. (75 cm)], including Washington's and New York City's greatest snowfalls (see section 5c).

3) CATEGORY 3 (NESIS = 4.0–5.99): MAJOR

A total of 16 cases occur in this category, including 11 of the 30 original cases, and this category includes

the mean of 4.80 for the 30 cases. This was the highest category attained by only two of the interior snowstorms and one moderate snowstorm because it was associated with a very widespread distribution of snow. This category encompasses the typical "major" Northeast snowstorm with large areas of 10-in. snows (generally between 50 and 150 × 100³ mi—roughly 1 to 3 times the size of the state of New York) with significant areas of 20-in. (50 cm) accumulations. An example of a category 3 snowfall is shown in Fig. 8-20b.

4) CATEGORY 4 (NESIS = 6.0–9.99): CRIPPLING

A total of 7 cases occur in this category, including 4 of the original 30 cases, 2 historical cases (the blizzards of 1888 and 1899), the February 1978 blizzard, and the recent Presidents' Day II snowstorm of 15–18 February 2003. These storms consist of some of the most widespread, heavy snows of the sample and can best be described as "crippling" to the northeast United States, with the impact to transportation and the economy felt throughout the United States. These storms encompass huge areas of 10-in. (25cm) snowfalls and each case is marked by large areas of snowfall accumulations greater than 20 in. (50 cm).

5) CATEGORY 5 (NESIS = 10.0+): EXTREME

Only two cases, the March 1993 Superstorm and the January Blizzard of 1996, occur in this category and represent the "extreme" storms that blanket large areas and population centers with heavy snowfall greater than 10, 20, and 30 in. (25, 50, and 75 cm). These are the only storms in which the 10-in. (25 cm) accumulations exceed 200 × 10³ mi²) and affect more than 60 million people (1999 census data). The March 1993 Superstorm derives the highest ranking given the largest area covered by greater than 10-in. (25 cm) snowfall in the entire sample, compounded by large areas of 20- (50) and 30-in. (75 cm) snowfall. The January 1996 snowstorm has similarities to other category 3 and 4 storms, except that this storm was accompanied by unusually large areas of snowfall greater than both 20 (50) and 30 in. (75 cm) that affected the large population centers within the entire Northeast urban corridor. The snowfall distribution of the January 1996 storm is shown in Fig. 8-20c as an example of a category 5 snowfall.

c. The effect of population change between 1900 and 1999 on NESIS values

Given the large population shifts that have occurred in the United States during the course of the 20th century, NESIS is evaluated utilizing census values from several periods between 1900 and 1999 to assess if the scale changes significantly as the population of the Northeast changes. In certain parts of the United States (especially portions of the southeast and the southwest),

TABLE 8-5. Seventy cases ranked from highest to lowest by NESIS value (using 1999 census data), as well as NESIS category (1–5) and description (see Table 8-4).

No.	Date	NESIS	Category	Description
1	12–14 Mar 1993	12.52	5	Extreme
2	6–8 Jan 1996	11.54	5	Extreme
3	15–18 Feb 2003	8.91	4	Crippling
4	11–14 Mar 1888	8.34	4	Crippling
5	11–14 Feb 1899	8.11	4	Crippling
6	2–5 Mar 1960	7.63	4	Crippling
7	10–12 Feb 1983	6.28	4	Crippling
8	5–7 Feb 1978	6.25	4	Crippling
9	2–5 Feb 1961	6.24	4	Crippling
10	14–17 Feb 1958	5.98	3	Major
11	19–21 Jan 1978	5.90	3	Major
12	11–14 Jan 1964	5.74	3	Major
13	25–28 Dec 1969	5.19	3	Major
14	29–31 Jan 1966	5.05	3	Major
15	21–23 Jan 1987	4.93	3	Major
16	7-8 Jan 1988	4.85	3	Major
17	8–12 Feb 1994	4.81	3	Major
18	11–13 Dec 1960	4.47	3	Major
19	22–23 Jan 1966	4.45	3	Major
20	17–19 Feb 1979	4.42	3	Major
21	24–25 Dec 2002	4.42	3	Major
22	18–20 Feb 1972	4.19	3	Major
23	14–15 Feb 1960	4.17	3	Major
24	16–18 Jan 1978	4.10	3	Major
25	22–28 Feb 1969	4.01	3	Major
26	18–21 Mar 1958	3.92	2	Significant
27	5–7 Feb 1967	3.82	2	Significant
28	23–25 Dec 1966	3.79	2	Significant
29	6–7 Apr 1982	3.75	2	Significant
30	3–5 Mar 1971	3.73	2	Significant
31	12–13 Mar 1959	3.64	2	Significant
32	27–29 Jan 1922	3.63	2	Significant
33	3–5 Mar 2001	3.53	2	Significant
34	2–4 Feb 1995	3.51	2	Significant
35	26–27 Dec 1947	3.50	2	Significant
36	18–21 Jan 1961	3.47	2	Significant
37	2–4 Mar 1994	3.46	2	Significant
38	8–10 Feb 1969	3.34	2	Significant
39	19–20 Dec 1995	3.32	2	Significant
40	22–23 Dec 1963	3.17	2	Significant
41	24–26 Jan 2000	3.14	2	Significant
42	10–12 Dec 1992	3.10	2	Significant
43	13–15 Jan 1982	3.08	2	Significant
44	16–17 Mar 1956	2.93	2	Significant
45	3–5 Jan 1994	2.87	2	Significant
46	6–7 Mar 1962	2.76	2	Significant
47	3–4 Jan 2003	2.65	2	Significant
48	30–31 Dec 2000	2.48	1	Notable
49	19–20 Feb 1964	2.39	1	Notable
50	31 Mar–1 Apr 197	2.37	1	Notable
51	25–27 Nov 1971	2.33	1	Notable
52	1–2 Jan 1987	2.26	1	Notable
53	18–19 Mar 1956	2.23	1	Notable
54	14–15 Mar 1999	2.20	1	Notable
55	16–17 Feb 1952	2.17	1	Notable
56	31 Dec 1970–1 Jan 1971	2.10	1	Notable
57	2–4 Feb 1996	2.03	1	Notable
58	4–5 Dec 2002	1.99	1	Notable
59	16–17 Jan 1965	1.95	1	Notable
60	28–29 Mar 1984	1.86	1	Notable
61	25–26 Jan 1987	1.70	1	Notable
62	16–17 Feb 1996	1.65	1	Notable
63	14–15 Feb 1962	1.59	1	Notable

TABLE 8-5. (Continued)

No.	Date	NESIS	Category	Description
64	26–27 Dec 1990	1.56	1	Notable
65	22–23 Feb 1987	1.46	1	Notable
66	23–25 Dec 1961	1.37	1	Notable
67	3–5 Dec 1957	1.32	1	Notable
68	8–9 Mar 1984	1.29	1	Notable
69	21–22 Mar 1967	1.20	1	Notable
70	6–7 Feb 2003	1.18	1	Notable

significant changes in population distribution during the 20th century would indicate that impact scales related to population density would change significantly over time. However, many of the metropolitan areas and population centers of the northeast United States were already established by 1900. Populations of some major metropolitan areas actually peaked in 1950 and decreased in subsequent years. The resulting shifts in population to suburban areas surrounding the cities in the Northeast corridor and the general shift of the populations from inland to coastal areas raises this question: Would the changing population distributions during the 20th century change the calibrated NESIS values in a fundamental way?

United States census data for 1900, 1950, 1960, 1970, 1980, 1990, and 1999 are used to compare the NESIS values of several storms given the change of population across the Northeast during the 20th century. Within the area called the Northeast urban corridor (Fig. 1-1), the population has increased approximately 70% since 1900 (from 14 to 45 million) while the remainder of the 13-state region between the Virginias and Maine has increased roughly 50% (from 10 to 20 million) in the same period. While there are some differences in the values for the 30 cases during the course of the century, the NESIS values remain relatively constant. Several cases increase slightly over time and others decrease slightly. While the March 1993 Superstorm is ranked first in 1900 (13.27) and 1999 (12.52), its value drops over the course of the century, while the NESIS values for the second-ranked storm, the January 1996 blizzard, remain relatively constant from 1900 (11.47) to 1999 (11.54). These differences reflect the decrease in population across some interior portions of the Northeast after 1950 (heaviest snows in the March 1993 storm occurred over the interior Northeast while the heaviest snows in January 1996 were centered over the Northeast urban corridor) and a slight general shift of the population toward the coastal areas during the same period. However, most of the values throughout the century fluctuate within 10% of the 1999 values for nearly all cases, indicating that the NESIS scale provides a consistent measure of impact for the northeast United States whose population *distribution* has shifted from the interior and city locations more toward the cities and the burgeoning suburbs during the 20th century. Given the population shifts

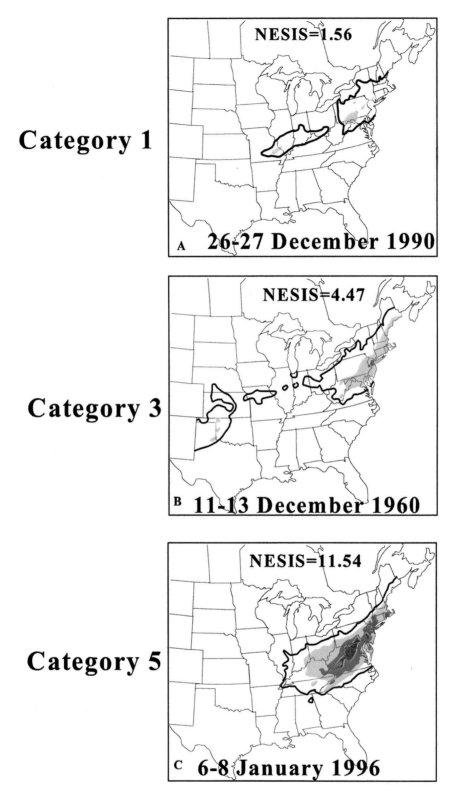

FIG. 8-20. Representative examples of snowfall distributions for (a) category 1, (b) category 3, and (c) category 5 storms.

NESIS Categories for 30 major snowstorms

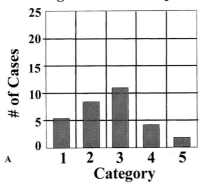

NESIS Categories for 70 cases

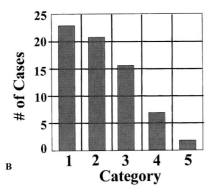

FIG. 8-21. Distribution of NESIS categories (1–5) for (top) 30 major snowstorms described in chapters 3 and 4; (bottom) 70 total cases (see text).

that will continue to occur in the 21st century, it is recommended that as new storms occur, their NESIS value should be computed with the most recent census count.

d. NESIS summary

A Northeast Snowfall Impact Scale (NESIS) is derived to convey a measure of the impact of snowstorms over the Northeast urban corridor as a function of the total snowfall distribution, snowfall amounts mapped directly to population density (based on 1999 census data). The scale is calibrated by the snowfall distributions of 30 high-impact snowstorms from a synoptic climatology provided in chapters 3 and 4 and applied to a total of 70 cases. The NESIS is an integrated measure of snowfall impact, rather than relying on instantaneous descriptions of a variety of parameters associated with the surface low. The scale also accounts for a greater impact associated with heavier snow amounts. Computed values are used to construct a categorical ranking from 1 to 5. NESIS provides a *simple* quantitative means to convey a measure of the impact of those storms in which the areal coverage of the snowfall upon

large population centers contributes to widespread human and economic disruption.

The NESIS is applied to 70 cases (Table 8-5), including 30 major snowstorms and 30 near-miss cases that occurred over the 50-yr period from 1950 to 2000 (as described in chapter 5), 4 historic cases, and 6 recent cases. NESIS differentiates limited moderate snowstorms that produce snows of mainly 4–10 in. (NESIS scores generally between 1 and 2; category 1) over a relatively small area over the Northeast, from interior snowstorms that miss the major cities but drop widespread heavy snowfall (NESIS scores generally between 2 and 4; categories 1–2). NESIS also differentiates major snowstorms with sizable areas of 10-in. snows over relatively large areas (NESIS scores between 4 and 6; category 3) from the rare "megastorm" that drops 10–20 in. (25–50 cm) and greater over large populated regions (NESIS values greater than 6; categories 4 and 5). For the 30 major snowstorms used to generate the scale, NESIS averages 4.8, with a distribution that is skewed toward the middle category (category 3; Fig. 8-21a). For the 70 cases, NESIS averages near 3.8, with a distribution that is skewed toward the first three categories (Fig. 8-21b).

Out of the 70 cases, 26 score greater than an NESIS value of 4 and include 3 interior snowstorms and 1 moderate snowstorm. The interior and moderate snowstorms that score greater than 4 are typically widespread snowstorms that affect the interior Northeast and other sections of the nation as the storm traverses the country toward the East Coast. Additional snowfall from areas outside the northeast United States can contribute as much as three points to the overall NESIS value when snowfall is widespread.

The occurrence of the three highest-ranked storms since 1990 represents an intriguing coincidence. Even when the population criteria are dropped from NESIS (not shown; the scale then simply becomes a measure of the snowfall distribution alone), the March 1993 Superstorm and the January 1996 blizzard stand alone as the most widespread snowfalls of the 70-case sample. The 2003 Presidents' Day Snowstorm moves down the ranking from third to fifth highest, scoring slightly less than the widespread snowstorms of March 1960 and the Blizzard of 1899. Therefore, while seasonal snowfall appears to be diminishing in recent years in the Northeast urban corridor as the number of storms producing significant snowfall decreases, it appears that for the storms that do produce significant snowfall, some of these storms may be producing *heavier, more widespread snows* than the storms that occurred during the first 90 years of the 20th century. Given uncertainties and temporal changes in data quality, changes in the density of the snow observation network, and subjectivity in contouring and spatial interpolation, more research is needed to quantify the significance of this finding.

At present, NESIS is best used as a relative assessment of storms that *have already occurred.* Furthermore, similar scales can also be developed and applied to other parts of the *country where synoptic climatologies of snowstorm snowfall are available.* Given the current difficulties in forecasting precipitation type and snowfall amounts and areal distribution associated with these events, we do not *yet* recommend the use of the NESIS in a *predictive* manner. As confidence and accuracy in predicting snowfall amounts and areal distribution increase, NESIS can provide an estimate of the upcoming impact of these storms, both in scope and population affected. The scale provides meteorologists, transportation officials, economists, planners, the media, and emergency managers the means to alert the public and business communities to take steps that can mitigate the impact of these storms.

APPENDIX

Seasonal Snowfall Statistics

Comprehensive snowfall statistics are included for Boston, Massachusetts; New York, New York; Philadelphia, Pennsylvania; Baltimore, Maryland; and Washington, D.C., through the 2006/07 winter season. The listing includes snowiest months of record, the 10 snowiest and least snowy seasons on record, and the 10 greatest snowstorms. Also included are listings of all seasonal snow totals and a summary of all snowstorms exceeding 25 cm. A less comprehensive listing of snowfall statistics is also provided for 29 other sites in the Northeast, including mean seasonal snowfall (1971–2000), the snowiest months, the 10 snowiest seasons, 10 least snowy seasons, and the 10 greatest snowstorms.

Boston snowfall
(1872/73–2006/07)

Snowiest months			**Snowiest seasons**		**Least snowy seasons**		**Biggest storms**	
In.	Month	Year	In.	Season	In.	Season	In.	Date
1.1	Oct	2005	107.6	1995/96	9.0	1936/37	27.6	17–18 Feb 2003
17.8	Nov	1898	96.4	1873/74	10.0	1875/76	27.1	6–7 Feb 1978
27.9	Dec	1970	96.3	1993/94	10.3	1972/73	26.3	24–27 Feb 1969
43.1	Jan	2005	89.2	1947/48	12.7	1979/80	25.4	31 Mar–1 Apr 1997
41.6	Feb	2003	86.6	2004/05	14.9	1994/95	22.5	22–23 Jan 2005
38.9	Mar	1993	85.1	1977/78	15.1	2001/02	21.5	20–21 Jan 1978
28.3	Apr	1873	83.9	1992/93	15.5	1988/89	19.8	3–5 Mar 1960
0.5	May	1977	79.2	1915/16	17.1	2006/07	19.4	16–17 Feb 1958
			73.4	1919/20	18.1	1985/86	18.7	8–10 Feb 1994
			73.1	1903/04	19.0	1888/89	18.2	20–22 Dec 1975
							18.2	7–8 Jan 1996

Seasonal snowfall

Season	In.	Season	In.	Season	In.	Season	In.
1869/70		1879/80	54.2	1889/90	39.1	1899/00	25.0
1870/71		1880/81	56.8	1890/91	59.3	1900/01	17.5
1871/72	44.3	1881/82	40.4	1891/92	43.5	1901/02	44.1
1872/73	49.1	1882/83	38.9	1892/93	67.3	1902/03	42.0
1873/74	96.4	1883/84	66.0	1893/94	64.0	1903/04	73.1
1874/75	50.0	1884/85	24.3	1894/95	47.2	1904/05	44.9
1875/76	10.0	1885/86	36.0	1895/96	38.9	1905/06	37.6
1876/77	42.3	1886/87	73.0	1896/97	43.2	1906/07	67.9
1877/78	27.7	1887/88	41.2	1897/98	51.9	1907/08	26.2
1878/79	43.4	1888/89	19.0	1898/99	71.1	1908/09	20.1
1909/10	37.0	1919/20	73.4	1929/30	31.4	1939/40	37.7
1910/11	40.6	1920/21	34.1	1930/31	40.8	1940/41	47.8
1911/12	31.6	1921/22	37.6	1931/32	24.2	1941/42	24.0
1912/13	19.4	1922/23	68.5	1932/33	40.6	1942/43	45.7
1913/14	39.4	1923/24	32.3	1933/34	62.7	1943/44	27.7
1914/15	22.3	1924/25	21.4	1934/35	45.4	1944/45	59.2
1915/16	79.2	1925/26	38.3	1935/36	29.5	1945/46	50.8
1916/17	54.2	1926/27	60.3	1936/37	9.0	1946/47	19.4
1917/18	45.7	1927/28	20.8	1937/38	45.7	1947/48	89.2
1918/19	21.1	1928/29	45.5	1938/39	40.2	1948/49	37.1
1949/50	31.9	1959/60	40.9	1969/70	48.8	1979/80	12.7
1950/51	29.7	1960/61	61.5	1970/71	57.3	1980/81	22.3
1951/52	39.6	1961/62	44.7	1971/72	47.5	1981/82	61.8
1952/53	29.8	1962/63	30.9	1972/73	10.3	1982/83	32.7
1953/54	23.6	1963/64	63.0	1973/74	36.9	1983/84	43.0
1954/55	25.1	1964/65	50.4	1974/75	27.6	1984/85	26.6
1955/56	60.9	1965/66	44.1	1975/76	46.6	1985/86	18.1
1956/57	52.0	1966/67	60.1	1976/77	58.5	1986/87	42.5
1957/58	44.7	1967/68	44.8	1977/78	85.1	1987/88	52.6
1958/59	34.1	1968/69	53.8	1978/79	27.5	1988/89	15.5
1989/90	39.2	1994/95	14.9	1999/2000	24.9	2004/05	86.6
1990/91	19.1	1995/96	107.6	2000/01	45.9	2005/06	39.9
1991/92	22.0	1996/97	51.9	2001/02	15.1	2006/07	17.1
1992/93	83.9	1997/98	25.6	2002/03	71.3		
1993/94	96.3	1998/99	36.4	2003/04	39.4		

Boston snowfall (1872/73–2006/07)

Snowstorms exceeding 10 in. (25 cm)

	1890/99			**1930/39**			**1970/79**
12.2	1–4 Mar 1892		10.5	11 Feb 1933		12.7	22–24 Dec 1970
10.2	18 Feb 1893		15.0	20 Feb 1934		10.9	9–10 Jan 1974
11.5	12–13 Feb 1894		12.4	23–24 Jan 1935		11.3	2–3 Feb 1974
14.7	28 Jan 1897		12.0	18–20 Jan 1936		18.2	20–22 Dec 1975
10.2	12–13 Feb 1897		11.0	13 Jan 1938		12.1	11–12 Jan 1976
10.0	1 Feb 1898		11.1	12–13 Mar 1939		11.3	29 Dec 1976
12.5	27–28 Nov 1898					13.8	7 Jan 1977
16.0	12–14 Feb 1899			**1940/49**		21.4	20–21 Jan 1978
			14.0	13–14 Feb 1940		27.1	6–7 Feb 1978
	1900/09		12.0	24–25 Jan 1941			
14.2	16–17 Feb 1903		13.3	27–29 Jan 1943			**1980/89**
11.8	2–3 Jan 1904		10.8	19–21 Mar 1944		12.9	5–6 Dec 1981
11.0	7–8 Jan 1904		14.0	8–9 Feb 1945		12.7	13–15 Jan 1982
10.5	24–26 Jan 1905		10.3	19–20 Dec 1945		13.3	6–7 Apr 1982
13.7	5–6 Feb 1907		10.6	24–25 Jan 1948		13.5	11–12 Feb 1983
12.3	26–27 Dec 1909						
				1950/59			**1990/99**
	1910/19		10.3	17–18 Feb 1952		12.8	13–14 Mar 1993
12.0	2–3 Feb 1916		11.7	7–9 Jan 1953		16.2	6–8 Jan 1994
15.0	11–13 Feb 1916		13.3	19–20 Mar 1956		18.7	8–10 Feb 1994
11.2	3–5 Mar 1917		19.4	16–17 Feb 1958		12.0	19–21 Dec 1995
						12.6	2–3 Jan 1996
	1920/29			**1960/69**		18.2	7–8 Jan 1996
10.1	23–25 Jan 1920		19.8	3–5 Mar 1960		12.4	5–8 Mar 1996
12.0	5 Feb 1920		13.0	12 Dec 1960		25.4	31 Mar–1 Apr 1997
16.5	20–21 Feb 1921		12.3	19–20 Jan 1961			
13.2	4–5 Feb 1926		14.4	4 Feb 1961			**2000–**
13.1	4–6 Dec 1926		10.0	24–25 Dec 1961		11.0	7 Feb 2003
11.3	19–21 Feb 1927		10.1	19–21 Feb 1964		27.6	17–18 Feb 2003
12.0	20–21 Feb 1929		10.9	5–7 Mar 1967		16.9	5–6 Dec 2003
			11.1	9–10 Feb 1969		22.5	22–23 Jan 2005
			26.3	24–27 Feb 1969		17.5	12 Feb 2006

Period	Mean snowfall (in.)
1870/99	42.5
1900/99	41.3
1870/99	47.8
1900/49	40.6
1950/99	42.0
1971–2000	41.6
Maximum	**107.6 (1995/96)**
Minimum	**9.0 (1936/37)**

New York City (Central Park) snowfall
(1869/70–2006/07)

Snowiest months			Snowiest seasons		Least snowy seasons		Biggest storms	
In.	Month	Year	In.	Season	In.	Season	In.	Date
0.8	Oct	1925	75.6	1995/96	2.8	1972/73	26.9	11–12 Feb 2006
19.0	Nov	1898	63.2	1947/48	3.5	2001/02	26.4	26–27 Dec 1947
29.6	Dec	1947	60.4	1922/23	3.8	1918/19	20.9	11–13 Mar 1888
27.4	Jan	1925	60.3	1872/73	5.3	1931/32	20.2	7–8 Jan 1996
27.9	Feb	1934	55.9	1898/99	5.5	1997/98	19.8	16–17 Feb 2003
30.5	Mar	1896	54.7	1960/61	8.1	1877/78	18.1	7–8 Mar 1941
13.5	Apr	1875	53.2	1993/94	8.1	1988/89	17.8	17–18 Feb 1893
T	May	1977*	53.2	1906/07	9.1	1900/01	17.7	6–7 Feb 1978
			52.0	1933/34	11.3	1941/42	17.6	11–12 Feb 1983
			51.5	1966/67	11.5	1954/55	17.5	22–24 Jan 1935

* In earlier years.

Seasonal snowfall

Season	In.	Season	In.	Season	In.	Season	In.
1869/70	27.8	1879/80	22.7	1889/90	24.3	1899/1900	13.4
1870/71	33.1	1880/81	35.5	1890/91	28.8	1900/01	9.1
1871/72	14.1	1881/82	31.4	1891/92	25.4	1901/02	30.0
1872/73	60.3	1882/83	44.0	1892/93	49.7	1902/03	28.7
1873/74	36.9	1883/84	43.1	1893/94	36.1	1903/04	32.2
1874/75	47.9	1884/85	34.2	1894/95	27.0	1904/05	48.1
1875/76	18.3	1885/86	20.8	1895/96	46.3	1905/06	20.0
1876/77	40.4	1886/87	32.9	1896/97	43.6	1906/07	53.2
1877/78	8.1	1887/88	45.6	1897/98	21.1	1907/08	33.4
1878/79	35.7	1888/89	16.5	1898/99	55.9	1908/09	20.3
1909/10	27.2	1919/20	47.6	1929/30	13.6	1939/40	25.7
1910/11	25.2	1920/21	18.6	1930/31	11.6	1940/41	39.0
1911/12	29.5	1921/22	27.8	1931/32	5.3	1941/42	11.3
1912/13	15.3	1922/23	60.4	1932/33	27.0	1942/43	29.5
1913/14	40.5	1923/24	27.5	1933/34	52.0	1943/44	23.8
1914/15	28.8	1924/25	29.6	1934/35	33.8	1944/45	27.1
1915/16	50.7	1925/26	32.4	1935/36	32.2	1945/46	31.4
1916/17	50.7	1926/27	22.3	1936/37	15.6	1946/47	30.6
1917/18	34.5	1927/28	14.5	1937/38	15.1	1947/48	63.2
1918/19	3.8	1928/29	13.8	1938/39	37.3	1948/49	46.6
1949/50	13.8	1959/60	39.2	1969/70	25.6	1979/80	12.8
1950/51	11.6	1960/61	54.7	1970/71	15.5	1980/81	19.4
1951/52	19.7	1961/62	18.1	1971/72	22.9	1981/82	24.6
1952/53	15.1	1962/63	16.3	1972/73	2.8	1982/83	29.2
1953/54	15.8	1963/64	44.7	1973/74	23.5	1983/84	25.4
1954/55	11.5	1964/65	24.4	1974/75	13.1	1984/85	24.1
1955/56	33.5	1965/66	21.4	1975/76	17.3	1985/86	13.0
1956/57	21.9	1966/67	51.5	1976/77	24.5	1986/87	23.1
1957/58	44.7	1967/68	19.5	1977/78	50.7	1987/88	19.1
1958/59	13.9	1968/69	30.2	1978/79	29.4	1988/89	8.1
1989/90	13.4	1994/95	11.8	1999/2000	16.3	2004/05	41.0
1990/91	24.9	1995/96	75.6	2000/01	35.0	2005/06	40.0
1991/92	12.6	1996/97	10.0	2001/02	3.5	2006/07	12.4
1992/93	24.5	1997/98	5.5	2002/03	49.3		
1993/94	53.4	1998/99	12.7	2003/04	42.6		

Snowstorms exceeding 10 in. (25 cm)

1880/89		**1920/29**		**1970/79**	
11.9	8 Jan 1886	17.5	4–7 Feb 1920	13.6	19–20 Jan 1978
20.9	11–13 Mar 1888	12.2	20 Feb 1921	17.7	6–7 Feb 1978
10.0	20–21 Mar 1889	10.0	2–3 Jan 1925	12.7	19 Feb 1979
		11.6	9–10 Feb 1926		
1890/99		**1930/39**		**1970/79**	
14.0	26–27 Dec 1890	10.0	26–27 Dec 1933	13.6	19–20 Jan 1978
15.4	16–18 Mar 1892	10.2	25–27 Feb 1934	17.7	6–7 Feb 1978
17.8	17–18 Feb 1893	17.5	22–24 Jan 1935	12.7	19 Feb 1979
14.0	12–13 Feb 1894				
15.2	25–27 Feb 1894	**1940/49**		**1980/89**	
10.5	15–16 Mar 1896	11.8	7–8 Mar 1941	17.6	11–12 Feb 1983
10.0	27–28 Jan 1897	10.1	11–12 Feb 1944		
15.5	12–14 Feb 1899	11.6	20–21 Feb 1947	**1990/99**	
		26.4	26–27 Dec 1947	10.8	12–13 Mar 1993
1900/09		16.7	19–20 Dec 1948	12.8	11 Feb 1994
10.9	16–17 Feb 1900			10.8	4 Feb 1995
10.5	15–17 Feb 1903	**1950/59**		20.2	7–8 Jan 1996
10.7	4–5 Feb 1907	13.5	18–19 Mar 1956	10.6	16–17 Feb 1996
10.2	23–24 Jan 1908	11.8	20–21 Mar 1958		
10.1	25–26 Dec 1909	13.7	21–22 Dec 1959	**2000–**	
				12.0	30 Dec 2000
1910/19		**1960/69**		19.8	16–17 Feb 2003
14.6	14–15 Jan 1910	14.2	3–4 Mar 1960	14.0	5–6 Dec 2003
11.8	24 Dec 1912	15.2	11–12 Dec 1960	10.3	27–28 Jan 2004
14.5	1–2 Mar 1914	17.4	3–4 Feb 1961	13.8	22–23 Jan 2005
10.2	3–4 Apr 1915	12.5	12–13 Jan 1964	26.9	11–12 Feb 2006
12.2	14–15 Dec 1916	15.2	6–7 Feb 1967		
		15.3	9–10 Feb 1969		

Period	Mean snowfall
1870/99	28.6
1900/99	27.0
1870/99	33.6
1900/49	29.2
1950/99	24.7
1971–2000	22.0
Maximum	**75.6 (1995/96)**
Minimum	**2.8 (1972/73)**

Philadelphia snowfall
(1884/85–2006/07)

Snowiest months			Snowiest seasons		Least snowy seasons		Biggest storms	
In.	Month	Year	In.	Season	In.	Season	In.	Date
2.2	Oct	1940	65.5	1995/96	T	1972/73	30.7	7–8 Jan 1996
13.4	Nov	1898	55.4	1898/99	0.8	1997/98	21.3	11–12 Feb 1983
22.4	Dec	1909	54.9	1977/78	1.9	1949/50	21.0	25–26 Dec 1909
34.6	Jan	1996	49.1	1960/61	3.3	2001/02	20.8	15–18 Feb 2003
31.5	Feb	1899	46.3	2002/03	4.1	1930/31	19.4	3–4 Apr 1915
15.2	Mar	1914	44.3	1966/67	4.5	1918/19	18.9	12–14 Feb 1899
19.4	Apr	1915	43.9	1917/18	4.6	1950/51	16.7	22–24 Jan 1935
T	May	1963	43.8	1904/05	4.7	1991/92	14.3	18–19 Feb 1979
			41.8	1957/58	5.1	1958/59	14.1	6–7 Feb 1978
			40.2	1978/79	6.2	1888/89	13.2	19–20 Jan 1961

Philadelphia snowfall (1884/85–2006/07)

Seasonal snowfall

Season	In.	Season	In.	Season	In.	Season	In.
		1889/90	7.4	1899/1900	20.6	1909/10	36.1
		1890/91	15.2	1900/01	10.6	1910/11	28.5
		1891/92	19.9	1901/02	28.8	1911/12	22.5
		1892/93	36.8	1902/03	16.8	1912/13	9.5
		1893/94	20.3	1903/04	29.4	1913/14	33.1
1884/85	27.0	1894/95	22.8	1904/05	43.8	1914/15	32.5
1885/86	20.2	1895/96	14.8	1905/06	20.5	1915/16	31.1
1886/87	25.8	1896/97	25.7	1906/07	38.6	1916/17	39.6
1887/88	30.0	1897/98	19.4	1907/08	25.8	1917/18	38.9
1888/89	6.2	1898/99	55.4	1908/09	18.1	1918/19	4.5
1919/20	23.2	1929/30	8.2	1939/40	22.3	1949/50	1.9
1920/21	11.7	1930/31	4.1	1940/41	31.5	1950/51	4.6
1921/22	28.0	1931/32	7.7	1941/42	10.3	1951/52	16.2
1922/23	19.5	1932/33	22.0	1942/43	16.3	1952/53	16.8
1923/24	21.8	1933/34	33.6	1943/44	15.7	1953/54	22.6
1924/25	12.1	1934/35	28.0	1944/45	21.1	1954/55	12.1
1925/26	19.1	1935/36	23.1	1945/46	18.7	1955/56	23.0
1926/27	11.8	1936/37	12.6	1946/47	23.7	1956/57	7.9
1927/28	15.5	1937/38	8.3	1947/48	31.7	1957/58	41.8
1928/29	11.9	1938/39	27.2	1948/49	19.3	1958/59	5.1
1959/60	21.8	1969/70	20.3	1979/80	20.9	1989/90	17.0
1960/61	49.1	1970/71	18.3	1980/81	15.4	1990/91	14.6
1961/62	29.2	1971/72	12.2	1981/82	25.4	1991/92	4.7
1962/63	20.5	1972/73	T	1982/83	35.9	1992/93	24.3
1963/64	32.9	1973/74	20.8	1983/84	21.6	1993/94	23.1
1964/65	26.2	1974/75	13.6	1984/85	16.5	1994/95	9.8
1965/66	27.4	1975/76	17.5	1985/86	16.4	1995/96	65.5
1966/67	44.3	1976/77	18.7	1986/87	25.7	1996/97	12.9
1967/68	15.9	1977/78	54.9	1987/88	15.0	1997/98	0.8
1968/69	23.7	1978/79	40.2	1988/89	11.2	1998/99	12.5
1999/00	18.1	2004/05	30.4				
2000/01	27.3	2005/06	19.5				
2001/02	3.3	2006/07	13.4				
2002/03	46.3						
2003/04	24.7						

Snowstorms exceeding 10 in. (25 cm)

	1880/89		1920/29		1960/69
				13.2	19–20 Jan 1961
				10.3	3–4 Feb 1961
10.5	12 Mar 1888	12.3	28–29 Jan 1922	12.7	24–25 Dec 1966
		10.4	28–29 Jan 1928		
	1890/99				1970/79
10.4	5–6 Feb 1893		1930/39	14.1	6–7 Feb 1978
11.4	5–8 Feb 1899	16.7	22–24 Jan 1935	14.3	18–19 Feb 1979
18.9	12–14 Feb 1899				
			1940/49		1980/89
	1900/09	10.1	28 Feb–1 Mar 1941	21.3	11–12 Feb 1983
11.3	16–18 Feb 1900	10.0	20–21 Feb 1947		
11.0	17 Feb 1902				1990/99
12.7	4–6 Feb 1907		1950/59	12.3	12–13 Mar 1993
21.0	25–26 Dec 1909	10.0	10–11 Jan 1954	30.7	7–8 Jan 1996
		13.0	15–16 Feb 1958		
	1910/19	11.4	19–21 Mar 1958		2000–
19.4	3–4 Apr 1915			20.8	15–18 Feb 2003
10.2	27–28 Jan 1918			12.6	22–23 Jan 2005
				12.0	11–12 Feb 2006

Period	Mean snowfall
1870/99	21.7
1900/99	21.9
1870/99	22.8
1900/49	21.8
1950/99	21.7
1971–2000	20.0
Maximum	**63.1 (1995/96)**
Minimum	**T (1972/73)**

Baltimore snowfall
(1883/84–2006/07)

Snowiest months			Snowiest seasons		Least snowy seasons		Biggest storms	
In.	Month	Year	In.	Season	In.	Season	In.	Date
2.5	Oct	1925	62.5	1995/96	0.7	1949/50	28.2	15–18 Feb 2003
9.7	Nov	1898	58.1	2002/03	1.2	1972/73	26.5	28–29 Jan 1922
20.4	Dec	1966	51.8	1963/64	2.3	2001/02	22.8	11–12 Feb 1983
31.3	Jan	1922	50.8	1891/92	2.4	1912/13	22.5	7–8 Jan 1996
40.5	Feb	2003	46.5	1960/61	3.2	1997/98	22.0	28–29 Mar 1942
25.6	Mar	1892	43.4	1966/67	4.0	1958/59	21.4	11–14 Feb 1899
9.4	Apr	1924	43.0	1957/58	4.1	1991/92	20.0	18–19 Feb 1979
T	May	1963	42.5	1978/79	4.6	1980/81	15.5	15–16 Feb 1958
			41.5	1898/99	6.2	1950/51	14.5	11–12 Dec 1960
			41.2	1903/04	7.8	1926/27	14.3	28–29 Jan 1928

Seasonal snowfall

1883/84	35.3	1889/90	15.5	1899/00	25.0	1909/10	31.8
1884/85	27.0	1890/91	25.3	1900/01	9.4	1910/11	22.0
1885/86	40.5	1891/92	50.8	1901/02	19.7	1911/12	27.3
1886/87	27.4	1892/93	27.1	1902/03	17.6	1912/13	2.4
1887/88	21.2	1893/94	20.7	1903/04	41.2	1913/14	25.5
1888/89	7.9	1894/95	15.1	1904/05	26.3	1914/15	17.8
		1895/96	23.8	1905/06	19.5	1915/16	24.7
		1896/97	8.0	1906/07	33.0	1916/17	27.3
		1897/98	18.2	1907/08	27.5	1917/18	24.3
		1898/99	41.5	1908/09	30.3	1918/18	9.1
1919/20	10.3	1929/30	19.1	1939/40	18.0	1949/50	0.7
1920/21	16.0	1930/31	12.5	1940/41	22.5	1950/51	6.2
1921/22	38.3	1931/32	18.2	1941/42	31.3	1951/52	14.1
1922/23	18.7	1932/33	21.2	1942/43	19.8	1952/53	11.8
1923/24	31.8	1933/34	40.0	1943/44	23.4	1953/54	22.1
1924/25	21.9	1934/35	40.7	1944/45	25.7	1954/55	10.1
1925/26	18.5	1935/36	21.2	1945/46	17.1	1955/56	19.1
1926/27	7.8	1936/37	23.6	1946/47	26.0	1956/57	15.4
1927/28	16.4	1937/38	15.7	1947/48	26.1	1957/58	43.0
1928/29	15.8	1938/39	15.3	1948/49	19.9	1958/59	4.0
1959/60	34.1	1969/70	21.0	1979/80	14.6	1989/90	17.3
1960/61	46.5	1970/71	13.0	1980/81	4.6	1990/91	9.4
1961/62	35.2	1971/72	14.0	1981/82	25.5	1991/92	4.1
1962/63	19.6	1972/73	1.2	1982/83	35.6	1992/93	24.4
1963/64	51.8	1973/74	17.1	1983/84	14.5	1993/94	17.3
1964/65	18.6	1974/75	12.2	1984/85	10.3	1994/95	8.0
1965/66	32.8	1975/76	11.5	1985/86	15.6	1995/96	62.5
1966/67	43.4	1976/77	11.1	1986/87	35.2	1996/97	15.3
1967/68	21.4	1977/78	34.3	1987/88	20.4	1997/98	3.2
1968/69	18.6	1978/79	42.5	1988/89	8.3	1998/99	15.2
1999/2000	26.1	2004/05	18.0				
2000/01	8.7	2005/06	19.6				
2001/02	2.3	2006/07	11.0				
2002/03	58.1						
2003/04	18.3						

Baltimore snowfall
(1883/84–2006/07)

Snowstorms exceeding 10 in. (25cm)

	1890/99		1930/39		1970/79
		10.0	16–17 Mar 1931		
16.0	15–18 Mar 1892	11.5	17 Dec 1932	20.0	18 Feb 1979
11.7	5–8 Feb 1899	11.5	23–24 Jan 1935		
21.4	11–14 Feb 1899				1980/89
	1900/09		1940/49	22.8	11–12 Feb 1983
12.0	16–18 Feb 1900	10.3	7–8 Mar 1941	12.3	21–23 Jan 1987
10.2	4 Mar 1909	22.0	28–29 Mar 1942	10.1	22–23 Feb 1987
10.0	25–26 Dec 1909	10.5	20–21 Feb 1947		
					1990/99
	1910/19		1950/59	11.9	12–13 Mar 1993
10.9	5–6 Dec 1910	15.5	15–16 Feb 1958	22.5	7–8 Jan 1996
10.5	26–28 Jan 1918				
			1960/69		2000–
	1920/29	10.5	2–4 Mar 1960	14.9	25 Jan 2000
26.5	28–29 Jan 1922	14.1	11–12 Dec 1960	28.2	15–18 Feb 2003
10.2	9–10 Feb 1926	10.7	3–4 Feb 1961	13.1	11–12 Feb 2006
14.3	28–29 Jan 1928	13.0	5–7 Mar 1962		
11.0	20–21 Feb 1929	11.5	21–22 Mar 1964		
		12.1	29–30 Jan 1966		
		10.6	6–7 Feb 1967		

Period	Mean snowfall
1870/99	21.2
1900/99	20.3
1870/99	25.3
1900/49	22.2
1950/99	18.6
1971–2000	18.0
Maximum	**62.5 (1995/96)**
Minimum	**1.2 (1972/73)**

Washington snowfall
(1884/85–2006/07)

Snowiest months			Snowiest seasons		Least snowy seasons		Biggest storms	
In.	Month	Year	In.	Season	In.	Season	In.	Date
2.2	Oct	1925	54.4	1898/99	0.1	1972/73	28.0	27–29 Jan 1922
11.5	Nov	1987	46.0	1995/96	0.1	1997/98	20.5	11–14 Feb 1899
16.2	Dec	1962	44.5	1921/22	2.2	1975/76	18.7	18–19 Feb 1979
31.5	Jan	1922	41.7	1891/92	2.5	1930/31	17.1	7–8 Jan 1996
35.2	Feb	1899	41.0	1904/05	2.7	2001/02	16.7	15–18 Feb 2003
19.3	Mar	1914	40.4	2002/03	3.3	1918/19	16.6	10–11 Feb 1983
5.5	Apr	1924	40.4	1957/58	3.4	1949/50	14.4	7 Feb 1936
T	May	1963	40.3	1960/61	4.5	1980/81	14.4	15–16 Feb 1958
			39.8	1910/11	4.9	1958/59	14.3	16–18 Feb 1900
			37.1	1890/91	5.0	1931/32	13.8	29–30 Jan 1966
							13.7	5–8 Feb 1899

Seasonal snowfall

		1889/90	6.5	1899/00	35.6	1909/10	20.0
		1890/91	37.1	1900/01	9.1	1910/11	39.8
		1891/92	41.7	1901/02	13.1	1911/12	21.8
		1892/93	31.0	1902/03	8.2	1912/13	8.7
		1893/94	25.4	1903/04	20.2	1913/14	28.6
1884/85	24.6	1894/95	24.8	1904/05	41.0	1914/15	14.5
1885/86	23.5	1895/96	9.3	1905/06	25.7	1915/16	17.4
1886/87	30.3	1896/97	16.2	1906/07	28.3	1916/17	18.8
1887/88	22.7	1897/98	11.0	1907/08	18.3	1917/18	36.4
1888/89	12.5	1898/99	54.4	1908/09	35.9	1918/19	3.3
1919/20	16.4	1929/30	18.1	1939/40	25.3	1949/50	3.4
1920/21	6.8	1930/31	2.5	1940/41	17.9	1950/51	10.2
1921/22	42.5	1931/32	5.0	1941/42	13.6	1951/52	10.2
1922/23	15.7	1932/33	23.8	1942/43	20.5	1952/53	8.3
1923/24	21.9	1933/34	30.7	1943/44	7.7	1953/54	18.0
1924/25	18.5	1934/35	31.4	1944/45	7.8	1954/55	6.6
1925/26	17.4	1935/36	33.0	1945/46	20.9	1955/56	11.3
1926/27	8.8	1936/37	20.2	1946/47	21.9	1956/57	14.2
1927/28	13.6	1937/38	5.4	1947/48	23.4	1957/58	40.4
1928/29	8.0	1938/39	15.1	1948/49	15.8	1958/59	4.9
1959/60	24.3	1969/70	14.0	1979/80	20.1	1989/90	15.3
1960/61	40.3	1970/71	11.7	1980/81	4.5	1990/91	8.1
1961/62	15.0	1971/72	16.8	1981/82	22.5	1991/92	6.6
1962/63	21.4	1972/73	0.1	1982/83	27.6	1992/93	11.7
1963/64	33.6	1973/74	16.7	1983/84	8.6	1993/94	13.2
1964/65	17.1	1974/75	12.8	1984/85	10.3	1994/95	10.1
1965/66	28.4	1975/76	2.2	1985/86	15.4	1995/96	46.0
1966/67	37.1	1976/77	11.1	1986/87	31.1	1996/97	6.7
1967/68	21.4	1977/78	22.7	1987/88	25.0	1997/98	0.1
1968/69	9.1	1978/79	27.7	1988/89	5.7	1998/99	11.6
1999/2000	15.4	2004/05	12.5				
2000/01	7.4	2005/06	13.6				
2001/02	2.7	2006/07	9.5				
2002/03	40.4						
2003/04	12.4						

Snowstorms exceeding 10 in. (25 cm)

	1880/89			**1920/29**			**1960/69**
12.4	3–4 Feb 1886		28.0	27–29 Jan 1922		13.8	29–30 Jan 1966
			10.5	28 Jan 1928		10.3	6–7 Feb 1967
	1890/99						
10.0	25 Dec 1890			**1930/39**			**1970/79**
12.0	27–28 Mar 1891		11.5	30 Jan 1930		10.2	16–17 Dec 1973
13.7	5–8 Feb 1899		12.0	17–18 Dec 1932		18.7	18–19 Feb 1979
20.5	11–14 Feb 1899		11.3	23–24 Jan 1935			
			14.4	7 Feb 1936			**1980/89**
	1900/09					16.6	11 Feb 1983
14.3	16–18 Feb 1900			**1940/49**		10.8	22 Jan 1987
10.0	15–16 Mar 1900		10.7	7–8 Mar 1941		10.3	22–23 Feb 1987
11.5	22–23 Dec 1908		11.5	29 Mar 1942		11.5	11 Nov 1987
			10.1	20–21 Feb 1947			
	1910/19						**1990/99**
	None			**1950/59**		17.1	6–8 Jan 1996
			11.4	3–4 Dec 1957			
			13.0	15–16 Feb 1958			**2000–**
						16.7	15–18 Feb 2003

Period	Mean snowfall
1870/99	18.2
1900/99	19.1
1870/99	24.7
1900/49	19.5
1950/99	16.9
1971–2000	14.7
Maximum	**54.4 (1898/99)**
Minimum	**0.1 (1972/73; 1997/98)**

Norfolk, VA
(1891/92–2006/07)

Snowiest months			Snowiest seasons		Least snowy seasons		Biggest storms	
Year	In.	Season	In.	Season	In.	Season	In.	Date
T	Oct	1910	41.9	1979/80	T	1996/97	18.6	27–28 Dec 1892
6.7	Nov	1891	37.7	1935/36	T	1997/98	15.4	17–19 Feb 1989
18.6	Dec	1892	33.3	1892/93	T	1991/92	13.7	1–2 Mar 1980
14.2	Jan	1966	24.9	1988/89	T	1990/91	13.5	11–13 Feb 1899
24.4	Feb	1989	24.0	1917/18	T	1975/76	12.4	6–7 Feb 1980
14.0	Mar	1914	22.9	1939/40	T	1920/21	12.4	9–10 Feb 1948
1.3	Apr	1940	22.8	1947/48	T	1897/98	11.4	11–12 Dec 1958
T	May	1951	21.7	1913/14	0.1	1902/03	11.0	7 Feb 1936
			21.3	1911/12	0.1	2006/07	11.0	2 Mar 1927
			20.1	1933/34	0.3	1980/81	11.0	2–3 Dec 1896

Roanoke, VA
(1901/02–1996/97)

Snowiest months			Snowiest seasons		Least snowy seasons		Biggest storms (1959/60–1995/96)	
In.	Month	Year	In.	Season	In.	Season	In.	Date
2.0	Oct	1925	72.9	1986/87	T	1919/20	24.9	6–8 Jan 1996
13.8	Nov	1968	62.7	1959/60	T	1918/19	18.6	10–11 Feb 1983
22.6	Dec	1966	55.9	1995/96	1.2	1990/91	17.4	2–3 Mar 1960
41.2	Jan	1966	50.3	1963/64	2.0	1937/38	16.4	25–26 Dec 1969
36.0	Feb	1987	49.9	1965/66	2.3	1975/76	16.0	12–13 Mar 1993
30.3	Mar	1960	44.0	1946/47	2.8	1990/91	14.2	7–8 Feb 1961
7.3	Apr	1971	41.7	1966/67	3.4	1934/35	13.9	25–26 Jan 1987
T	May	1990	41.7	1913/14	3.7	1956/57	13.7	22–23 Jan 1966
			40.8	1960/61	4.0	1996/97	12.3	29–30 Jan 1966
			39.1	1968/69	5.0	1991/92	11.7	13–14 Jan 1968

Lynchburg, VA
(1893/94–1997/98)

Snowiest months			Snowiest seasons		Least snowy seasons		Biggest storms	
In.	Month	Year	In.	Season	In.	Season	In.	Date
2.9	Oct	1925	58.4	1986/87	1.3	1931/32	21.4	6–8 Jan 1996
11.6	Nov	1968	57.8	1995/96	1.6	1918/19	20.2	27–28 Jan 1922
17.9	Dec	1966	45.2	1959/60	1.9	1975/76	17.9	5–7 Mar 1962
31.8	Jan	1966	38.2	1965/66	2.1	1902/03	14.8	25–26 Feb 1934
26.8	Feb	1987	37.2	1966/67	2.5	1997/98	14.6	10–11 Feb 1983
24.9	Mar	1960	35.4	1917/18	2.9	1924/25	14.5	31 Jan–1 Feb 1948
4.8	Apr	1971	35.0	1981/82	3.1	1991/92	13.7	28 Feb–1 Mar 1969
T	May	1923	34.9	1968/69	3.1	1937/38	13.0	12–13 Mar 1993
			33.0	1979/80	3.3	1944/45	12.8	13–14 Feb 1960
			27.7	1970/71	4.1	1990/91	12.7	25–26 Dec 1969

Richmond, VA
(1897/98–2006/07)

Snowiest months			Snowiest seasons		Least snowy seasons		Biggest storms	
In.	Month	Year	In.	Season	In.	Season	In.	Date
T	Oct	1979	38.9	1961/62	T	1918/19	21.6	23–24 Jan 1940
9.2	Nov	1938	38.6	1979/80	T	1920/21	17.7	27–29 Jan 1922
17.2	Dec	1908	35.8	1966/67	0.2	1998/99	17.7	10–11 Feb 1983
28.5	Jan	1940	34.2	1939/40	0.5	1944/45	17.2	22–23 Dec 1908
21.4	Feb	1983	33.7	1917/18	0.7	1950/51	16.3	11–14 Feb 1899
19.7	Mar	1960	32.3	1959/60	0.9	1991/92	15.2	5–7 Mar 1962
10.0	Apr	1915	30.4	1995/96	1.0	1980/81	14.9	4–5 Jan 1980
T	May	1950	30.2	1898/1899	1.0	1920/21	14.8	25–27 Jan 1966
			29.4	1982/83	1.1	1955/56	13.0	1–2 Mar 1980
			29.2	1965/66	1.2	1975/76	12.6	6–7 Feb 1936

Washington Dulles Airport
(1958/59–2006/07)

Snowiest months			Snowiest seasons		Least snowy seasons		Biggest storms	
In.	Month	Year	In.	Season	In.	Season	In.	Date
1.3	Oct	1979	61.9	1995/96	2.2	1972/73	24.5	6–8 Jan 1996
11.4	Nov	1967	50.1	2002/03	2.6	2001/02	22.8	11–12 Feb 1983
24.2	Dec	1966	44.4	1966/67	4.4	1980/81	22.4	15–18 Feb 2003
30.9	Jan	1996	42.7	1986/87	5.9	1997/98	16.3	18 Feb 1979
34.9	Feb	2003	40.6	1978/79	6.7	1991/92	15.4	31 Dec 1970–1 Jan 1971
15.5	Mar	1993	39.2	1982/83	9.1	1975/76	13.9	13 Mar 1993
4.0	Apr	1990	30.6	1965/66	9.9	1988/89	12.1	25–26 Dec 1969
T	May	1963	30.5	1967/68	10.6	1976/77	12.0	22–23 Feb 1987
			30.3	1992/93	13.8	2005/06	11.8	6–7 Feb 1967
			30.2	1969/70	14.6	1990/91	11.4	30 Nov 1967

Wilmington, DE
(1950/51–2006/07)

Snowiest months			Snowiest seasons		Least snowy seasons		Biggest storms	
In.	Month	Year	In.	Season	In.	Season	In.	Date
2.5	Oct	1979	55.9	1995/96	T	1931/32	22.2	15–18 Feb 2003
11.9	Nov	1953	49.5	1957/58	T	1997/98	22.0	7–8 Jan 1996
21.5	Dec	1966	46.0	2002/03	1.0	1949/50	17.9	19–21 Mar 1958
26.2	Jan	1996	45.6	1977/78	1.2	1972/73	16.5	18–19 Feb 1979
31.6	Feb	2003	44.7	1933/34	2.4	2001/02	14.5	5–7 Feb 1978
20.3	Mar	1958	44.7	1906/07	3.5	1991/92	14.4	11–12 Feb 2006
2.6	Apr	1982	44.2	1978/79	4.3	1918/19	13.7	13–14 Mar 1993
T	May	1963	44.1	1960/61	4.8	1950/51	13.6	11 Feb 1983
			43.5	1966/67	5.7	1930/31	12.4	24 Dec 1966
			40.0	1986/87	7.3	1958/59	12.1	22 Jan 1987

Harrisburg, PA
(1889/90–2006/07)

Snowiest months			Snowiest seasons		Least snowy seasons		Biggest storms	
In.	Month	Year	In.	Season	In.	Season	In.	Date
2.1	Oct	1925	81.3	1960/61	8.8	1937/38	25.0	11–12 Feb 1983
15.4	Nov	1953	75.9	1993/94	9.8	1949/50	22.2	6–7 Jan 1996
22.1	Dec	1960	74.7	1963/64	10.6	2001/02	21.0	15–16 Jan 1945
38.9	Jan	1996	70.6	1977/78	10.9	1930/31	20.8	18–20 Feb 1964
30.3	Feb	1893	67.5	1995/96	11.4	1997/98	20.4	13–14 Mar 1993
22.8	Mar	1993	60.6	1969/70	12.9	1991/92	20.3	15–18 Feb 2003
18.0	Apr	1894	58.5	1981/82	13.3	1972/73	18.7	19–20 Jan 1961
T	May	1966	57.7	2002/03	14.6	1979/80	18.1	7–8 Mar 1941
			53.8	1893/94	15.3	1931/32	18.1	12–13 Jan 1964
			52.2	1892/93	15.3	1912/13	18.0	27–28 Mar 1891
			52.0	1904/05				

Reading, PA
(1904/05–1968/69)

Snowiest months			Snowiest seasons		Least snowy seasons		Biggest storms	
In.	Month	Year	In.	Season	In.	Season	In.	Date
1.1	Oct	1925	62.1	1904/05	6.8	1918/19	20.3	25–26 Dec 1909
13.4	Nov	1938	58.8	1960/61	7.1	1930/31	18.7	15–16 Feb 1958
27.1	Dec	1966	58.2	1966/67	10.8	1949/50	17.3	3–4 Feb 1961
36.8	Jan	1925	53.2	1957/58	12.0	1941/42	16.0	19–20 Mar 1958
22.3	Feb	1926	51.8	1940/41	12.4	1950/51	15.5	2 Jan 1925
21.5	Mar	1958	51.1	1963/64	12.6	1931/32	15.2	6–8 Mar 1915
10.0	Apr	1924	47.1	1906/07	13.4	1900/01	14.9	28 Jan 1928
T	May	1966	47.0	1916/17	13.8	1902/03	14.8	7–9 Mar 1941
			45.2	1917/18	14.7	1967/68	14.5	22–24 Jan 1935
			45.0	1909/10	15.4	1937/38	14.2	14–15 Feb 1940

Allentown, PA
(1922/23–2006/07)

Snowiest months			Snowiest seasons		Least snowy seasons		Biggest storms (1943/44–1995/96)	
In.	Month	Year	In.	Season	In.	Season	In.	Date
2.2	Oct	1925	75.2	1993/94	5.0	1931/32	25.6	6–8 Jan 1996
15.0	Nov	1938	71.8	1995/96	7.4	1972/73	25.2	11–12 Feb 1983
28.4	Dec	1966	67.2	1967/68	9.2	1937/38	20.3	19–21 Mar 1958
43.2	Jan	1925	65.1	1960/61	9.7	2001/02	20.1	16–18 Feb 2003
29.5	Feb	1983	63.6	1957/58	11.0	1930/31	17.6	13–14 Mar 1993
30.5	Mar	1958	55.6	1977/78	11.4	1949/50	17.3	3–4 Feb 1961
13.4	Apr	1982	54.9	2002/03	12.2	1988/89	16.0	19–20 Jan 1961
T	May	1977	54.7	1963/64	13.9	1928/29	15.8	15–16 Feb 1958
			50.7	1947/48	15.6	1991/92	15.2	11–12 Feb 2006
			47.2	2001/02	15.7	1954/55	14.4	3–5 Mar 1960

Scranton, PA
(1901/02–1995/96)

Snowiest months			Snowiest seasons		Least snowy seasons		Biggest storms	
In.	Month	Year	In.	Season	In.	Season	In.	Date
4.4	Oct	1962	98.3	1995/96	7.3	1988/89	21.5	7–8 Jan 1996
22.5	Nov	1971	90.4	1993/94	13.9	1952/53	21.4	13–14 Mar 1993
33.9	Dec	1969	88.6	1904/05	16.7	1928/29	20.5	24–25 Nov 1971
42.3	Jan	1994	82.8	1915/16	18.5	1943/44	20.0	19–20 Jan 1936
27.9	Feb	1914	76.8	1969/70	20.8	1954/55	19.6	18–20 Apr 1983
38.0	Mar	1916	75.3	1966/67	21.1	1912/13	17.3	13–14 Feb 1914
26.7	Apr	1983	74.7	1963/64	22.8	1929/30	16.5	16–17 Jan 1994
2.4	May	1977	73.7	1960/61	22.9	1953/54	15.5	3–4 Feb 1926
			73.5	1977/78	23.2	1972/73	15.0	9 Feb 1904
			66.7	1907/08	24.5	1991/92	15.0	1–2 Mar 1914

Atlantic City, NJ
(1883/84–2006/07)

Snowiest months			Snowiest seasons		Least snowy seasons		Biggest storms	
In.	Month	Year	In.	Season	In.	Season	In.	Date
T	Oct	1990	51.2	1898/99	0.7	1949/50	21.5	15–18 Feb 2003
16.7	Nov	1898	46.9	1966/67	1.5	1918/19	18.0	16–17 Feb 1902
17.5	Dec	1935	43.1	1978/79	2.4	1936/37	17.1	18–19 Feb 1979
22.3	Jan	1905	42.3	2002/03	2.6	2001/02	16.3	25–26 Jan 1987
35.2	Feb	1967	40.4	1904/05	2.7	1980/81	16.2	28 Feb–1 Mar 1941
23.6	Mar	1914	38.1	1963/64	3.0	1997/98	16.0	25 Jan 1905
6.7	Apr	1915	37.3	1933/34	3.0	1943/44	15.1	11–13 Feb 1899
			34.5	1892/93	3.2	1991/92	14.7	12–13 Jan 1964
			33.7	1901/02	3.3	1980/81	14.0	13–15 Feb 1885
			33.5	1957/58	3.3	1941/42	14.0	26–27 Nov 1898

Trenton, NJ
(1865/66–1981/82)

Snowiest months			Snowiest seasons		Least snowy seasons		Biggest storms	
In.	Month	Year	In.	Season	In.	Season	In.	Date
2.5	Oct	1972	63.0	1867/68	2.0	1918/19	22.0	12–13 Feb 1899
14.0	Nov	1898	61.0	1898/99	3.3	1972/73	21.0	12–13 Mar 1888
21.5	Dec	1965	54.0	1966/67	3.8	1949/50	18.0	26 Dec 1872
20.8	Jan	1978	51.7	1957/58	5.1	1930/31	17.8	19–21 Mar 1958
34.0	Feb	1899	51.3	1977/78	5.2	1931/32	16.6	11–12 Dec 1960
22.5	Mar	1888	48.5	1960/61	5.6	1950/51	16.1	6–7 Feb 1978
16.0	Apr	1915	42.5	1887/88	7.3	1937/38	16.0	3–4 Apr 1915
T	May	1977	40.9	1922/23	7.6	1905/06	14.0	25 Jan 1905
			40.1	1933/34	8.7	1929/30	13.8	20–22 Feb 1922
			38.8	1955/56	8.9	1941/42	13.8	6–7 Feb 1967

Newark, NJ
(1842/43–2006/07)

Snowiest months			Snowiest seasons		Least snowy seasons		Biggest storms	
In.	Month	Year	In.	Season	In.	Season	In.	Date
0.3	Oct	1952	78.4	1995/96	1.9	1972/73	27.8	7–8 Jan 1996
14.2	Nov	1938	73.5	1873/74	3.5	2001/02	26.0	26–27 Dec 1947
29.1	Dec	1947	64.9	1993/94	6.5	1997/98	23.1	16–18 Feb 2003
31.6	Jan	1996	64.5	1947/48	7.3	1931/32	22.6	3–4 Feb 1961
33.4	Feb	1994	62.5	1977/78	7.5	1988/89	21.3	11–12 Feb 2006
26.0	Mar	1956	58.3	1992/93	10.7	1949/50	20.4	11–12 Dec 1960
13.8	Apr	1982	57.3	1915/16	10.9	1950/51	20.0	4–5 Feb 1845
T	May	1977	52.3	2002/03	11.0	1936/37	20.0	5–6 Jan 1856
			43.2	1919/20	11.1	1937/38	20.0	20 Dec 1875
			42.1	1903/04	12.8	1998/99	20.0	4–5 Feb 1907
							19.6	6–7 Feb 1978

New York LaGuardia Airport
(1943/44–2006/07)

Snowiest months			Snowiest seasons		Least snowy seasons		Biggest storms	
In.	Month	Year	In.	Season	In.	Season	In.	Date
1.2	Oct	1962	78.4	1995/96	1.9	1972/73	25.4	11–12 Feb 2006
6.1	Nov	1989	64.2	1947/48	3.4	2001/02	23.8	7–8 Jan 1996
26.8	Dec	1947	58.5	1993/94	7.1	1997/98	23.6	26–27 Dec 1947
27.6	Jan	1996	56.5	1960/61	10.1	1994/95	22.0	11–12 Feb 1983
26.4	Feb	1983	51.0	2002/03	10.3	1979/80	19.0	3–4 Feb 1961
18.9	Mar	1958	46.7	1948/49	10.8	1988/89	17.6	16–18 Feb 2003
8.2	Apr	1982	43.5	1977/78	11.3	1954/55	16.1	9–10 Feb 1969
T	May	1961	43.4	1966/67	11.7	1974/75	15.9	19–20 Dec 1948
			42.2	2000/01	13.5	1952/53	15.0	3–5 Mar 1960
			35.5	1946/47	14.3	1991/92	14.9	19–20 Dec 1995
							14.4	6–7 Feb 1978

New York Kennedy Airport
(1959/60–2006/07)

Snowiest months			Snowiest seasons		Least snowy seasons		Biggest storms	
In.	Month	Year	In.	Season	In.	Season	In.	Date
0.5	Oct	1962	69.0	1995/96	1.6	1972/73	26.0	16–18 Feb 2003
3.7	Nov	1989	58.4	1960/61	3.6	1997/98	24.0	3–4 Feb 1961
16.4	Dec	1960	56.2	2002/03	4.5	2001/02	21.7	11–12 Feb 1983
23.0	Jan	1996	48.5	1977/78	8.2	1988/89	20.7	7–8 Jan 1996
32.1	Feb	2003	47.0	1966/67	9.6	1989/90	20.2	9–10 Feb 1969
21.1	Mar	1960	45.2	1993/94	10.0	1994/95	16.7	11–12 Feb 2006
8.2	Apr	1982	34.4	1963/64	10.5	1991/92	14.3	3–5 Mar 1960
T	May	1967	34.2	1959/60	10.5	1974/75	14.2	19–20 Jan 1978
			34.0	2000/01	11.0	1979/80	13.9	12–14 Jan 1964
			32.1	1982/83	14.9	1971/72	13.7	6–7 Feb 1978

New York (The Battery), NY
(1884/85–1959/60)

Snowiest months			Snowiest seasons		Least snowy seasons		Biggest storms	
In.	Month	Year	In.	Season	In.	Season	In.	Date
0.4	Oct	1925	77.6	1892/93	3.5	1918/19	25.8	26–27 Dec 1947
14.0	Nov	1898	61.5	1947/48	5.1	1931/32	20.9	12–13 Mar 1888
29.0	Dec	1947	58.3	1898/99	9.1	1952/53	17.8	17–18 Feb 1893
26.2	Jan	1925	57.8	1904/05	9.2	1900/01	17.5	4–7 Feb 1920
37.2	Feb	1894	56.1	1893/94	9.7	1930/31	17.5	22–24 Jan 1935
28.5	Mar	1896	55.4	1919/20	10.2	1941/42	17.0	11–12 Dec 1960
10.2	Apr	1915	55.2	1922/23	10.4	1952/53	16.7	19–20 Dec 1948
T	May	1946	53.1	1933/34	10.9	1954/55	15.5	12–14 Feb 1899
			52.4	1906/07	10.9	1950/51	15.2	25–27 Feb 1894
			51.6	1915/16	11.9	1936/37	14.6	14–15 Jan 1910

Brookhaven Lab, Upton, NY
(1947/48–2005/06)

\multicolumn{3}{Snowiest months}			\multicolumn{2}{Snowiest seasons}		\multicolumn{2}{Least snowy seasons}		\multicolumn{2}{Biggest storms}
In.	Month	Year	In.	Season	In.	Season	In.	Date
1.0	Oct	1962	90.8	1995/96	4.5	1997/98	23.0	6–7 Feb 1978
7.5	Nov	1989	74.5	1966/67	5.5	2001/02	19.0	6–8 Jan 1996
23.7	Dec	1948	66.7	1947/48	5.8	1972/73	19.0	26–27 Dec 1947
29.0	Jan	2005	62.4	1977/78	6.5	1994/95	17.3	3–5 Mar 1960
32.1	Feb	1967	62.0	2002/03	11.0	1979/80	17.2	6–7 Feb 1967
31.5	Mar	1967	61.1	1957/58	11.8	1952/53	17.0	3–4 Feb 1961
16.0	Apr	1996	60.2	2003/04	13.4	1953/54	17.0	16–17 Feb 1958
T	May	1977	57.5	1960/61	15.2	1958/59	16.5	28 Feb–1 Mar 1949
			55.0	1993/94	15.3	1950/51	16.0	21–22 Mar 1967
			54.9	1948/49	15.7	1956/57	16.0	20–22 Mar 1958

Bridgehampton, NY
(1959/60–1995/96)

In.	Month	Year	In.	Season	In.	Season	In.	Date
T	Oct	1971	—	1995/96	3.0	1994/95	21.0	3–5 Mar 1960
10.0	Nov	1989	68.9	1966/67	9.4	1972/73	18.5	6–7 Feb 1978
17.4	Dec	1963	59.1	1977/78	10.0	1990/91	18.0	15–16 Feb 1958
27.4	Jan	1961	58.0	1960/61	10.8	1971/72	16.0	22 Mar 1967
30.2	Feb	1967	56.3	1955/56	11.3	1979/80	15.3	19–20 Jan 1961
29.0	Mar	1956	—	1993/94	11.5	1988/89	14.4	19–20 Mar 1956
6.5	Apr	1982	46.8	1963/64	12.2	1952/53	14.0	4 Feb 1961
T	May	1977	44.4	1959/60	14.1	1962/63	14.0	19–20 Jan 1978
			43.2	1957/58	14.3	1965/66	13.5	6–7 Feb 1967
			36.5	1986/87	14.5	1984/85	13.0	25 Feb 1990

Albany, NY
(1884/85–2006/07)

In.	Month	Year	In.	Season	In.	Season	In.	Date
6.5	Oct	1987	112.5	1970/71	13.8	1912/13	46.7	11–14 Mar 1888
24.6	Nov	1972	110.4	1887/88	19.0	1988/89	26.6	13–14 Mar 1993
57.5	Dec	1969	105.4	2002/03	24.8	1929/30	26.4	25–28 Dec 1969
47.8	Jan	1987	99.6	1890/91	26.7	1918/19	24.7	13–15 Dec 1915
40.7	Feb	1893	97.1	1981/82	27.4	1979/80	24.5	15–16 Jan 1983
50.9	Mar	1888	94.7	1915/16	28.1	1889/90	23.5	14 Feb 1914
17.7	Apr	1982	94.2	1992/93	28.4	1936/37	23.5	18–22 Dec 1887
5.4	May	1945	94.2	1886/87	28.7	1990/91	22.5	24–25 Nov 1971
			92.4	1977/78	28.7	1914/15	21.0	25–26 Dec 2002
			90.0	1947/48	28.9	1896/97	20.8	2–3 Jan 2003

Bridgeport, CT
(1930/31–2006/07)

\multicolumn{3}{Snowiest months}			\multicolumn{2}{Snowiest seasons}		\multicolumn{2}{Least snowy seasons}		\multicolumn{2}{Biggest storms (after 1951/52)}
In.	Month	Year	In.	Season	In.	Season	In.	Date
0.5	Oct	1987	76.8	1995/96	8.2	1972/73	17.7	9–10 Feb 1969
14.1	Nov	1938	71.3	1933/34	8.5	2001/02	17.3	17–18 Feb 2003
25.8	Dec	1945	65.7	1947/48	8.7	1952/53	16.7	19–20 Jan 1978
30.3	Jan	1923	61.6	1966/67	8.7	1931/32	15.5	5–7 Mar 2001
47.0	Feb	1934	59.7	1922/23	9.2	1997/98	15.0	7–8 Jan 1996
21.8	Mar	1967	55.5	2002/03	9.6	1979/80	15.0	15–16 Feb 1958
8.1	Apr	1924	55.3	2000/01	9.9	1954/55	14.9	6–7 Feb 1967
T	May	1977	55.0	1993/94	11.5	1980/81	13.5	11–12 Feb 2006
			52.7	1977/78	13.7	1950/51	13.3	6–7 Feb 1978
			49.3	1948/49	14.2	1958/59	12.9	3–4 Feb 1961

New Haven, CT
(1873/74–1974/75)

Snowiest months (1872/73–1974/75)			Snowiest seasons		Least snowy seasons		Biggest storms (1872/73–1974/75)	
In.	Month	Year	In.	Season	In.	Season	In.	Date
2.0	Oct	1876	76.0	1915/16	7.3	1918/19	44.7	11–14 Mar 1888
23.0	Nov	1898	74.2	1892/93	11.3	1877/78	23.2	19–20 Feb 1934
27.3	Dec	1904	72.4	1933/34	13.4	1900/01	20.0	28 Jan 1897
31.2	Jan	1923	71.6	1947/48	13.6	1952/53	17.5	30–31 Mar 1881
46.3	Feb	1934	69.5	1880/81	14.5	1936/37	17.2	15–16 Feb 1958
44.9	Mar	1888	67.4	1896/97	15.6	1954/55	16.8	25–27 Nov 1898
10.0	Apr	1874	66.9	1898/99	16.5	1928/29	16.1	28 Feb–1 Mar 1949
1.2	May	1876	66.2	1887/88	17.5	1888/89	15.3	20–21 Feb 1947
			65.8	1893/94	18.5	1941/42	15.0	19–20 Dec 1948
			65.3	1906/07	18.6	1958/59	15.0	9 Feb 1969

Hartford, CT
(1904/05–2006/07)

Snowiest months			Snowiest seasons		Least snowy seasons		Biggest storms	
In.	Month	Year	In.	Season	In.	Season	In.	Date
1.7	Oct	1979	115.2	1995/96	12.0	2001/02	21.9	11–12 Feb 2006
15.6	Nov	1938	84.9	1993/94	14.7	1936/37	21.0	11–12 Feb 1983
45.3	Dec	1945	82.8	1966/67	14.9	1988/89	19.2	28 Feb–1 Mar 1949
40.1	Jan	1923	80.2	1960/61	16.4	1979/80	18.2	19–20 Dec 1945
32.7	Feb	1926	79.9	1945/46	16.6	1941/42	17.4	19–20 Feb 1934
43.3	Mar	1956	77.7	1947/48	17.2	1918/19	17.0	20–21 Feb 1921
14.3	Apr	1982	77.1	1915/16	17.3	1931/32	16.9	26–27 Dec 1947
1.3	May	1977	76.1	1955/56	17.7	1980/81	16.9	6–7 Feb 1978
			74.5	2002/03	19.0	1924/25	16.3	5 Feb 2001
			70.6	1906/07	19.7	1912/13	16.2	4 Feb 1926

Block Island, RI
(1880/81–1977/78)

Snowiest months			Snowiest seasons		Least snowy seasons		Biggest storms	
In.	Month	Year	In.	Season	In.	Season	In.	Date
0.3	Oct	1925	51.2	1977/78	2.2	1888/89	22.0	19–20 Jan 1978
7.5	Nov	1898	47.8	1881/82	3.2	1918/19	21.2	4–5 Feb 1882
19.9	Dec	1904	45.3	1933/34	5.1	1890/91	16.9	3–4 Feb 1961
30.0	Jan	1978	43.9	1886/87	5.8	1931/32	14.0	15–17 Mar 1967
22.2	Feb	1899	43.3	1898/99	6.5	1905/06	13.2	12–14 Feb 1899
24.1	Mar	1956	42.4	1955/56	6.7	1972/73	12.5	3–5 Mar 1960
13.2	Apr	1887	38.6	1966/67	6.8	1912/13	12.0	22–24 Jan 1935
T	May	1941	38.3	1960/61	6.9	1920/21	11.6	17–18 Feb 1902
			37.9	1896/97	8.0	1952/53	11.2	1–2 Feb 1887
			36.8	1895/96	8.2	1970/71	11.1	11 Jan 1927

Providence, RI
(1905/06–2006/07)

Snowiest months			Snowiest seasons		Least snowy seasons		Biggest storms	
In.	Month	Year	In.	Season	In.	Season	In.	Date
2.5	Oct	1979	106.1	1995/96	4.3	1997/98	28.6	6–7 Feb 1978
10.2	Nov	1945	75.6	1947/48	10.2	2001/02	24.0	7–8 Jan 1996
26.7	Dec	1945	72.2	2004/05	10.9	1988/89	23.4	22–23 Jan 2005
37.4	Jan	1996	71.4	1906/07	11.3	1972/73	18.9	14–16 Feb 1962
30.9	Feb	1962	70.2	1977/78	11.8	1936/37	18.3	4 Feb 1961
31.6	Mar	1956	63.5	1993/94	12.2	1979/80	18.0	31 Mar–1 Apr 1997
18.0	Apr	1997	62.7	1960/61	12.5	1994/95	17.7	3–5 Mar 1960
7.0	May	1977	58.1	1966/67	13.7	1918/19	16.0	27–29 Jan 1943
			56.7	1944/45	15.1	2006/07	15.9	24–27 Feb 1969
			56.2	1989/90	15.8	1914/15	15.0	17–18 Feb 2003
					16.0	1912/13		

Nantucket, MA
(1887/88–1969/70)

Snowiest months			Snowiest seasons		Least snowy seasons		Biggest storms	
In.	Month	Year	In.	Season	In.	Season	In.	Date
0.2	Oct	1925	82.0	1903/04	3.8	1952/53	31.3	3–5 Mar 1960
8.1	Nov	1904	73.1	1963/64	5.4	1931/32	21.4	27–29 Feb 1952
24.7	Dec	1963	72.2	1904/05	5.8	1936/37	21.4	25–26 Jan 1905
39.5	Jan	1904	62.1	1960/61	6.5	1920/21	19.2	13–14 Jan 1964
36.4	Feb	1952	59.4	1959/60	8.9	1908/09	16.0	2–3 Jan 1904
40.2	Mar	1960	57.6	1966/67	11.3	1949/50	16.0	19–20 Jan 1961
9.5	Apr	1955	54.3	1901/02	11.4	1912/13	15.9	3–4 Feb 1951
T	May	1917	54.1	1951/52	12.0	1953/54	15.7	14–15 Jan 1910
			53.1	1964/65	14.9	1914/15	15.6	2–4 Mar 1916
			49.2	1906/07	15.0	1965/66	15.5	12 Dec 1960

Blue Hill Observatory, MA
(1885/86–2006/07)

Snowiest months			Snowiest seasons		Least snowy seasons		Biggest storms	
In.	Month	Year	In.	Season	In.	Season	In.	Date
6.8	Oct	1979	144.4	1995/96	12.4	1936/37	38.7	24–28 Feb 1969
23.0	Nov	1889	136.0	1947/48	17.9	1994/95	30.1	3–4 Mar 1960
45.2	Dec	1945	119.4	2004/05	22.6	1900/01	30.1	6–7 Feb 1978
56.3	Jan	1948	113.1	2002/03	24.0	1888/89	30.0	31 Mar–1 Apr 1997
65.4	Feb	1969	111.5	1915/16	24.2	1972/73	24.7	17–18 Feb 2003
52.0	Mar	1956	109.6	1966/67	25.1	1918/19	24.3	5–7 Dec 2003
24.2	Apr	1996	106.8	1955/56	25.3	1979/80	23.0	5–7 Mar 2001
7.8	May	1977	103.4	1903/04	26.4	1985/86	22.2	16–17 Feb 1958
			102.5	1922/23	27.2	1991/92	21.0	11–12 Dec 1960
			101.1	1993/94	27.6	2006/07	21.0	9–10 Feb 1969

Worcester, MA
(1892/93–2006/07)

Snowiest months			Snowiest seasons		Least snowy seasons		Biggest storms	
In.	Month	Year	In.	Season	In.	Season	In.	Date
7.5	Oct	1979	132.9	1995/96	21.2	1954/55	33.0	31 Mar–1 Apr 1997
20.7	Nov	1971	120.1	1992/93	23.0	1997/98	32.1	11–12 Dec 1992
37.0	Dec	1992	111.8	2002/03	24.6	1936/37	24.8	14–16 Feb 1962
50.7	Jan	2005	111.5	2004/05	24.9	1994/95	24.1	22–23 Jan 2005
55.0	Feb	1893	104.3	1960/61	26.6	1979/80	24.0	7–8 Mar 2001
44.1	Mar	1993	100.2	1993/94	28.1	1988/89	22.1	3–5 Mar 1960
24.0	Apr	1997	99.3	1971/72	31.2	1953/54	20.8	17–18 Feb 2003
12.7	May	1977	98.4	1947/48	31.3	1943/44	20.2	6–7 Feb 1978
			97.5	1957/58	33.8	1973/74	20.1	13–14 Mar 1993
			94.2	1966/67	35.5	1941/42	19.6	24–27 Feb 1969

Concord, NH
(1903/04–2006/07)

Snowiest months			Snowiest seasons		Least snowy seasons		Biggest storms	
In.	Month	Year	In.	Season	In.	Season	In.	Date
3.0	Oct	1884	113.2	1995/96	27.0	1979/80	22.0	5–6 Dec 2003
25.0	Nov	1873	100.0	1971/72	29.1	1988/89	21.5	19–21 Feb 1929
43.0	Dec	1876	98.1	1992/93	29.3	1912/13	21.2	24–27 Feb 1969
46.7	Jan	1935	95.8	1970/71	31.5	1929/30	18.7	6 Jan 1944
59.0	Feb	1893	93.9	1916/17	33.5	1948/49	18.3	12–13 Apr 1933
38.3	Mar	1956	92.6	1951/52	33.6	1990/91	17.0	13–14 Mar 1993
35.0	Apr	1874	90.0	1981/82	34.6	1950/51	16.9	27–29 Dec 1946
5.0	May	1945	89.1	1955/56	34.8	2001/02	15.8	20–22 Jan 1978
			88.8	2002/03	35.1	1952/53	15.6	7–10 Mar 1916
			88.1	1947/48	35.2	1991/92	15.5	7 Jan 1977

Portland, ME
(1882/83–2006/07)

Snowiest months			Snowiest seasons		Least snowy seasons		Biggest storms	
In.	Month	Year	In.	Season	In.	Season	In.	Date
3.8	Oct	1969	141.5	1970/71	27.5	1979/80	27.1	17–18 Jan 1979
24.3	Nov	1921	137.5	1905/06	29.4	1936/37	26.9	24–27 Feb 1969
54.8	Dec	1970	133.9	1897/98	30.1	1912/13	25.3	17–18 Feb 1952
62.4	Jan	1979	125.5	1886/87	30.9	1988/89	22.8	28–29 Dec 1946
61.2	Feb	1969	125.0	1922/23	31.3	1952/53	22.8	17–18 Dec 1970
49.0	Mar	1993	123.0	1995/96	32.4	1990/91	22.7	26–27 Feb 1934
20.5	Apr	1906	119.5	1906/07	32.6	2001/02	21.6	13–14 Mar 1939
7.0	May	1945	119.1	1933/34	38.7	1994/95	21.5	9–10 Feb 1969
			118.3	1883/84	38.8	1980/81	21.0	26 Jan 1888
			116.5	1955/56	40.0	1891/92		

REFERENCES

Achtor, T. H., and L. H. Horn, 1986: Spring season Colorado cyclones. Part I: Use of composites to relate upper and lower tropospheric wind fields. *J. Climate Appl. Meteor.,* **25,** 732–743.

Alpert, P., M. Tsidulko, and U. Stein, 1995: Can sensitivity studies yield absolute comparisons for the effects of several processes? *J. Atmos. Sci.,* **52,** 597–601.

Ambrose, K., 1993: *Blizzards and Snowstorms of Washington, D.C.* Historical Enterprises, 115 pp.

——, 1994: *Great Blizzards of New York City.* Historical Enterprises, 123 pp.

Anthes, R. A., and D. Keyser, 1979: Tests of a fine-mesh model over Europe and the United States. *Mon. Wea. Rev.,* **107,** 963–984.

Atlas, R., 1987: The role of oceanic fluxes and initial data in the numerical prediction of an intense coastal storm. *Dyn. Atmos. Oceans,* **10,** 359–388.

Austin, J. M., 1941: Favorable conditions for cyclogenesis near the Atlantic coast. *Bull. Amer. Meteor. Soc.,* **22,** 270.

——, 1951: Mechanisms of pressure change. Compendium *of Meteorology,* T. F. Malone, Ed., Amer. Meteor. Soc., 630–638.

Bailey, R. E., 1960: Forecasting of heavy snowstorms associated with major cyclones. *Weather Forecasting for Aeronautics,* J. J. George, Ed., Academic Press, 468–475.

Baker, D. G., 1970: A study of high pressure ridges to the east of the Appalachian Mountains. Ph.D. dissertation, Massachusetts Institute of Technology, 127 pp.

Baker, R., 1996: What a lovely blizzard. *New York Times,* 9 January, p 31.

Ballentine, R. J., 1980: A numerical investigation of New England coastal frontogenesis. *Mon. Wea. Rev.,* **108,** 1479–1497.

——, A. J. Stamm, E. E. Chermack, G. P. Byrd, and D. Schleede, 1998: Mesoscale model simulation of the 4–5 January 1995 lake-effect snowstorm. *Wea. Forecasting,* **13,** 893–920.

Barnes, S. L., 1964: A technique for maximizing details in numerical weather map analysis. *J. Appl. Meteor.,* **3,** 396–409.

——, and B. R. Colman, 1993: Quasigeostrophic diagnosis of cyclogenesis associated with a cutoff extratropical cyclone—The Christmas 1987 storm. *Mon. Wea. Rev.,* **121,** 1613–1634.

Barnston, A. G., and R. E. Livezey, 1987: Classification, seasonality and persistence of low-frequency atmospheric circulation patterns. *Mon. Wea. Rev.,* **115,** 1083–1126.

Beckman, S. K., 1987: Use of enhanced IR/visible satellite imagery to determine heavy snow areas. *Mon. Wea. Rev.,* **115,** 2060–2087.

Bell, G. D., and L. F. Bosart, 1988: Appalachian cold-air damming. *Mon. Wea. Rev.,* **116,** 137–161.

——, and ——, 1989: The large-scale atmospheric structures accompanying New England coastal frontogenesis and associated North American east coast cyclogenesis. *Quart. J. Roy. Meteor. Soc.,* **115,** 1133–1146.

——, and M. S. Halpert, 1998: Climate assessment for 1997. *Bull. Amer. Meteor. Soc.,* **79** (5), S1–S49.

Bennetts, D. A., and B. J. Hoskins, 1979: Conditional symmetric instability—A possible explanation of frontal rainbands. *Quart. J. Roy. Meteor. Soc.,* **105,** 945–962.

Bjerknes, J. 1919: On the structure of moving cyclones. *Geofys. Publ., Norske Videnskaps-Akad. Oslo,* **1** (1), 1–8.

——, 1951: Extratropical cyclones. *Compendium of Meteorology,* T. F. Malone, Ed., Amer. Meteor. Soc., 577–598.

——, 1954: The diffluent upper trough. *Arch. Meteor. Geophys. Bioklimatol.,* **A7,** 41–46.

——, 1969: Atmospheric teleconnections from the equatorial Pacific. *Mon. Wea. Rev.,* **97,** 163–172.

——, and H. Solberg, 1922: Life cycle of cyclones and the polar front theory of atmospheric circulation. *Geofys. Publ., Norske Videnskaps-Akad. Oslo,* **3** (1), 1–18.

——, and J. Holmboe, 1944: On the theory of cyclones. *J. Meteor.,* **1,** 1–22.

Bleck, R., 1973: Numerical forecasting experiments based on the conservation of potential vorticity on isentropic surfaces. *J. Appl. Meteor.,* **12,** 737–752.

——, 1974: Short-range prediction in isentropic coordinates with filtered and unfiltered numerical models. *Mon. Wea. Rev.,* **102,** 813–829.

——, 1977: Numerical simulation of lee cyclogenesis in the Gulf of Genoa. *Mon. Wea. Rev.,* **116,** 137–161.

——, and C. Mattocks, 1984: A preliminary analysis of the role of potential vorticity in Alpine lee cyclogenesis. *Beitr. Phys. Atmos.,* **57,** 357–368.

Bonner, W., 1965: Statistical and kinematical properties of the low-level jet stream. Satellite and Mesometeorology Research Project Research Paper, 38, University of Chicago, 54 pp.

——, 1989: NMC overview: Recent progress and future plans. *Wea. Forecasting,* **4,** 275–285.

Bosart, L. F., 1975: New England coastal frontogenesis. *Quart. J. Roy. Meteor. Soc.,* **101,** 957–978.

——, 1981: The Presidents' Day snowstorm of 18–19 February 1979: A subsynoptic-scale event. *Mon. Wea. Rev.,* **109,** 1542–1566.

——, 1999: Observed cyclone life cycles. *The Life Cycles of Extratropical Cyclones,* M. A. Shapiro and S. Grønås, Eds., Amer. Meteor. Soc., 187–213.

——, 2003a: Whither the weather analysis and forecast process? *Wea. Forecasting,* **18,** 520–529.

——, 2003b: Tropopause folding, upper-level frontogenesis, and beyond. *A Half Century of Progress in Meteorology: A Tribute to Richard J. Reed, Meteor. Monogr.,* Amer. Meteor. Soc., **31,** 13–47.

——, and J. P. Cussen Jr., 1973: Gravity wave phenomena accompanying East Coast cyclogenesis. *Mon. Wea. Rev.,* **101,** 5445–5454.

——, and S. C. Lin, 1984: A diagnostic analysis of the Presidents' Day storm of February 1979. *Mon. Wea. Rev.,* **112,** 2148–2177.

——, and F. Sanders, 1986: Mesoscale structure in the Megalopolitan snowstorm of 11–12 February 1983. Part III: A large amplitude gravity wave. *J. Atmos. Sci.,* **43,** 924–939.

——, and A. Seimon, 1988: A case study of an unusually intense atmospheric gravity wave. *Mon. Wea. Rev.,* **116,** 1857–1886.

——, and J. A. Bartlo, 1991: Tropical storm formation in a baroclinic environment. *Mon. Wea. Rev.,* **119,** 1979–2013.

——, and F. Sanders, 1991: An early-season coastal storm: Conceptual success and model failure. *Mon. Wea. Rev.,* **119,** 2831–2851.

——, and W. E. Bracken, 1996: Coastal cyclonic development in the northwest periphery of a larger parent cyclone. Preprints, *14th Conf. on Weather Analysis and Forecasting,* Dallas, TX, Amer. Meteor. Soc., 551–554.

——, C. J. Vaudo, and J. H. Helsdon Jr., 1972: Coastal frontogenesis. *J. Appl. Meteor.,* **11,** 1236–1258.

——, G. J. Hakim, K. R. Tyle, M. A. Bedrick, W. E. Bracken, M. J. Dickinson, and D. M. Schultz, 1996: Large-scale antecedent conditions associated with the 12–14 March 1993 cyclone ("Superstorm '93") over eastern North America. *Mon. Wea. Rev.,* **124,** 1865–1891.

——, W. E. Bracken, and A. Seimon, 1998: A study of cyclone mesoscale structure with emphasis on a large-amplitude inertia–gravity wave. *Mon. Wea. Rev.,* **126,** 1497–1527.

Boucher, R. J., and R. J. Newcomb, 1962: Synoptic interpretation of some TIROS vortex patterns: A preliminary cyclone model. *J. Appl. Meteor.,* **1,** 127–136.

Boyle, J. S., and L. F. Bosart, 1983: A cyclone/anticyclone couplet over North America: An example of anticyclone evolution. *Mon. Wea. Rev.,* **111,** 1025–1045.

——, and ——, 1986: Cyclone–anticyclone couplets over North America. Part II: Analysis of a major cyclone event over the eastern United States. *Mon. Wea. Rev.,* **114,** 2432–2465.

Brandes, E. A., and J. Spar, 1971: A search for necessary conditions for heavy snow on the East Coast. *J. Appl. Meteor.,* **11,** 397–409.

Branick, M. L., 1997: A climatology of significant winter-type weather events in the contiguous United States 1982–94. *Wea. Forecasting,* **12,** 193–207.

Brill, K. F., L. W. Uccellini, R. P. Burkhart, T. T. Warner, and R. A. Anthes, 1985: Numerical simulations of a transverse indirect circulation and low-level jet in the exit region of an upper-level jet. *J. Atmos. Sci.,* **42,** 1306–1320.

——, ——, J. Manobianco, P. J. Kocin, and J. H. Homan, 1991: The use of successive dynamic initialization by nudging to simulate cyclogenesis during GALE IOP1. *Meteor. Atmos. Phys.,* **45,** 15–40.

Bristor, C. L., 1951: The Great Storm of November 1950. *Weatherwise,* **4,** 10–16.

Brooks, C. F., 1914: The distribution of snowfall in cyclones of the eastern United States. *Mon. Wea. Rev.,* **42,** 318–329.

Brown, H. E., and D. A. Olson, 1978: Performance of NMC in forecasting a record-breaking winter storm, 6–7 February 1978. *Bull. Amer. Meteor. Soc.,* **59,** 562–575.

Browne, R. F., and R. J. Younkin, 1970: Some relationships between 850-millibar lows and heavy snow occurrences over the central and eastern United States. *Mon. Wea. Rev.,* **98,** 399–401.

Browning, K. A., 1971: Radar measurements of air motion near fronts. *Weather,* **26,** 320–340.

——, 1986: Conceptual models of precipitation systems. *Wea. Forecasting,* **1,** 23–41.

——, 1990: Organization of clouds and precipitation in extratropical cyclones. *Extratropical Cyclones: The Erik Palmén Memorial Volume,* C. W. Newton and E. O. Holopainen, Eds., Amer. Meteor. Soc., 129–153.

——, and T. W. Harrold, 1969: Air motion and precipitation growth in a wave depression. *Quart. J. Roy. Meteor. Soc.,* **95,** 288–309.

Brunk, I., 1949: The pressure pulsation of 11 April 1944. *J. Meteor.,* **6,** 181–187.

Buizza, R., and P. Chessa, 2002: Prediction of the U.S. storm of 24–26 January 2000 with the ECMWF Ensemble Prediction System. *Mon. Wea. Rev.,* **130,** 1531–1551.

Burnham, G. H., 1922: Economic effects of New England's unprecedented ice storm of November 25–29, 1921. *J. Geogr.,* **21,** 161–168.

Buzzi, A., and S. Tibaldi, 1978: Cyclogenesis in the lee of the Alps: A case study. *Quart. J. Roy. Meteor. Soc.,* **104,** 271–287.

Byrd, G. P., R. A. Anstett, J. E. Heim, and D. M. Usinski, 1991: Mobile sounding observations of lake-effect snowbands in western and central New York. *Mon. Wea. Rev.,* **119,** 2323–2332.

Cahir, J. J. 1971: Implications of circulations in the vicinity of jet streaks at subsynoptic scales. Ph.D. thesis, The Pennsylvania State University, 170 pp.

Caplan, P., 1995: The 12–14 March 1993 Superstorm: Performance of the global model. *Bull. Amer. Meteor. Soc.,* **76,** 201–212.

Caplovich, J. 1987: *Blizzard! The Great Storm of '88.* Vero Publishing Co., 242 pp.

Carlson, T. N., 1961: Lee-side frontogenesis in the Rocky Mountains. *Mon. Wea. Rev.,* **89,** 163–172.

——, 1980: Airflow through midlatitude cyclones and the comma cloud pattern. *Mon. Wea. Rev.,* **108,** 1498–1509.

——, 1991: *Mid-latitude Weather Systems.* HarperCollins Academic, 507 pp.

Carpenter, D. M., 1993: The lake effect of the Great Salt Lake: Overview and forecast problems. *Wea. Forecasting,* **8,** 181–193.

Chang, C. B., D. J. Perkey, and C. W. Kreitzberg, 1982: A numerical case study of the effects of latent heating on a developing wave cyclone. *J. Atmos. Sci.,* **39,** 1555–1570.

Charney, J. G., 1947: The dynamics of long waves in a baroclinic westerly current. *J. Meteor.,* **4,** 135–162.

——, and N. A. Phillips, 1953: Numerical integration of the quasi-geostrophic equations for barotropic and simple baroclinic flows. *J. Meteor.,* **10,** 71–99.

Colucci, S. J., 1976: Winter cyclone frequencies over the eastern United States and adjacent western Atlantic, 1963–1973. *Bull Amer. Meteor. Soc.,* **57,** 548–553.

——, 1985: Explosive cyclogenesis and large-scale circulation changes: Implications for atmospheric blocking. *J. Atmos. Sci.,* **42,** 2701–2717.

Cressman, G., 1970: Public forecasting: Present and future. *A Century of Weather Progress,* J. E. Caskey Jr., Ed., Amer. Meteor. Soc., 71–77.

Crum, T. D., and R. L. Alberty, 1993: The WSR-88D and the WSR-88D Operational Support Facility. *Bull. Amer. Meteor. Soc.,* **74,** 1669–1687.

Cunningham, R. M., and F. Sanders, 1987: Into the teeth of the gale: The remarkable advance of a cold front at Grand Manan. *Mon. Wea. Rev.,* **115,** 2450–2462.

Danard, M. B., 1964: On the influence of released latent heat on cyclone development. *J. Appl. Meteor.,* **3,** 27–37.

——, and G. E. Ellenton, 1980: Physical influences on East Coast cyclogenesis. *Atmos.–Ocean,* **18,** 65–82.

Danielsen, E. F., 1966: Research in four-dimensional diagnosis of cyclone storm cloud systems. Air Force Cambridge Research Lab. Rep. 66–30, Bedford, MA, 53 pp. [NTIS-AD-632668.]

——, 1968: Stratospheric–tropospheric exchange based upon radioactivity, ozone and potential vorticity. *J. Atmos. Sci.,* **25,** 502–518.

Day, P. C., and S. P. Fergusson, 1922: The great snowstorm of January 27–29, 1922 over the Atlantic Coast states. *Mon. Wea. Rev.,* **50,** 21–24.

Dean, D. B., and L. F. Bosart, 1996: Northern Hemisphere 500-hPa trough merger and fracture: A climatology and case study. *Mon. Wea. Rev.,* **124,** 2644–2671.

DeGaetano, A. T., 2000: Climatic perspective and impacts of the 1998 northern New York and New England ice storm. *Bull. Amer. Meteor. Soc.,* **81,** 237–254.

Dickinson, M. J., L. F. Bosart, W. E. Bracken, G. J. Hakim, D. M. Schultz, M. A. Bedrick, and K. R. Tyle, 1997: The March 1993 Superstorm cyclogenesis: Incipient phase synoptic- and convective-scale flow interaction and model performance. *Mon. Wea. Rev.,* **125,** 3041–3072.

Dines, W. H., 1925: The correlation between pressure and temperature in the upper air with a suggested explanation. *Quart. J. Roy. Meteor. Soc.,* **51,** 31–38.

Dirks, R. A., J. P. Kuettner, and J. A. Moore, 1988: Genesis of Atlantic Lows Experiment (GALE): An overview. *Bull. Amer. Meteor. Soc.,* **69,** 148–160.

Doesken, N. J., and A. Judson, 1996: *The Snow Booklet: A Guide to the Science, Climatology and Measurement of Snow in the United States.* Colorado State University, 85 pp.

Dole, R., 1986: Persistent anomalies of the extratropical Northern Hemisphere wintertime circulation: Structure. *Mon. Wea. Rev.,* **114,** 178–207.

——, 1989: Life cycles of persistent anomalies. Part I: Evolution of 500 mb height fields. *Mon. Wea. Rev.,* **117,** 177–211.

——, and N. D. Gordon, 1983: Persistent anomalies of the extratropical Northern Hemisphere wintertime circulation: Geographical distribution and regional persistence characteristics. *Mon. Wea. Rev.,* **111,** 1567–1586.

Doyle, J. D., and T. T. Warner, 1993a: A three-dimensional numerical investigation of a Carolina coastal low-level jet during GALE IOP2. *Mon. Wea. Rev.,* **121,** 1030–1047.

——, and ——, 1993b: A numerical investigation of coastal frontogenesis and mesoscale cyclogenesis during GALE IOP2. *Mon. Wea. Rev.,* **121,** 1048–1077.

——, and ——, 1993c: Nonhydrostatic simulations of coastal mesobeta-scale vortices and frontogenesis. *Mon. Wea. Rev.,* **121,** 3371–3392.

Dunn, L. B., 1987: Cold-air damming by the Front Range of the Colorado Rockies and its relationship to locally heavy snows. *Wea. Forecasting,* **2,** 177–189.

——, 1988: Vertical motion evaluation of a Colorado snowstorm from a synoptician's perspective. *Wea. Forecasting,* **3,** 261–272.

——, 1992: Evidence of ascent in a sloped barrier jet and an associated heavy-snow band. *Mon. Wea. Rev.,* **120,** 914–924.

Eady, E. T., 1949: Long waves and cyclone waves. *Tellus,* **1,** 33–52.

Egger, J., 1974: Numerical experiments on lee cyclogenesis. *Mon. Wea. Rev.,* **102,** 847–860.

Eliassen, A., and E. Kleinschmidt, 1957: Dynamic meteorology. *Handbuch der Physik,* S. Flugge, Ed., Vol. 48, Springer-Verlag, 1–154.

Emanuel, K., 1979: Inertial instability and mesoscale convective systems. Part I: Linear theory of inertial instability in rotating viscous fields. *J. Atmos. Sci.,* **36,** 2425–2449.

——, 1983: The Lagrangian parcel dynamics of moist symmetric instability. *J. Atmos. Sci.,* **40,** 2368–2376.

——, 1985: Frontogenesis in the presence of low moist symmetric stability. *J. Atmos. Sci.,* **42,** 1062–1071.

Evans, M. S., D. Keyser, L. F. Bosart, and G. M. Lackmann, 1994: A satellite-derived classification scheme for rapid maritime cyclogenesis. *Mon. Wea. Rev.,* **122,** 1381–1416.

Farrell, B., 1984: Modal and nonmodal baroclinic waves. *J. Atmos. Sci.,* **41,** 668–673.

——, 1985: Transient growth of damped baroclinic waves. *J. Atmos. Sci.,* **42,** 2718–2727.

Ferber, G. K., C. F. Mass, G. M. Lackmann, and M. W. Patnoe, 1993: Snowstorms over the Puget Sound lowlands. *Wea. Forecasting,* **8,** 481–504.

Ferretti, R., F. Einaudi, and L. W. Uccellini, 1988: Wave disturbances associated with the Red River Valley severe weather outbreak of 10–11 April 1979. *Meteor. Atmos. Phys.,* **39,** 132–168.

Forbes, G. S., R. A. Anthes, and D. W. Thomson, 1987: Synoptic and mesoscale aspects of an Appalachian ice storm associated with cold air damming. *Mon. Wea. Rev.,* **115,** 564–591.

Foster, J. L., and R. J. Leffler, 1979: The extreme weather of February 1979 in the Baltimore–Washington area. *Natl. Wea. Dig.,* **4,** 16–21.

Friday, E. W., 1994: The modernization and associated restructuring of the National Weather Service: An overview. *Bull. Amer. Meteor. Soc.,* **75,** 43–52.

Fujita, T. T., 1971: Proposed characterization of tornadoes and hurricanes by area and intensity. Satellite and Meteorology Research Paper 91, The University of Chicago, Chicago, IL, 42 pp.

Gall, R., 1976: Structural changes of growing baroclinic waves. *J. Atmos. Sci.,* **33,** 374–390.

Garriott, E. B., 1899: Forecasts and warnings. *Mon. Wea. Rev.,* **27,** 41–44.

Gaza, R. S., and L. F. Bosart, 1990: Trough merger characteristics over North America. *Wea. Forecasting,* **5,** 314–331.

Gedzelman, S. D., and E. Lewis, 1990: Warm snowstorms: A forecaster's dilemma. *Weatherwise,* **43,** 265–270.

Gilhousen, D. B., 1994: The value of NDBC observations during March 1993's "Storm of the Century." *Wea. Forecasting,* **9,** 255–264.

Glickman, T., Ed., 2000: *Glossary of Meteorology.* 2d ed. Amer. Meteor. Soc., 855 pp.

Godev, N., 1971a: The cyclogenetic properties of the Pacific coast: Possible source of errors in numerical prediction. *J. Atmos. Sci.,* **28,** 968–972.

——, 1971b: Anticyclonic activity over south Europe and its relation to orography. *J. Appl. Meteor.,* **10,** 1097–1102.

Goree, P. A., and R. J. Younkin, 1966: Synoptic climatology of heavy snowfall over the central and eastern United States. *Mon Wea. Rev.,* **94,** 663–668.

Graves, C. E., J. T. Moore, M. J. Singer, and S. Ng, 2003: Band on the run: Chasing the physical processes associated with heavy snowfall. *Bull. Amer. Meteor. Soc.,* **84,** 990–995.

Green, J. S. A., F. H. Ludlam, and J. F. R. McIlveen, 1966: Isentropic relative-flow analysis and the parcel theory. *Quart. J. Roy. Meteor. Soc.,* **92,** 210–219.

Grumm, R. H., 1993: Characteristics of surface cyclone forecasts in the aviation run of the global spectral model. *Wea. Forecasting,* **8,** 87–112.

——, R. J. Oravec, and A. L. Siebers, 1992: Systematic model forecast errors of surface cyclones in NMC's Nested-Grid Model, December 1988 through November 1990. *Wea. Forecasting,* **7,** 65–87.

Gurka, J. J., E. P. Auciello, A. F. Gigi, J. S. Waldstreicher, K. K. Keeter, S. Businger, and L. G. Lee, 1995: Winter weather forecasting throughout the eastern United States. Part II. An operational perspective of cyclogenesis. *Wea. Forecasting,* **10,** 21–41.

Gyakum, J. R., 1983a: On the evolution of the *QE II* storm. I: Synoptic aspects. *Mon. Wea. Rev.,* **111,** 1137–1155.

——, 1983b: On the evolution of the *QE II* storm. II: Dynamic and thermodynamic structure. *Mon. Wea. Rev.,* **111,** 1156–1173.

——, 1987: On the evolution of a surprise snowfall in the United States Midwest. *Mon. Wea. Rev.,* **115,** 2322–2345.

——, and E. S. Barker, 1988: A case study of explosive subsynoptic-scale cyclogenesis. *Mon. Wea. Rev.,* **116,** 2225–2253.

——, and P. J. Roebber, 2001: The 1998 ice storm—Analysis of a planetary-scale event. *Mon. Wea. Rev.,* **129,** 2983–2997.

——, J. R. Anderson, R. H. Grumm, and E. L Gruner, 1989: North Pacific cold-season surface cyclone activity. *Mon. Wea. Rev.,* **117,** 1141–1155.

Hadlock, R., and C. W. Kreitzberg, 1988: The Experiment on Rapidly Intensifying Cyclones over the Atlantic (ERICA) field study: Objectives and plans. *Bull. Amer. Meteor. Soc.,* **69,** 1309–1320.

Hakim, G. J., and L. W. Uccellini, 1992: Diagnosing coupled jet-streak circulations for a northern plains snowband from the operational Nested Grid Model. *Wea. Forecasting,* **7,** 26–48.

——, L. F. Bosart, and D. Keyser, 1995: The Ohio Valley wave-merger cyclogenesis event of 25–26 January 1978. Part 1: Multiscale case study. *Mon. Wea. Rev.,* **123,** 2663–2692.

——, D. Keyser, and L. F. Bosart, 1996: The Ohio Valley wave-merger cyclogenesis event of 25–26 January 1978: Part II: Diagnosis using quasigeostrophic potential vorticity inversion. *Mon. Wea. Rev.,* **124,** 2176–2205.

Halpert, M. S., and C. F. Ropelewski, 1992: Surface temperature patterns associated with the Southern Oscillation. *J. Climate,* **5,** 577–593.

——, and G. D. Bell, 1997: Climate assessment for 1996. *Bull. Amer. Meteor. Soc.,* **78** (5), S1–S49.

Harlin, B. W., 1952: The great southern glaze storm of 1951. *Weatherwise,* **5,** 10–13.

Harrold, T. W., 1973: Mechanisms influencing the distribution of precipitation within baroclinic disturbances. *Quart. J. Roy. Meteor. Soc.,* **99,** 232–251.

Hart, R. E., and R. H. Grumm, 2001: Using normalized climatological anomalies to rank synoptic-scale events objectively. *Mon. Wea. Rev.,* **129,** 2426–2442.

Hayden, B. P., 1981: Secular variations in Atlantic coast extratropical cyclones. *Mon. Wea. Rev.,* **109,** 159–167.

Hayden, E., 1888: *The Great Storm off the Atlantic Coast of the United States March 11–14, 1888. Nautical Monogr.,* No. 5, U.S. Government Printing Office, 65 pp.

Hibbard, W., D. Santek, L. Uccellini, and K. Brill, 1989: Application of the 4-D McIDAS to a model diagnostic study of the Presidents' Day cyclone. *Bull. Amer. Meteor. Soc.* **70,** 1394–1403.

Hjelmfelt, M. R., 1990: Numerical study of the influence of environmental conditions on lake effect snowstorms over Lake Michigan. *Mon. Wea. Rev.,* **118,** 138–150.

——, 1992: Orographic effects in simulated lake-effect snowstorms over Lake Michigan. *Mon. Wea. Rev.,* **120,** 373–377.

Holton, J. R., 1979: *An Introduction to Dynamic Meteorology.* 2d ed. Academic Press, 391 pp.

Homan, J., and L. W. Uccellini, 1987: Winter forecast problems associated with light to moderate snow events in the mid-Atlantic states on 14 and 22 February 1986. *Wea. Forecasting,* **2,** 206–228.

Hooke, W. H., 1986: Gravity waves. *Mesoscale Meteorology and Forecasting,* P. Ray, Ed., Amer. Meteor. Soc., 272–288.

Hoskins, B. J., M. E. McIntyre, and A. W. Robertson, 1985: On the use and significance of isentropic potential vorticity maps. *Quart. J. Roy. Meteor. Soc.,* **111,** 877–946.

Houghton, J. T., Y. Ding, D. J. Griggs, M. Noguer, P. J. van der Linden, and D. Xiaosu, Eds., 2001: Climate Change 2001: The Scientific Basis. Contribution of Working Group I to the Third Assessment Report of the Intergovernmental Panel on Climate Change. Cambridge University Press, 881 pp.

Houze, R. A., Jr., 1993: *Cloud Dynamics.* Academic Press, 573 pp.

Hovanec, R. D., and L. H. Horn, 1975: Static stability and the 300 mb isotach field in the Colorado cyclogenetic area. *Mon. Wea. Rev.,* **103,** 628–638.

Howard, K. W., and E. I. Tollerud, 1988: The structure and evolution of heavy-snow-producing Colorado cyclones. Preprints, *Palmén Symp. on Extratropical Cyclones and Their Role in the General Circulation,* Helsinki, Finland, Amer. Meteor. Soc., 168–171.

Huffman, G. J., and G. A. Norman Jr., 1988: The supercooled warm rain process and the specification of freezing precipitation. *Mon. Wea. Rev.,* **116,** 2172–2182.

Hughes, P., 1976: *American Weather Stories.* NOAA, 116 pp.

——, 1981: The blizzard of '88. *Weatherwise,* **34,** 250–256.

Huo, Z., D.-L. Zhang, and J. R. Gyakum, 1995: A diagnostic analysis of the Superstorm of March 1993. *Mon. Wea. Rev.,* **123,** 1740–1761.

——, ——, and ——, 1998: An application of potential vorticity inversion to improving the numerical prediction of the March 1993 Superstorm. *Mon. Wea. Rev.,* **126,** 424–436.

——, ——, and ——, 1999a: Interaction of portential vorticity anomalies in extratropical cyclogenesis. Part I: Static piecewise inversion. *Mon. Wea. Rev.,* **127,** 2546–2561.

——, ——, and ——, 1999b: Interaction of potential vorticity anomalies in extratropical cyclogenesis. Part II: Sensitivity to initial perturbations. *Mon. Wea. Rev.,* **127,** 2563–2575.

Hurell, J., 1995: Decadal trends in the North Atlantic Oscillation: Regional temperatures and precipitation. *Science,* **269,** 676–679.

Johnson, D. R., and W. K. Downey, 1976: The absolute angular momentum budget of an extratropical cyclone: Quasi-Lagrangian diagnostics. *Mon. Wea. Rev.,* **104,** 3–14.

Junker, N. W., J. E. Hoke, and R. H. Grumm, 1989: Performance of NMC's regional models. *Wea. Forecasting,* **4,** 368–390.

Kain, J. S., S. M. Goss, and M. E. Baldwin, 2000: The melting effect as a factor in precipitation-type forecasting. *Wea. Forecasting,* **15,** 700–714.

Kalnay, E., and Coauthors, 1996: The NCEP/NCAR 40-Year Reanalysis Project. *Bull. Amer. Meteor. Soc.,* **77,** 437–471.

Kaplan, M. L., J. W. Zack, V. C. Wong, and J. J. Tuccillo, 1982: A sixth-order mesoscale atmospheric simulation system applicable to research and real-time forecasting problems. *Proc. CIMMS 1982 Symp.,* Norman, OK,? ?sponsor??, 38–84.

Karl, T., Ed.,1996: *Long-Term Climate Monitoring by the Global Climate Observing System.* Kluwer Academic, 648 pp.

Keeter, K. K., and J. W. Cline, 1991: The objective use of observed and forecast thickness values to predict precipitation type in North Carolina. *Wea. Forecasting,* **6,** 456–469.

——, S. Businger, L. G. Lee, and J. S. Waldstreicher, 1995: Winter weather forecasting throughout the eastern United States. Part III: The effects of topography and the variability of winter weather in the Carolinas and Virginia. *Wea. Forecasting,* **10,** 42–60.

Keshishian, L. G., and L. F. Bosart, 1987: A case study of extended East Coast frontogenesis. *Mon. Wea. Rev.,* **115,** 100–117.

——, ——, and W. E. Bracken, 1994: Inverted troughs and cyclogenesis over interior North America: A limited regional climatology and case studies. *Mon. Wea. Rev.,* **122,** 565–607.

Keyser, D., and D. R. Johnson, 1984: Effects of diabatic heating on the ageostrophic circulation of an upper tropospheric jet streak. *Mon. Wea. Rev.,* **112,** 1709–1724.

——, and M. Shapiro, 1986: A review of the structure and dynamics of upper-level frontal zones. *Mon. Wea. Rev.,* **114,** 452–499.

——, B. D. Schmidt, and D. G. Duffy, 1989: A technique for representing three-dimensional vertical circulations in baroclinic disturbances. *Mon. Wea. Rev.,* **117,** 2463–2494.

Klein, W. J., and J. S. Winston, 1958: Geographical frequency of troughs and ridges on mean 700 mb charts. *Mon. Wea. Rev.,* **86,** 344–358.

Kleinschmidt, E., 1950: On the structure and origin of cyclones (Part 1). *Meteor. Rundsch.,* **3,** 1–6.

——, 1957: Cyclones and anticyclones. *Handbuch der Physik,* S. Flugge, Ed., Vol. 48, Springer-Verlag, 1–154.

Koch, S. E., and P. B. Dorian, 1988: A mesoscale gravity wave event observed during CCOPE. Part III: Wave environment and probable source mechanisms. *Mon. Wea. Rev.,* **116,** 2570–2592.

——, and R. E. Golus, 1988: A mesoscale gravity wave event observed during CCOPE. Part I: Multiscale statistical analysis of wave characteristics. *Mon. Wea. Rev.,* **116,** 2527–2544.

——, and C. O'Handley, 1997: Operational forecasting and detection of mesoscale gravity waves. *Wea. Forecasting,* **12,** 253–281.

——, M. L. des Jardins, and P. J. Kocin, 1983: An interactive Barnes objective map analysis scheme for use with satellite and conventional data. *J. Climate Appl. Meteor.,* **22,** 1487–1503.

Kocin, P. J., 1983: An analysis of the "Blizzard of '88." *Bull. Amer. Meteor. Soc.,* **64,** 1258–1272.

——, and L. W. Uccellini, 1990: *Snowstorms along the Northeastern Coast of the United States: 1955 to 1985. Meteor. Monogr.,* No. 44, Amer. Meteor. Soc., 280 pp.

——, and ——, 2004a: *Overview.* Vol. 1. *Northeast Snowstorms. Meteor. Monogr.,* No. 54, Amer. Meteor. Soc., in press.

——, and ——, 2004b: *The Cases.* Vol. 2. *Northeast Snowstorms. Meteor. Monogr.,* No. 54, Amer. Meteor. Soc., in press.

——, and ——, 2004c: A northeast snowfall impact scale. *Bull. Amer. Meteor. Soc.,* **85,** 177–194.

——, ——, J. W. Zack, and M. L. Kaplan, 1985: A mesoscale numerical forecast of an intense convective snowburst along the East Coast. *Bull. Amer. Meteor. Soc.,* **66,** 1412–1424.

——, L. W. Uccellini, and R. A. Petersen, 1986: Rapid evolution of a jet stream circulation in a pre-convective environment. *Meteor. Atmos. Phys.,* **35,** 103–138.

——, A. D. Weiss, and J. J. Wagner, 1988: The great Arctic outbreak and East Coast blizzard of February 1899. *Wea. Forecasting,* **3,** 305–318.

——, P. N. Schumacher, R. F. Morales Jr., and L. W. Uccellini, 1995: Overview of the 12–14 March 1993 Superstorm. *Bull. Amer. Meteor. Soc.,* **76,** 165–182.

——, L. W. Uccellini, K. F. Brill, and M. Zika, 1998: Northeast snowstorms: An update. Preprints, *16th Conf. on Weather Analysis and Forecasting,* Phoenix, AZ, Amer. Meteor. Soc., 421–423.

Koppel, L. L., L. F. Bosart, and D. Keyser, 2000: A 25-yr climatology of large-amplitude hourly surface pressure changes over the conterminous United States. *Mon. Wea. Rev.,* **128,** 51–68.

Krishnamurti, T. N., 1968: A study of a developing wave cyclone. *Mon. Wea. Rev.,* **96,** 208–217.

Kristovich, D. A. R., and Coauthors, 2000: The Lake-Induced Convection Experiments and the Snowband Dynamics Project. *Bull. Amer. Meteor. Soc.,* **81,** 519–542.

Kuo, Y.-H., and S. Low-Nam, 1990: Prediction of nine explosive cyclones over the western Atlantic Ocean with a regional model. *Mon. Wea. Rev.,* **118,** 3–25.

Lackmann, G. M., L. F. Bosart, and D. Keyser, 1996: Planetary- and synoptic-scale characteristics of explosive wintertime cyclogenesis over the western North Atlantic Ocean. *Mon. Wea. Rev.,* **124,** 2672–2702.

——, ——, and ——, 1997: A characteristic life cycle of upper-tropospheric cyclogenetic precursors during the Experiment on Rapidly Intensifying Cyclones over the Atlantic (ERICA). *Mon. Wea. Rev.,* **125,** 2729–2758.

——, ——, and ——, 1999: Energetics of an intensifying jet streak during the Experiment on Rapidly Intensifying Cyclones over the Atlantic (ERICA). *Mon. Wea. Rev.,* **127,** 2777–2795.

Lai, C.-C., and L. F. Bosart, 1988: A case study of trough mergers in split westerly flow. *Mon. Wea. Rev.,* **116,** 1838–1856.

Langland, R. J., M. A. Shapiro, and R. Gelaro, 2002: Initial condition sensitivity and error growth in forecasts of the 25 January 2000 East Coast snowstorm. *Mon. Wea. Rev.,* **130,** 957–974.

LaPenta, W. M., and N. L. Seaman, 1990: A numerical investigation of East Coast cyclogenesis during the cold-air damming event of 27–28 February 1982: Part I: Dynamic and thermodynamic structure. *Mon. Wea. Rev.,* **118,** 2668–2695.

——, and ——, 1992: A numerical investigation of East Coast cyclogenesis during the cold-air damming event of 27–28 February 1982: Part II: Importance of physical mechanisms. *Mon. Wea. Rev.,* **120,** 52–76.

Lau, K.-M., and H. Weng, 1999: Interannual, decadal–interdecadal, and global warming signals in sea surface temperature during 1955–97. *J. Climate,* **12,** 1257–1267.

Leathers, D. J., D. R. Kluck, and S. Kroczynski, 1998: The severe flooding event of January 1996 across north-central Pennsylvania. *Bull. Amer. Meteor. Soc.,* **79,** 785–797.

Leese, J. A., 1962: The role of advection in the formation of vortex cloud patterns. *Tellus,* **14,** 409–421.

Lindzen, R. S., and K.-K. Tung, 1976: Banded convective activity and ducted gravity waves. *Mon. Wea. Rev.,* **104,** 1602–1617.

Livezey, R. E., and T. M. Smith, 1999: Covariability of aspects of North American climate with global sea surface temperatures on interannual to interdecadal timescales. *J. Climate,* **12,** 289–302.

Loughe, A., C.-C. Lai, and D. Keyser, 1995: A technique for diagnosing three-dimensional circulations in baroclinic disturbances on limited-area domains. *Mon. Wea. Rev.,* **123,** 1476–1504.

Ludlum, D. M., 1956: The great Atlantic low. *Weatherwise,* **9,** 64–65.

——, 1966: *Early American Winters I 1604–1820.* Amer. Meteor. Soc., 198 pp.

——, 1968: *Early American Winters II 1821–1870.* Amer. Meteor. Soc., 257 pp.

——, 1971: *The Weather Record Book.* Weatherwise, Inc., 98 pp.

——, 1976: *The Country Journal New England Weather Book.* Houghton Mifflin, 148 pp.

——, 1982: *The American Weather Book.* Houghton Mifflin, 296 pp.

——, 1983: *The New Jersey Weather Book.* Rutgers University Press, 256 pp.

Maddox, R. A., D. J. Perkey, and J. M. Fritsch, 1981: Evolution of upper tropospheric features during the development of a mesoscale convective complex. *J. Atmos. Sci.,* **38,** 1664–1674.

Maglaras, G. J. F. Waldstreicher, P. J. Kocin, A. F. Gigi, and R. A. Marine, 1995: Winter weather forecasting throughout the eastern United States. Part I: An overview. *Wea. Forecasting,* **10,** 5–20.

Mahoney, J. L., J. M. Brown, and E. I. Tollerud, 1995: Contrasting meteorological conditions associated with winter storms at Denver and Colorado Springs. *Wea. Forecasting,* **10,** 245–260.

Mailhot, J., and C. Chouinard, 1989: Numerical forecasts of explosive winter storms: Sensitivity experiments with a meso-scale model. *Mon. Wea. Rev.,* **117,** 1311–1343.

Manney, G. L., J. D. Farrara, and C. R. Mechoso, 1994: Simulations of the February 1979 stratospheric sudden warming: Model comparisons and three-dimensional evolution. *Mon. Wea. Rev.,* **122,** 1115–1140.

Manobianco, J., L. W. Uccellini, K. F. Brill, and Y.-H. Kuo, 1992: The impact of dynamic data assimilation on the numerical simulations of the *QE II* cyclone and an analysis of the jet streak influencing the precyclogenetic environment. *Mon. Wea. Rev.,* **120,** 1973–1996.

Marks, F. D., Jr. and P. M. Austin, 1979: Effects of the New England coastal front on the distribution of precipitation. *Mon. Wea. Rev.,* **107,** 53–67.

Martin, J. E., 1998: The structure and evolution of a continental winter cyclone. Part II: Frontal forcing of an extreme snow event. *Mon. Wea. Rev.,* **126,** 329–348.

——, 1999: Quasigeostrophic forcing of ascent in the occluded sector of cyclones and the Trowal airstream. *Mon. Wea. Rev.,* **127,** 70–88.

Marwitz, J. D., and J. Toth, 1993: A case study of heavy snowfall in Oklahoma. *Mon. Wea. Rev.,* **121,** 648–660.

Mass, C. F., and D. M. Schultz, 1993: The structure and evolution of a simulated midlatitude cyclone over land. *Mon. Wea. Rev.,* **121,** 889–917.

Mather, J. R., H. Adams, and G. A. Yoshioka, 1964: Coastal storms of the eastern United States. *J. Appl. Meteor.,* **3,** 693–706.

Mattocks, C., and R. Bleck, 1986: Jet streak dynamics and geostrophic adjustment processes during the initial stages of lee cyclogenesis. *Mon. Wea. Rev.,* **114,** 2033–2056.

McKelvey, B., 1995: *Snow in the Cities: A History of America's Urban Response.* University of Rochester Press, 202 pp.

McQueen, H. R., and H. C. Keith, 1956: The ice storm of January 7–10, 1956 over the northeastern United States. *Mon. Wea. Rev.,* **84,** 35–45.

Miller, J. E., 1946: Cyclogenesis in the Atlantic coastal region of the United States. *J. Meteor.,* **3,** 31–44.

Mook, C. P., 1956: The "Knickerbocker" snowstorm of January 1922 at Washington, D.C. *Weatherwise,* **9,** 188–191.

——, and K. S. Norquest, 1956: The heavy snowstorm of 18–19 March 1956. *Mon. Wea. Rev.,* **84,** 116–125.

Moore, J. T., and P. D. Blakely, 1988: The role of frontogenetical forcing and conditional symmetrical instability in the Midwest snowstorm of 30–31 January 1982. *Mon. Wea. Rev.,* **116,** 2153–2176.

——, and V. Vanknowe, 1992: The effect of jet-streak curvature on kinematic fields. *Mon. Wea. Rev.,* **120,** 2429–2441.

Mote, T. L., D. W. Gamble, S. J. Underwood, and M. L. Bentley, 1997: Synoptic-scale features common to heavy snowstorms in the southeast United States. *Wea. Forecasting,* **12,** 5–23.

Mullen, S. L., and B. B. Smith, 1990: An analysis of sea-level cyclone errors in NMC's Nested Grid Model (NGM) during the 1987–1988 winter season. *Wea. Forecasting,* **5,** 433–447.

Murray, R., and S. M. Daniels, 1953: Transverse flow at entrance and exit to jet streams. *Quart. J. Roy. Meteor. Soc.,* **79,** 236–241.

Naistat, R. J., and J. A. Young, 1973: A linear model of boundary layer flow applied to the St. Patrick's Day storm of 1965. *J. Appl. Meteor.,* **12,** 1151–1162.

Namias, J., and P. F. Clapp, 1949: Confluence theory of the high tropospheric jet stream. *J. Meteor.,* **6,** 330–336.

NCDC, cited 2003: Billion dollar weather disasters. [Available online at http://www.ncdc.noaa.gov/oa/reports/billionz.html#LIST.]

Newton, C. W., 1954: Frontogenesis and frontolysis as a three-dimensional process. *J. Meteor.,* **11,** 449–461.

——, 1956: Mechanisms of circulation change during a lee cyclogenesis. *J. Meteor.,* **13,** 528–539.

——, and A. V. Persson, 1962: Structural characteristics of the subtropical jet stream and certain lower-stratospheric wind systems. *Tellus,* **14,** 221–241.

——, and A. Trevisan, 1984: Clinogenesis and frontogenesis in jet stream waves. Part II: Channel model numerical experiments. *J. Atmos. Sci.,* **41,** 2735–2755.

Nicosia, D. J., and R. H. Grumm, 1999: Mesoscale band formation in three major northeastern United States snowstorms. *Wea. Forecasting,* **14,** 346–368.

Nielsen, J. W., 1989: The formation of New England coastal fronts. *Mon. Wea. Rev.,* **117,** 1380–1401.

——, and P. P. Neilley, 1990: The vertical structure of New England coastal fronts. *Mon. Wea. Rev.,* **118,** 1793–1807.

Niziol, T. A., 1987: Operational forecasting of lake effect snow in western and central New York. *Wea. Forecasting,* **2,** 310–321.

——, W. R. Snyder, and J. S. Waldstreicher, 1995: Winter weather forecasting throughout the eastern United States. Part IV: Lake effect snow. *Wea. Forecasting,* **10,** 61–77.

NOAA, 1971: *Storm Data.* Vol. 13, No. 11.

O'Handley, C., and L. F. Bosart, 1996: The impact of the Appalachian

Mountains on cyclonic weather systems. Part I: A climatology. *Mon Wea. Rev.,* **124,** 1353–1373.

Oravec, R. J., and R. H. Grumm, 1993: The prediction of rapidly deepening cyclones by NMC's Nested Grid Model: Winter 1989—autumn 1991. *Wea. Forecasting,* **8,** 248–270.

Orlanski, I., 1975: A rational subdivision of scales for atmospheric processes. *Bull. Amer. Meteor. Soc.,* **56,** 527–530.

——, and K. M. Chang, 1993: Ageostrophic geopotential fluxes in downstream and upstream development of baroclinic waves. *J. Atmos. Sci.,* **50,** 212–225.

Palmén, E., 1951: The aerology of extratropical disturbances. *Compendium of Meteorology,* T. F. Malone, Ed., Amer. Meteor. Soc., 599–620.

——, and C. W. Newton, 1969: *Atmospheric Circulation Systems.* Academic Press, 603 pp.

Petterssen, S., 1955: A general survey of factors influencing development at sea-level. *J. Meteor.,* **12,** 36–42.

——, 1956: *Weather Analysis and Forecasting.* Vol. 1. McGraw-Hill, 428 pp.

——, D. L. Bradbury, and K. Pedersen, 1962: The Norwegian cyclone models in relation to heat and cold sources. *Geofys. Publ., Norske Viderskaps-Akad. Oslo,* **24,** 243–280.

Phillips, N. A., 1951: A simple three-dimensional model for the study of large-scale extratropical flow patterns. *J. Meteor.,* **8,** 381–394.

——, 2000: A review of theoretical question in the early days of NWP. *50th Anniversary of Numerical Weather Prediction Commemorative Symposium,* A. Spekat, Ed., Deutsche Meteorologische Gesellschaft, 13–28.

Platzman, G. W., 1952: Some remarks on high-speed computers and their use in meteorology. *Tellus,* **4,** 168–178.

——, 1979: The ENIAC computations of 1950—Gateway to numerical weather prediction. *Bull. Amer. Meteor. Soc.,* **60,** 302–312.

Pokrandt, P. J., G. J. Tripoli, and D. D. Houghton, 1996: Processes leading to the formation of mesoscale waves in the Midwest cyclone of 15 December 1987. *Mon. Wea. Rev.,* **124,** 2726–2752.

Powers, J. G., and R. J. Reed, 1993: Numerical simulation of the large-amplitude mesoscale gravity-wave event of 15 December 1987 in the central United States. *Mon. Wea. Rev.,* **121,** 2285–2308.

Ramamurthy, M. K., R. M. Rauber, B. P. Collins, M. T. Shields, P. C. Kennedy, and W. L. Clark, 1991: UNIWIPP: A University of Illinois field experiment to investigate the structure of mesoscale precipitation in winter storms. *Bull. Amer. Meteor. Soc.,* **72,** 764–776.

Rasmussen, R., and Coauthors, 1992: Winter Icing and Storms Project (WISP). *Bull. Amer. Meteor. Soc.,* **73,** 951–974.

Rauber, R. M., M. K., Ramamurthy, and A. Tokay, 1994: Synoptic and mesoscale structure of severe freezing rain event: The St. Valentine's Day ice storm. *Wea. Forecasting,* **9,** 183–208.

Reed, R. J., 1955: A study of a characteristic type of upper-level frontogenesis. *J. Meteor.,* **12,** 226–237.

——, and F. Sanders, 1953: An investigation of the development of a mid-tropospheric frontal zone and its associated vorticity field. *J. Meteor.,* **10,** 338–349.

——, and E. F. Danielsen, 1959: Fronts in the vicinity of the tropopause. *Arch. Meteor. Geophys. Bioklim.,* **A11,** 1–17.

——, M. T. Stoelinga, and Y.-H. Kuo, 1992: A model-aided study of the origin and evolution of the anomalously high potential vorticity in the inner region of a rapidly deepening marine cyclone. *Mon. Wea. Rev.,* **120,** 893–913.

——, Y.-H. Kuo, and S. Low-Nam, 1994: An adiabatic simulation of the ERICA IOP 4 storm: An example of quasi-ideal frontal cyclone development. *Mon. Wea. Rev.,* **122,** 2688–2708.

Reitan, C. H., 1974: Frequencies of cyclones and cyclogenesis for North America, 1950–1970. *Mon. Wea. Rev.,* **102,** 861–868.

Reiter, E. R., 1963: *Jet Stream Meteorology.* The University of Chicago Press, 515 pp.

——, 1969: Tropospheric circulations and jet streams. *Climate of the Free Atmosphere,* D. F. Rex, Ed., Vol. 4, *World Survey of Climatology,* Elsevier Science, 85–203.

Richwein, B. A., 1980: The damming effect of the southern Appalachians. *Natl. Wea. Dig.,* **5,** 2–12.

Riehl, H., and Coauthors, 1952: *Forecasting in the Middle Latitudes.* Meteor. Monogr., No. 5, Amer. Meteor. Soc., 80 pp.

Riordan, A. J., 1990: Examination of the mesoscale features of the GALE coastal front of 24–25 January, 1986. *Mon. Wea. Rev.,* **118,** 258–282.

Roebber, P. J., 1984: Statistical analysis and updated climatology of explosive cyclones. *Mon. Wea. Rev.,* **112,** 1577–1589.

——, 1993: A diagnostic case study of self-development as an antecedent conditioning process in explosive cyclogenesis. *Mon. Wea. Rev.,* **121,** 976–1006.

——, J. R. Gyakum, and D. N. Trat, 1994: Coastal frontogenesis and precipitation during ERICA IOP 2. *Wea. Forecasting,* **9,** 21–44.

Ropelewski, C. F., and M. S. Halpert, 1986: North American precipitation and temperature patterns associated with the El Niño/Southern Oscillation (ENSO). *Mon. Wea. Rev.,* **114,** 2352–2362.

——, and ——, 1987: Global and regional scale precipitation patterns associated with the El Niño/Southern Oscillation (ENSO). *Mon. Wea. Rev.,* **115,** 1606–1626.

——, and ——, 1989: Precipitation patterns associated with the high index phase of the Southern Oscillation. *J. Climate,* **2,** 268–284.

——, and ——, 1996: Quantifying Southern Oscillation–precipitation relationships. *J. Climate,* **9,** 1043–1059.

Rosenblum, H. S., and F. Sanders, 1974: Meso-analysis of a coastal snowstorm in New England. *Mon. Wea. Rev.,* **102,** 433–442.

Saffir, H. S., 1977: Design and construction requirements for hurricane resistant construction. American Society of Civil Engineers Preprint 2830, 20 pp.

Salmon, E., and P. J. Smith, 1980: A synoptic analysis of the 25–26 January 1978 blizzard cyclone in the central United States. *Bull. Amer. Meteor. Soc.,* **61,** 453–460.

Sanders, F., 1986a: Explosive cyclogenesis in the west-central North Atlantic Ocean, 1981–84. Part I: Composite structure and mean behavior. *Mon. Wea. Rev.,* **114,** 1781–1794.

——, 1986b: Frontogenesis and symmetric stability in a major New England snowstorm. *Mon. Wea. Rev.,* **114,** 1847–1862.

——, 1987: Skill of NMC operational dynamical models in prediction of explosive cyclogenesis. *Wea. Forecasting,* **2,** 322–336.

——, 1990: Surface analysis over the oceans—Searching for sea truth. *Wea. Forecasting,* **5,** 596–612.

——, 1992: Skill of operational dynamical models in cyclone prediction out to five-days range during ERICA. *Wea. Forecasting,* **7,** 3–25.

——, and J. R. Gyakum, 1980: Synoptic-dynamic climatology of the "bomb." *Mon. Wea. Rev.,* **108,** 1589–1606.

——, and ——, 1983a: On the evolution of the *QE II* storm. Part I: Synoptic aspects. *Mon. Wea. Rev.,* **111,** 1137–1155.

——, and L. F. Bosart, 1985a: Mesoscale structure in the Megalopolitan snowstorm of 11–12 February 1983. Part I: Frontogenetical forcing and symmetric instability. *J. Atmos. Sci.,* **42,** 1050–1061.

——, and ——, 1985b: Mesoscale structure in the Megalopolitan snowstorm, 11–12 February 1983. Part II: Doppler radar study of the New England snowband. *J. Atmos. Sci.,* **42,** 1398–1407.

——, and E. Auciello, 1989: Skill in prediction of explosive cyclogenesis over the western North Atlantic Ocean, 1987/1988: A forecast checklist and NMC dynamical models. *Wea. Forecasting,* **4,** 157–172.

Sanderson, A. N., and R. B. Mason Jr., 1958: Behavior of two East Coast storms, 13–24 March 1958. *Mon. Wea. Rev.,* **86,** 109–115.

Santer, B. D., K. E. Taylor, J. E. Penner, T. M. L. Wigley, U. Cubasch, and P. D. Jones, 1996: Towards the detection and attribution of an anthropogenic effect on climate. *Climate Dyn.,* **12,** 77–100.

Scherhag, R., 1937: Bermerkurgen über die bedeutung der konvergenzen und divergenzen du geschwindigkeitsfeldes fur die Druckänderungen. *Beitr. Phys. Atmos.,* **24,** 122–129.

Schneider, R. S., 1990: Large-amplitude mesoscale wave disturbances within the intense Midwest extratropical cyclone of 15 December 1987. *Wea. Forecasting, 5,* 533–558.

Schultz, D. M., 1999: Lake-effect snowstorms in northern Utah and western New York with and without lightning. *Wea. Forecasting,* **14,** 1023–1031.

——, 2001: Reexamining the cold conveyor belt. *Mon. Wea. Rev.,* **129,** 2205–2225.

——, and C. Mass, 1993: The occlusion process in a midlatitude cyclone over land. *Mon. Wea. Rev.,* **121,** 918–940.

——, and P. N. Schumacher, 1999: The use and misuse of conditional symmetric instability. *Mon. Wea. Rev.,* **127,** 2709–2732.

Seimon, A., L. F. Bosart, W. E. Bracken, and W. R. Snyder, 1996: Large-amplitude inertia–gravity waves. Part II: Structure of an extreme gravity wave event over New England on 4 January 1994 revealed by WSR-88D radar and mesoanalysis. Preprints, *14th Conf. on Weather Analysis Forecasting,* Dallas, TX, Amer. Meteor. Soc., 434–441.

Shapiro, M. A., and P. J. Kennedy, 1981: Research aircraft measurements of jet stream geostrophic and ageostrophic winds. *J. Atmos. Sci.,* **38,** 2642–2652.

——, and D. Keyser, 1990: Fronts, jet streams and the tropopause. *Extratropical Cyclones: The Erik Palmén Memorial Volume,* C. W. Newton and E. O. Holopainen, Eds., Amer. Meteor. Soc., 167–193.

——, H. Wernli, N. A. Bond, and R. Langland, 2000: The influence of the 1997–1999 ENSO on extratropical baroclinic life cycles over the eastern North Pacific. *Quart. J. Roy. Meteor. Soc.,* **126,** 1–20.

Shields, M. T., R. M. Rauber, and M. K. Ramamurthy, 1991: Dynamical forcing and mesoscale organization of precipitation bands in a Midwest witner cyclonic storm. *Mon. Wea. Rev.,* **119,** 936–964.

Shuman, F. G., 1989: History of numerical weather prediction at the National Meteorological Center. *Wea. Forecasting, 4,* 286–296.

Simmons, A. J., and B. J. Hoskins, 1979: The downstream and upstream development of unstable baroclinic waves. *J. Atmos. Sci.,* **36,** 1239–1260.

Sinclair, M. R., and R. L. Elsberry, 1986: A diagnostic study of baroclinic disturbances in polar air streams. *Mon. Wea. Rev.,* **114,** 1957–1983.

Smith, B. B., and S. L. Mullen, 1993: An evaluation of sea-level cyclone forecasts produced by NMC's Nested-Grid Model and Global Spectral Model. *Wea. Forecasting, 8,* 37–56.

Smith, C. D., Jr., 1950: The destructive storm of November 25–27, 1950. *Mon. Wea. Rev., 78,* 204–209.

Smith, R. B., 1979: The influence of the mountains on the atmosphere. *Advances in Geophysics,* Vol. 21, Academic Press, 87–230.

——, 1984: A theory of lee cyclogenesis. *J. Atmos. Sci.,* **41,** 1159–1168.

Smith, S. R., and J. J. O'Brien, 2001: Regional snowfall distributions associated with ENSO: Implications for seasonal forecasting. *Bull. Amer. Meteor. Soc., 82,* 1179–1191.

Snook, J. S., and R. A. Pielke, 1995: Diagnosing a Colorado heavy snow event with a nonhydrostatic mesoscale numerical model structured for operational use. *Wea. Forecasting,* **10,** 261–285.

Spiegler, D. B., and G. E. Fisher, 1971: A snowfall prediction method for the Atlantic seaboard. *Mon. Wea. Rev., 99,* 311–325.

Staley, D. O., 1960: Evaluation of potential-vorticity changes near the tropopause and the related vertical motions, vertical advection of vorticity, and transfer of radioactive debris from stratosphere to troposphere. *J. Meteor.,* **17,** 591–620.

Stauffer, D. R., and T. T. Warner, 1987: A numerical study of Appalachian cold-air damming and coastal frontogenesis. *Mon. Wea. Rev.,* **115,** 799–821.

Stein, U., and P. Alpert, 1993: Factor separation in numerical simulations. *J. Atmos. Sci.,* **50,** 2107–2115.

Stewart, G. R., 1941: *Storm.* Random House, 349 pp. [Reprinted 2003, Heyday Books, 352 pp.]

Stewart, R. E., 1985: Precipitation types in winter storms. *Pure Appl. Geophys.,* **123,** 597–609.

——, 1992: Precipitation types in the transition region of winter storms. *Bull. Amer. Meteor. Soc.., 73,* 287–296.

——, and P. King, 1987a: Freezing precipitation in winter storms. *Mon. Wea. Rev.,* **115,** 1270–1279.

——, and ——, 1987b: Rain–snow boundaries over southern Ontario. *Mon. Wea. Rev.,* **115,** 1894–1907.

——, R. W. Shaw, and G. A. Isaac, 1987: Canadian Atlantic Storms program: The meteorological field project. *Bull. Amer. Meteor. Soc.,* **68,** 338–345.

——, J. D. Marwitz, and R. E. Carbone, 1984: Characteristics through the melting layer of stratiform clouds. *J. Atmos. Sci.,* **41,** 3227–3237.

Stokols, P. M., J. P. Gerrity, and P. J. Kocin, 1991: Improvements at NMC in numerical weather prediction and their effect on winter storm forecasts. Preprints, *First Int. Symp. on Winter Storms,* New Orleans, LA, Amer. Meteor. Soc., 15–19.

Suckling, P. W., 1991: Spatial and temporal climatology of snowstorms in the Deep South. *Phys. Geogr.,* **12,** 124–139.

Sutcliffe, R. C., 1939: Cyclonic and anticyclonic development. *Quart. J. Roy. Meteor. Soc.,* **65,** 518–524.

——, 1947: A contribution to the problem of development. *Quart. J. Roy. Meteor. Soc.,* **73,** 370–383.

——, and A. G. Forsdyke, 1950: The theory and use of upper air thickness patterns in forecasting. *Quart. J. Roy. Meteor. Soc.,* **76,** 189–217.

Tepper, M., 1954: Pressure jump lines in midwestern United States, January–August 1951. U.S. Weather Bureau Research Paper 376, 70 pp.

Thorncroft, C. D., B J. Hoskins, and M. E. McIntyre, 1993: Two paradigms of baroclinic-wave life-cycle behavior. *Quart. J. Roy. Meteor. Soc.,* **119,** 17–56.

Tracton, M. S., 1993: On the skill and utility of NMC's medium-range central guidance. *Wea. Forecasting, 8,* 147–153.

——, and E. Kalnay, 1993: Operational ensemble prediction at the National Meteorological Center: Practical aspects. *Wea. Forecasting, 8,* 379–398.

Trenberth, K. E., 1976: Spatial and temporal variations of the Southern Oscillation. *Quart. J. Roy. Meteor. Soc.,* **102,** 639–653.

——, 1997: The definition of El Niño. *Bull. Amer. Meteor. Soc., 78,* 2771–2778.

Uccellini, L. W., 1975: A case study of apparent gravity wave initiation of severe convective storms. *Mon. Wea. Rev.,* **103,** 497–513.

——, 1984: Comments on "Comparative diagnostic case study of East Coast secondary cyclogenesis under weak versus strong synoptic forcing." *Mon. Wea. Rev.,* **112,** 2540–2541.

——, 1990: Processes contributing to the rapid development of Extratropical Cylcones. *Extratropical Cyclones: The Erik Palmén Memorial Volume,* C. W. Newton and E. O. Holopainen, Eds., Amer. Meteor. Soc., 81–105.

——, and D. R. Johnson, 1979: The coupling of upper- and lower-tropospheric jet streaks and implications for the development of severe convective storms. *Mon. Wea. Rev.,* **107,** 682–703.

——, and S. E. Koch, 1987: The synoptic setting and possible energy sources for mesoscale wave disturbances. *Mon. Wea. Rev.,* **115,** 763–786.

——, and P. J. Kocin, 1987: An examination of vertical circulations associated with heavy snow events along the East Coast of the United States. *Wea. Forecasting, 2,* 289–308.

——, ——, R. A. Petersen, C. H. Wash, and K. F. Brill, 1984: The Presidents' Day cyclone of 18–19 February 1979: Synoptic overview and analysis of the subtropical jet streak influencing the precyclogenetic period. *Mon. Wea. Rev.,* **112,** 31–55.

——, D. Keyser, K. F. Brill, and C. H. Wash, 1985: The Presidents' Day cyclone of 18–19 February 1979: Influence of upstream trough amplification and associated tropopause folding on rapid cyclogenesis. *Mon. Wea. Rev.,* **115,** 2227–2261.

——, R. A. Petersen, K. F. Brill, P. J. Kocin, and J. J. Tuccillo, 1987:

Synergistic interactions between an upper-level jet streak and diabatic processes that influence the development of a low-level jet and a secondary coastal cyclone. *Mon. Wea. Rev.,* **115,** 2227–2261.

——, P. J. Kocin, R. S. Schneider, P. M. Stokols, and R. A. Dorr, 1995: Forecasting the 12–14 March 1993 Superstorm. *Bull. Amer. Meteor. Soc.,* **76,** 183–199.

——, J. M. Sienkiewicz, and P. J. Kocin, 1999: Advances in forecasting extratropical cyclogenesis at the National Meteorological Center. *The Life Cycles of Extratropical Cyclones,* M. A. Shapiro and S. Grønås, Eds., Amer. Meteor. Soc., 317–336.

Upton, W., 1888: The storm of March 11–14, 1888. *Amer. Meteor. J.,* **5,** 19–37.

van Loon, H., and J. C. Rogers, 1978: The seesaw in winter temperatures between Greenland and northern Europe. Part I: General description. *Mon. Wea. Rev.,* **106,** 296–310.

Wagner, A. J., 1957: Mean temperature from 1000 mb to 500 mb as a predictor of precipitation type. *Bull. Amer. Meteor. Soc.,* **38,** 584–590.

Walker, G. T., and E. W. Bliss, 1932: World weather V. *Mem. Roy. Meteor. Soc.,* **4,** 53–84.

Wallace, J. M., and D. S. Gutzler, 1981: Teleconnections in the geopotential height field during the Northern Hemisphere winter. *Mon. Wea. Rev.,* **109,** 784–812.

Wash, C. H., J. E. Peak, W. E. Calland, and W. A. Cook, 1988: Diagnostic study of explosive cyclogenesis during FGGE. *Mon. Wea. Rev.,* **116,** 431–451.

Weisman, R. A., 1996: The Fargo snowstorm of 6–8 January 1989. *Wea. Forecasting,* **11,** 198–215.

Weismuller, J. L., and S. M. Zubrick, 1998: Evaluation application of conditional symmetric instability, equivalent potential vortic-

ity, and frontogenetic forcing in an operational environment. *Wea. Forecasting,* **13,** 84–101.

Weldon, R. B., 1979: Cloud patterns and upper air wind field. Part IV. Satellite training course notes. AWZ/TR-79/0003, United States Air Force.

Werstein, I., 1960: *The Blizzard of '88.* Thomas Y. Crowell Company, 157 pp.

Wesley, D. A., R. M. Rasmussen, and B. J. Bernstein, 1995: Snowfall associated with a terrain-generated convergence zone during the Winter Icing and Storm Project. *Mon. Wea. Rev.,* **123,** 2957–2977.

Wexler, R., R. J. Reed, and J. Honig, 1954: Atmospheric cooling by melting snow. *Bull. Amer. Meteor. Soc.,* **35,** 48–51.

Whitaker, J. S., L. W. Uccellini, and K. F. Brill, 1988: A model-based diagnostic study of the explosive development phase of the Presidents' Day cyclone. *Mon. Wea. Rev.,* **116,** 2337–2365.

Whittier, J. G., 1961: Snowbound. *The Family Book of Verse,* L. Gannett, Ed., Harper and Row, 302–303.

Widger, W. K., Jr., 1964: A synthesis of interpretations of extratropical vortex patterns as seen by TIROS. *Mon. Wea. Rev.,* **92,** 263–282.

Wolfsberg, D. G., K. A. Emanuel, and R. E. Passarelli, 1986: Band formation in a New England snowstorm. *Mon. Wea. Rev.,* **114,** 1552–1569.

Younkin, R. J., 1968: Circulation patterns associated with heavy snowfall over the western United States. *Mon. Wea. Rev.,* **96,** 851–853.

Zhang, F., C. Snyder, and R. Rotunno, 2002: Mesoscale predictability of the "surprise" snowstorm of 24–25 January 2000. *Mon. Wea. Rev.,* **130,** 1617–1632.

Zielinski, G., 2002: A classification scheme for winter storms in the eastern and central United States with an emphasis on nor'easters. *Bull. Amer. Meteor. Soc.,* **83,** 37–51.